Diseases of Economically Important Horticultural Crops

First published 2027
by CRC Press
4 Park Square, Milton Park, Abingdon, Oxon, OX14 4RN

and by CRC Press
2385 NW Executive Center Drive, Suite 320, Boca Raton FL 33431

CRC Press is an imprint of Informa UK Limited

British Library Cataloguing-in-Publication Data
A catalogue record for this book is available from the British Library

ISBN: 9781041349044 (hbk)
ISBN: 9781041387268 (pbk)
ISBN: 9781042014460 (ebk)

DOI: 10.1201/9781042014460

Typeset in Times New Roman
by Content Vibes

CV CONTENT VIBES
PUBLISHERS • PRINTERS • AGGREGATORS

Diseases of Economically Important Horticultural Crops

About the Editors

Mr. Deepak Mourya (Silver Medalist): Mr. Deepak Mourya pursued his B.Sc.(Hons) Horticulture degree from Banda University of Agriculture and Technology, Banda, U.P. in 2018, his M.Sc. (Ag.) Plant Pathology degree from Department of Plant Protection, Aligarh Muslim University, Aligarh, U.P. in 2020, and is currently pursuing his Ph.D. (Plant Pathology) at Shri Khushal Das University in Hanumangarh, Rajasthan. He was awarded the Vice-Chancellor Silver Medal in the UG Program. He has attended numerous national and international conferences and symposiums, where he has garnered best poster presentation prizes. He has contributed to numerous publications, including 5 research articles, 5 book chapters, 21 popular articles, and 3 abstracts in national and international journals.

Mr. Pankaj Singh: is Currently pursuing Doctor of Philosophy in Department of Plant Pathology at Sardar Vallabhbhai Patel University of Agriculture & Technology, Meerut (U.P.). He completed his Bachelor of Science (Agriculture) Honours from Chatrapati Sahu Ji Maharaj University, Kanpur (U.P.) and Master of Science (Agriculture) in Plant Pathology from Acharya Narendra Dev University of Agriculture & Technology, Ayodhya (U.P.). He has participated in several national and international conferences and symposiums and received best poster presentation awards. He has also published several research papers, book chapters, articles and abstracts.

Diseases of Economically Important Horticultural Crops

Edited by:
Deepak Mourya
Pankaj Singh

CONTENT VIBES
PUBLISHERS • PRINTERS • AGGREGATORS

CRC Press
Taylor & Francis Group
Boca Raton London New York

CRC Press is an imprint of the
Taylor & Francis Group, an **informa** business

First published 2027
by CRC Press
4 Park Square, Milton Park, Abingdon, Oxon, OX14 4RN

and by CRC Press
2385 NW Executive Center Drive, Suite 320, Boca Raton FL 33431

CRC Press is an imprint of Informa UK Limited

British Library Cataloguing-in-Publication Data
A catalogue record for this book is available from the British Library

ISBN: 9781041349044 (hbk)
ISBN: 9781041387268 (pbk)
ISBN: 9781042014460 (ebk)

DOI: 10.1201/9781042014460

Typeset in Times New Roman
by Content Vibes

CV CONTENT VIBES
PUBLISHERS • PRINTERS • AGGREGATORS

College of Agriculture

(S.K.N. Agriculture University, Jobner)
Bhusawar -321406, Bharatpur (Rajasthan), India
Mobile No.: 9414741234
Email: dean.coabhusawer@sknau.ac.in

Dr. Udai Bhan Singh
Professor and Dean

Foreword

I was honored to foreword the book "Diseases of Economically Important Horticultural Crops" edited by Deepak Mourya and Pankaj Singh and published by Publish My Work (P.M.W.) at Pitam Pura, New Delhi, India.

Horticultural crop production is critical to global agriculture, not just for nutrition and food security, but also for contributing significantly to global economics. These crops include fruits, vegetables, ornamental plants, and herbs are vital to the livelihoods of millions of farmers and critical to the economic stability of many regions.

This edited collection "Diseases of Economically Important Horticultural Crops" brings together the most recent research and ideas from top specialists in the area. This book provides valuable information on different aspects of the above-given area through twenty- two chapters, such as diseases and the impact of essential horticulture crops, as well as their epidemiology, diagnostic procedures, and creative management practices. Each chapter is intended to give useful information for academics, practitioners, and policymakers, thereby bridging the gap between scientific advances and practical applications.

In presenting this collection, we hope to provide stakeholders with the knowledge they need to protect horticulture output and preserve the sector's long-term viability and profitability. This book shows a joint effort to address the difficulties faced by horticulture crops and secure a healthy, productive future for global agriculture.

We believe that this book is a valuable resource that motivates additional study and innovation in the fight against horticultural crop diseases.

Date: 01-07-2024
Place: Bhusawar, Bharatpur

Udai Bhan Singh

Preface

Plant pathology is an important branch of Agriculture due to the losses caused by the plant diseases. Plant diseases are one of the important constraints of crop production causing tremendous losses to crop yields to the tune of millions of money besides causing death to thousands of people died due to hunger worldwide. Considering its importance plant pathology is being taught as a compulsory subject to the students of Agriculture. Through this book attempts have been made to provide most recent information about the diseases of horticultural crops to the students of Agriculture doing under-graduate courses as per the 5th Deans committee recommendations.

Discussion has been made on the importance, economic losses due to the disease, symptoms, etiology, disease cycle, factors favorable and management of most of the plant diseases in brief so that it can give a clear cut concept on the diseases. The major diseases as of considered now are discussed elaborately as compared to the minor diseases.

Number of books, research paper, manuals and other published articles on the diseases of horticultural crops are studied and drawn information from those and presented in this book. I wish to express my gratitude to the authors of various book and scientific literature as those were the basic resource while preparing this book. I sincerely acknowledge the help, inspiration and advice of all the well-wishers and friends while writing this book. I also express my thankfulness to Publish My Work for publishing the book to place in the hands of the students, teachers and other readers. Last but not the least acknowledges the encouragement and constant support of my lovely parents.

2024

Deepak Mourya
Pankaj Singh

Contents

1

Economic Importance of Horticultural Crops in India's GDP

Amit Kumar[1], Arvind Kumar[2], Vibhu Pandey[3] and Shalini Singh[4]

[1]Department of Fruit Science, Sardar Vallabhbhai Patel University of Agriculture and Technology Meerut, Uttar Pradesh
[2]Department of Fruit Science, Sardar Vallabhbhai Patel University of Agriculture and Technology
Meerut-250110, Uttar Pradesh
[3]Defence Institue of High Altitude Research, Leh (Ladakh)
[4]Department of Agriculture Science, Maharishi Arvind University Jaipur, Rajasthan

Abstract

Horticultural crops constitute a vital sector within India's agriculture, contributing significantly to the nation's Gross Domestic Product (GDP). This study conducts a thorough analysis of the economic significance of horticultural crops in the Indian economy, focusing on their pivotal role in driving economic growth, generating employment, and enhancing nutritional security. The cultivation of horticultural crops encompasses a diverse array of fruits, vegetables, flowers, spices, plantation crops, and medicinal and aromatic plants. This sector constitutes a substantial portion of India's agricultural GDP, with a contribution of approximately 30%. Moreover, horticulture serves as a major employer, especially for smallholder farmers and laborers, providing critical livelihood opportunities across the country. In addition to domestic consumption, horticultural crops play a pivotal role in international trade, with India emerging as a prominent exporter of fruits, vegetables, and spices. This export orientation has not only bolstered India's global trade position but also led to increased foreign exchange earnings, further reinforcing the economic importance of this sector. Beyond economic gains, horticultural crops play a crucial role in ensuring nutritional security. Rich in essential vitamins, minerals, and dietary fiber, they form a cornerstone of a balanced and healthy diet, contributing to improved public health outcomes. Furthermore, the cultivation of horticultural crops aligns with sustainable agricultural practices, promoting ecological balance and

DOI: 10.1201/9781042014460-1

environmental resilience. This underscores the imperative of recognizing and fostering the horticultural sector's continued growth. By implementing policies that support research, technology adoption, and market access, India can unlock the full potential of horticultural crops for sustained economic development and food security.

Keywords: *Indian economy, GDP contribution, Employment generation, Nutritional security*

1. Introduction

Horticultural crops have emerged as a cornerstone of India's agricultural sector, wielding substantial influence over the nation's economic landscape. Comprising an extensive range of fruits, vegetables, ornamental plants, plantation crops, spices, and medicinal herbs, this sector has witnessed remarkable growth and transformation in recent years. The significance of horticultural crops extends beyond sheer economic value, permeating various facets of Indian agriculture and socio-economic development (Ravichandra & Ravichandra 2014).The unique agro-climatic diversity of India allows for the cultivation of an unparalleled variety of horticultural produce, contributing to its global prominence as a horticultural powerhouse. From the verdant orchards of Himachal Pradesh to the spice gardens of Kerala, the country's geographical expanse fosters an environment conducive to diverse horticultural practices.

Horticulture is a linchpin of the Indian economy, providing crucial employment opportunities and raw materials for food processing industries. It also drives higher profitability and boosts export earnings. In recent decades, it has gained prominence, significantly contributing to the GDP, becoming a primary force in augmenting income, alleviating poverty, and generating jobs. Over 7 percent of cultivated land, spanning 27.4 million hectares, is dedicated to horticultural crops(NHB, 2020-21). India leads in Okra production and ranks second in vegetables like potatoes, cauliflower, and brinjal. It also tops the charts in fruit cultivation, including mango, banana, guava, lemon, and papaya. The cultivation of fruits and vegetables has overshadowed that of food grains. This sector offers diverse employment avenues, from running nurseries to turf grass and florist operations. Making up 6% of the GDP and contributing a third of the total agricultural output, horticulture plays a crucial role in India's economic growth. The Horticulture Division of the Indian Council of Agricultural Research (ICAR) actively furthers its objectives, resulting in substantial progress in various Indian states. The Indian horticulture sector yields around 334 million tons of produce from an area of 27.4 million hectares (NHB 2020-21), contributing approximately 33% of the total gross value addition (GVA). With the right allocation of resources, infrastructure development, technological advancements, and a robust policy framework, India has the potential to become a major cultivator and exporter in the horticulture sector.

The economic importance of horticultural crops in India's GDP is unquestionable(Mitra& Panda 2020). However, it is not without its challenges. One of the significant challenges associated with the economic importance of horticultural crops in India's GDP is the prevalence of diseases. Plant diseases can have a devastating impact on horticultural yields, leading to substantial economic losses for farmers. Factors such as inadequate knowledge of disease management practices, limited access to disease-resistant varieties, and a lack of effective pest control measures contribute to the vulnerability of horticultural crops. Collaborative efforts between agricultural institutions, government agencies, and the private sector are essential in developing and disseminating effective disease management strategies to safeguard the economic significance of horticultural crops in India's GDP (Nabi *et al.* 2017).

The challenges in India's horticultural sector are also diverse and impactful. Limited access to credit and insurance inhibits investment in modern practices. Inadequate post-harvest infrastructure leads to significant produce losses. Market instability and price fluctuations create uncertainty for farmers.

2. Purpose and Objectives

The primary purpose of this paper is to provide a comprehensive and updated assessment of the economic significance of horticultural crops in the Indian GDP. To achieve this, the study will:

Quantify Contribution to GDP: Evaluate the exact economic contribution of horticultural crops to the overall GDP of the Indian economy, providing a clear understanding of its relative importance.

Analyze Employment Generation: Investigate the role of horticultural crops in creating employment opportunities, particularly for smallholder farmers and laborers, and assess their socio-economic impact.

Examine Export Dynamics: Scrutinize the export trends and dynamics of horticultural produce, exploring how they bolster India's global trade position and contribute to foreign exchange earnings.

Assess Nutritional Significance: Investigate the nutritional value and public health implications of horticultural consumption, emphasizing their role in enhancing dietary diversity and overall well-being.

3. Horticultural Crops in India

India boasts a rich tapestry of horticultural crops, reflecting its agricultural diversity and cultural heritage. From the iconic mango orchards of Ratnagiri to the sprawling vineyards of Nashik, the country's agricultural landscape is adorned with an extensive array of fruits, vegetables, flowers, spices, and medicinal plants.

Diverse Range of Crops

Fruits hold a prominent position in Indian horticulture, with varieties ranging from the ubiquitous banana, guava, and citrus fruits, to region-specific gems like Alphonso mangoes and Kinnow oranges. Additionally, India is a treasure trove of diverse vegetables, encompassing staples like potatoes, onions, and tomatoes, alongside a kaleidoscope of regional specialties such as Okra (Bhindi), pointed gourd (Parwal), and Yam (Suran). The aromatic spices like cardamom, pepper, and cinnamon, which are integral to Indian cuisine, also form a significant part of horticultural cultivation.

Cultural and Economic Significance

Beyond their economic utility, horticultural crops hold a profound cultural significance in India. Festivals and celebrations often feature specific fruits and flowers as symbols of prosperity, auspiciousness, and religious devotion. For instance, the marigold flower is a ubiquitous presence in Indian festivals, adorning homes, temples, and public spaces alike.

Economically, horticultural crops contribute substantially to rural livelihoods, supporting the income of millions of smallholder farmers across the country. Their cultivation has facilitated economic empowerment, particularly for women engaged in activities like vegetable vending and flower trade. Additionally, the export of horticultural produce has positioned India as a major player in the global market, generating substantial revenue and bolstering the country's trade profile.

4. Contribution to Agricultural GDP

Horticultural crops wield substantial influence over India's agricultural Gross Domestic Product (GDP), playing a pivotal role in shaping the economic contours of the nation. Data and statistics underscore the magnitude of their impact, highlighting a sustained and upward trajectory in their contribution.

Share in Agricultural GDP

Recent data reveals that horticultural crops account for approximately 30% of India's total agricultural GDP. This sizeable share not only signifies the sheer volume of production but also underscores the critical role played by horticulture in the overall agricultural landscape. This contribution has been steadily increasing over the years, reflecting a shift in agricultural patterns towards high-value and commercially viable crops (Kondal 2014).

Trends and Growth Patterns

The growth patterns of horticultural crops in India are characterized by several noteworthy trends:

Increasing Acreage: There has been a discernible expansion in the acreage dedicated to horticultural cultivation. Farmers are increasingly diversifying their crops, incorporating a wider range of fruits, vegetables, and ornamental plants into their agricultural practices.

Technological Adoption: The horticulture sector has witnessed a surge in the adoption of modern technologies, including precision farming, greenhouse cultivation, and drip irrigation. These advancements have not only enhanced productivity but also facilitated year-round cultivation.

Market-Oriented Production: The growing demand for horticultural produce, both domestically and in international markets, has incentivized farmers to adopt market-oriented production practices. This has led to a shift from subsistence farming to commercial horticulture.

Varietal Improvement: Research and development efforts have resulted in the introduction of high-yielding and disease-resistant varieties of horticultural crops. This has been instrumental in augmenting productivity and quality.

Value Addition and Processing: The horticulture sector has witnessed a surge in value-addition activities, including post-harvest processing, packaging, and export-oriented marketing. This has added a new dimension to the economic significance of horticultural crops.

The trends and growth patterns observed in the horticulture sector indicate a dynamic and evolving landscape. The increasing contribution of horticultural crops to India's agricultural GDP not only signifies their economic importance but also underscores their potential for continued growth and development. The next section will delve into the role of horticultural crops in generating employment opportunities, a critical aspect of their economic impact.

5. Employment Generation

The horticulture sector in India stands as a formidable generator of employment, wielding significant socio-economic impact. Its ability to provide livelihoods, especially in rural areas, has been instrumental in alleviating poverty and enhancing economic resilience for millions of individuals.

Role in Livelihoods

Horticultural crops serve as a reliable source of livelihood for a diverse spectrum of individuals. In rural areas, where the majority of horticultural production is concentrated, smallholder farmers and laborers find substantial employment opportunities. The sector also empowers women, who play a pivotal role in tasks such as seedling transplantation, weeding, and post-harvest handling (Nath 2011).

Stages of the Supply Chain

The horticultural supply chain encompasses various stages, each offering distinct employment opportunities:

Cultivation and Pre-harvest Operations: This stage involves activities like land preparation, planting, irrigation, pest management, and fertilization. Smallholder farmers, agricultural laborers, and skilled technicians are key players in this phase.

Harvesting and Post-harvest Handling: The harvesting phase requires a substantial workforce for tasks like picking, sorting, and packaging. Additionally, grading, washing, and storage activities create employment opportunities in both rural and urban areas.

Transportation and Marketing: Moving horticultural produce from farms to markets necessitates a network of logistics and transportation services. This involves truck drivers, loaders, and market intermediaries.

Retail and Distribution: Retailers, vendors, and market stall owners form an essential part of the supply chain, ensuring the accessibility of horticultural produce to consumers.

Value Addition and Processing: Processing units for activities such as canning, juicing, and packaging create employment, particularly in semi-urban and urban settings.

Export and International Trade: International trade in horticultural produce necessitates a range of professionals, including exporters, quality controllers, and logistics experts.

6. Export Revenue and Global Market

The export of horticultural crops stands as a cornerstone of India's global trade portfolio, playing a pivotal role in generating foreign exchange earnings (Kaul 1997). This dynamic sector has not only strengthened India's position in the international market but has also contributed significantly to the nation's economic resilience and stability.

Export Dynamics

India's prowess in horticultural exports is evident in its diverse array of commodities that find favor in markets worldwide. Fruits like mangoes, renowned for their exceptional quality and flavor, hold a prominent place in international trade. Additionally, the export of grapes, pomegranates, and citrus fruits has surged, establishing India as a key player in the global fruit market.

Vegetables, particularly onions and potatoes, have also witnessed a substantial increase in export volumes. These staples form the backbone of India's horticultural exports, meeting the demand of countries across the Middle East and Southeast Asia. Furthermore, spices such as cardamom, pepper, and turmeric, valued for their aromatic and culinary significance, continue to be sought-after commodities in the global spice trade.

Foreign Exchange Earnings

The revenue generated from horticultural exports contributes significantly to India's foreign exchange reserves. The sector has consistently registered positive growth in export earnings, serving as a reliable source of foreign currency. This influx of revenue not only strengthens India's economic stability but also provides a cushion against external economic shocks.

Market Diversification

India's horticultural exports have seen a positive trend in market diversification. While traditional markets in the Middle East remain crucial, there has been a notable expansion into regions like Europe, North America, and East Asia. This diversification mitigates risks associated with over-reliance on specific markets and broadens the scope for further growth.

Quality Assurance and Global Standards

Maintaining and improving the quality of horticultural produce is paramount in ensuring continued success in the global market. Adhering to international standards and certifications for safety, hygiene, and quality control is imperative. This not only fosters consumer trust but also opens doors to premium markets with discerning consumers.

Income Diversification for Farmers

The cultivation of horticultural crops stands as a transformative strategy for Indian farmers, offering them the opportunity to diversify their income streams and mitigate risks associated with traditional agriculture. This shift towards horticulture has not only bolstered economic resilience but has also empowered farmers to explore new avenues of prosperity.

Diversification through Horticulture

Traditionally, Indian agriculture has been dominated by the cultivation of staple crops like rice and wheat. However, the cultivation of horticultural crops presents an alternative, and often more profitable, avenue for farmers. By incorporating fruits, vegetables, spices, and ornamental plants into their agricultural practices, farmers can stagger their harvesting seasons, leading to a more consistent and steady income throughout the year.

Year-round Cultivation

One of the key advantages of horticultural crops is the potential for year-round cultivation. Unlike traditional field crops, which are typically limited to one or two seasons, many horticultural crops can be cultivated throughout the year. This not only ensures a more consistent income for farmers but also reduces their vulnerability to seasonal fluctuations and crop failures.

Case Studies in Income Diversification

Numerous case studies highlight the success stories of farmers who have embraced horticulture as a means of income diversification. For instance, in the state of Maharashtra, a group of smallholder farmers transitioned from traditional rice cultivation to the cultivation of high-value crops like grapes and pomegranates. This shift not only led to a significant increase in their income but also transformed the socio-economic landscape of their village.

Similarly, in Tamil Nadu, a cooperative of women farmers focused on the cultivation of organic vegetables. Through effective marketing strategies and direct-to-consumer sales, they were able to command premium prices for their produce, resulting in higher profits and improved livelihoods.

Skill Enhancement and Technological Adoption

Cultivating horticultural crops often requires a higher level of skill and knowledge compared to traditional agriculture. This has prompted many farmers to invest in training and education, leading to a more skilled and empowered farming community. Additionally, the adoption of modern technologies and practices, such as drip irrigation and precision farming, has further enhanced productivity and profitability.

7. Nutritional Security and Public Health

The cultivation and consumption of horticultural crops play a pivotal role in enhancing nutritional security and promoting public health in India (Keatinge *et al.* 2011). These crops, rich in essential vitamins, minerals, and dietary fiber, form a cornerstone of a balanced and healthy diet, contributing to improved overall well-being.

Nutritional Abundance

Horticultural crops are a treasure trove of essential nutrients, offering a wide spectrum of vitamins, including A, C, and K, as well as minerals like potassium, calcium, and folate. Additionally, they are often high in dietary fiber, which aids digestion and helps prevent various chronic diseases. The diversity of horticultural produce ensures that individuals have access to a wide range of nutrients, contributing to a well-rounded and nutritionally balanced diet.

Addressing Malnutrition

India faces significant challenges related to malnutrition, with a substantial portion of its population experiencing deficiencies in key nutrients. The incorporation of horticultural crops into diets can play a critical role in alleviating this issue. For example, the inclusion of fruits like papaya and guava, which are rich in vitamin C and A, can help combat deficiencies commonly seen in children and pregnant women.

Promoting Public Health

The consumption of horticultural crops has far-reaching implications for public health. Regular intake of fruits and vegetables has been associated with a reduced risk of chronic diseases such as heart disease, diabetes, and certain types of cancer. Additionally, the presence of antioxidants in horticultural produce helps protect cells from damage, further contributing to overall health and well-being.

Culinary Diversity and Traditional Medicine

Horticultural crops form the foundation of India's diverse culinary heritage. Spices like turmeric, ginger, and garlic are not only used for their flavor-enhancing properties but also for their medicinal benefits. Additionally, the use of medicinal herbs in traditional Indian medicine (Ayurveda) underscores the therapeutic potential of horticultural biodiversity.

Community and Household Gardens

The cultivation of horticultural crops in community and household gardens has gained traction as a means of promoting local food security and enhancing dietary diversity. These initiatives empower communities to take charge of their nutrition, reducing their dependence on external food sources.

8. Market Demand and Consumer Preferences

Analyzing Changing Consumer Preferences

In recent years, there has been a discernible shift in consumer preferences towards diverse and exotic horticultural produce in India. This transformation is driven by factors such as rising income levels, urbanization, and an increasing awareness of health and nutrition. Consumers are now seeking a wider variety of fruits, vegetables, and specialty crops, including those that were previously considered exotic.

Exotic fruits like avocados, kiwis, and dragon fruits, once rare in Indian markets, have gained popularity. Similarly, there is a growing demand for unique varieties of vegetables, herbs, and spices, driven by a desire for culinary diversity and heightened awareness of their health benefits.

Meeting Evolving Market Demands: To cater to this evolving market, it is imperative for the horticulture sector to adopt strategies that align with changing consumer preferences:

Diversification of Cultivation: Farmers can diversify their crops by introducing a wider range of horticultural produce, including exotic varieties. This not only taps into emerging market trends but also reduces the risk associated with over-reliance on a single crop.

Adoption of High-Value Crops: High-value crops, such as organic produce and specialty herbs, are gaining traction in the market. Farmers can explore the cultivation of these crops to tap into premium segments and command higher prices.

Value Addition through Processing: Processing activities like canning, juicing, and preservation can enhance the marketability of horticultural produce. Processed products not only have an extended shelf life but also cater to the convenience preferences of modern consumers.

Market Intelligence and Research: Staying attuned to market trends and consumer preferences is critical. Regular market research and intelligence can provide insights into emerging demands, allowing farmers to make informed planting decisions.

Building Sustainable Supply Chains: Establishing efficient and sustainable supply chains ensures that produce reaches consumers in a timely and quality-assured manner. This involves close collaboration between farmers, processors, distributors, and retailers.

Engaging in Direct-to-Consumer Models: Platforms such as farmers' markets, community-supported agriculture (CSA), and e-commerce channels provide opportunities for farmers to directly connect with consumers, bypassing traditional middlemen.

Adoption of Technology: Leveraging technology for precision farming, greenhouse cultivation, and post-harvest management can improve productivity and quality, meeting the exacting standards of discerning consumers (Ugochukwu & Phillips 2018).

By adopting these strategies, the horticulture sector can not only meet evolving market demands but also position itself as a dynamic and responsive player in the agricultural landscape. Additionally, such initiatives promote economic sustainability by diversifying income sources for farmers and fostering a more resilient agricultural sector.

9. Climate Resilience and Sustainable Agriculture

The Role of Horticultural Crops in Climate-Smart Agriculture

Horticultural crops play a pivotal role in climate-smart agriculture, which focuses on sustainable practices that address the challenges posed by climate change. Their inherent characteristics make them well-suited to adapt to changing climatic conditions and contribute to the resilience of agricultural systems.

Shorter Growth Cycles: Many horticultural crops have shorter growth cycles compared to traditional field crops. This allows for quicker adjustments in response to sudden weather fluctuations, reducing the risk of crop failure.

Crop Diversity: The wide variety of horticultural crops available allows farmers to diversify their plantings. This not only spreads risk but also ensures that even if one crop is affected by adverse weather, others may thrive.

Drought Resistance: Certain varieties of horticultural crops, such as succulents and xerophytes, are naturally adapted to arid conditions. These plants require less water, making them more resilient in regions prone to drought.

Adaptability to Protected Cultivation: Greenhouse and polyhouse cultivation of horticultural crops provide a controlled environment that protects plants from extreme weather events. This allows for year-round cultivation and reduces the impact of adverse weather.

Varieties Resilient to Changing Climate

Drought-Tolerant Varieties: Crops like drought-tolerant varieties of tomatoes, millets, and pulses have been developed to withstand water stress, making them suitable for regions with irregular rainfall patterns.

Heat-Tolerant Varieties: Certain varieties of fruits like pomegranates and grapes, bred for heat tolerance, can thrive in regions with high temperatures, where other crops might struggle.

Salinity-Tolerant Varieties: In coastal areas where salinity in soil and water is a concern, horticultural crops like specific varieties of mangoes and coconuts have been developed to tolerate higher salt concentrations.

Frost-Resistant Varieties: Varieties of fruits like apples and strawberries that are less susceptible to frost damage have been bred for regions with colder climates.

Pest and Disease-Resistant Varieties: Breeding programs have focused on developing horticultural varieties that are naturally resistant to common pests and diseases, reducing the need for chemical interventions.

Early-Maturing Varieties: Early-maturing varieties of horticultural crops allow for a quicker turnaround between planting and harvest, providing a buffer against unpredictable weather events.

By prioritizing the cultivation of climate-resilient horticultural varieties, farmers can enhance their capacity to adapt to changing climatic conditions. Additionally, research and development efforts in breeding programs and agronomic practices should continue to identify and promote varieties that demonstrate increased resilience to the challenges posed by climate change.

10. Value Addition and Processing

Opportunities for Value Addition

Value addition through processing activities is a crucial dimension of the horticulture sector, presenting opportunities to enhance the economic returns of horticultural produce. Various processing methods, including canning, juicing, and preservation, enable the extension of shelf life and the creation of innovative products that cater to diverse consumer preferences (Sharangi & Datta 2015).

Canning and Preservation: Canning involves sealing horticultural products in airtight containers after heat treatment, preserving their freshness for extended periods. This technique is particularly valuable for fruits and vegetables that are susceptible to spoilage. Similarly, techniques like pickling, drying, and freezing allow for the preservation of produce, reducing post-harvest losses.

Juicing and Beverage Production: The extraction of juices from fruits and vegetables offers a popular and versatile avenue for value addition. Additionally, the production of beverages like fruit juices, smoothies, and concentrates provides consumers with convenient and nutritious options.

Processed Foods and Snacks: The transformation of horticultural produce into processed foods and snacks, such as jams, jellies, sauces, and dehydrated fruits, adds value and caters to changing consumer lifestyles and dietary preferences.

Herbal and Medicinal Products: The processing of medicinal herbs and plants from the horticulture sector into products like herbal teas, essential oils, and natural remedies taps into the growing market for wellness and alternative medicine.

Economic Benefits of Processing Horticultural Products

Increased Market Value: Processed horticultural products command higher prices in the market compared to raw produce. This value addition translates to increased income for farmers and other stakeholders in the supply chain.

Reduction of Post-Harvest Losses: Processing activities help extend the shelf life of horticultural produce, reducing the likelihood of spoilage and minimizing post-harvest losses. This is especially significant in regions where transportation and storage infrastructure may be less developed (Ridolfi *et al.* 2018).

Generation of Employment Opportunities: The processing sector provides employment opportunities in areas such as food processing units, packaging, transportation, and marketing. This contributes to rural development and economic diversification (Singh 2013).

Market Diversification: Processed horticultural products open up new markets and distribution channels, both domestically and internationally. They cater to a wider consumer base and enable access to premium segments.

Promotion of Sustainable Practices: Processing can also contribute to sustainability efforts by utilizing imperfect or surplus produce that may not meet the aesthetic standards of fresh markets. This reduces food wastage and promotes a more efficient use of resources.

By encouraging and supporting value addition through processing activities, India's horticulture sector can unlock significant economic potential. Government policies, infrastructure development, research, and capacity-building initiatives are integral in enabling the growth and competitiveness of the processed horticultural products industry.

11. Government Policies and Support

Existing Policies and Initiatives

India has implemented several policies and initiatives aimed at promoting horticulture and realizing its economic potential. Some notable ones include:

National Horticulture Mission (NHM): Launched in 2005, NHM focuses on holistic development of horticulture by providing assistance for production, post-harvest management, and marketing. It encompasses various sub-schemes targeting fruits, vegetables, spices, and other horticultural crops (Singh *et al.* 2022).

Mission for Integrated Development of Horticulture (MIDH): This mission, initiated in 2014, amalgamated several horticultural schemes, including NHM, to provide comprehensive support for horticulture development. It encompasses areas like research and development, capacity-building, and market access (Jaiswal *et al.* 2023).

Pradhan Mantri Kisan Sampada Yojana (PMKSY): Introduced in 2016, PMKSY aims to create modern infrastructure for post-harvest management and processing of horticultural produce. It focuses on reducing post-harvest losses and enhancing value addition.

Rashtriya Krishi Vikas Yojana (RKVY): Although not exclusively for horticulture, RKVY supports various agricultural activities, including horticulture, through state-specific projects aimed at increasing agricultural productivity.

Effectiveness and Potential Improvements

While these policies have made commendable strides in promoting horticulture, there are areas where improvements can be considered:

Access to Credit and Insurance: Ensuring that horticulturists, especially smallholder farmers, have easy access to credit and insurance schemes is crucial. This provides a safety net in case of crop failures or adverse weather conditions.

Market Linkages and Infrastructure: Strengthening market linkages and creating modern infrastructure, including cold storage facilities and agro-processing units, can significantly reduce post-harvest losses and enhance the value chain.

Research and Development: Continued investment in research and development is essential for the introduction of new and climate-resilient varieties. This can help farmers adapt to changing climatic conditions and consumer preferences.

Training and Capacity Building: Training programs that equip farmers with modern techniques, sustainable practices, and entrepreneurial skills are vital. This empowers them to make informed decisions and tap into emerging opportunities.

Promotion of Organic and Sustainable Practices: Encouraging organic farming and sustainable horticultural practices not only benefits the environment but also caters to the growing demand for organic produce in both domestic and international markets.

Market Intelligence and Price Stabilization: Implementing measures to provide farmers with timely market intelligence and ensuring fair prices for their produce can help stabilize incomes and reduce market risks.

Promotion of Agro-Processing Industries: Supporting the establishment of agro-processing units, especially in rural areas, can add value to horticultural produce and generate employment opportunities.

By evaluating and refining existing policies, India can further leverage the economic potential of its horticulture sector. Additionally, a collaborative approach involving government, agricultural experts, research institutions, and farmers themselves is crucial in ensuring the effectiveness and impact of these policies.

12. Challenges and Future Prospects

Challenges Faced by the Horticulture Sector

Despite its significant contributions, the horticulture sector in India encounters several challenges that require strategic interventions:

Limited Access to Credit and Insurance: Many smallholder farmers lack access to formal credit and insurance facilities, making them vulnerable to financial risks associated with crop failures or adverse weather conditions (Singh & Agrawal 2020).

Post-Harvest Losses: Inadequate post-harvest infrastructure and practices contribute to substantial losses, particularly in perishable horticultural produce. This reduces overall profitability and income for farmers (Gajanana *et al.* 2011).

Market Instability and Price Fluctuations: Farmers often face market uncertainties, including volatile prices and inconsistent demand. This can lead to income instability and affect long-term planning.

Lack of Market Information: Access to timely and accurate market information is crucial for farmers to make informed decisions regarding crop selection, timing of harvest, and pricing strategies.

Climate Change and Environmental Factors: Erratic weather patterns, changing climate conditions, and increasing incidences of extreme weather events pose a significant threat to horticulture. Farmers need strategies to adapt to these challenges (Malhotra 2017).

Pest and Disease Management: Effective pest and disease management is critical in horticulture. Inadequate knowledge and access to safe and sustainable pest control methods can lead to crop losses (Matthiessen & Kirkegaard 2006).

Infrastructure Gaps: Insufficient infrastructure for storage, transportation, and processing hampers the sector's growth potential, particularly in remote and rural areas.

Strategies for Sustainable Growth

To address these challenges and ensure the sustainable growth of the horticulture sector, several strategies can be considered:

Promoting Farmer Producer Organizations (FPOs): FPOs can enhance the collective bargaining power of smallholder farmers, providing access to credit, technology, and markets.

Investing in Post-Harvest Infrastructure: Building and upgrading storage facilities, cold chains, and processing units will help reduce post-harvest losses and improve the overall value chain.

Enhancing Market Intelligence: Implementing systems for real-time market information and price forecasting empowers farmers to make informed decisions regarding planting and marketing strategies.

Integrating Technology and Innovation: Encouraging the adoption of modern technologies, such as precision farming, drip irrigation, and greenhouse cultivation, can significantly enhance productivity and resilience.

Promoting Organic and Sustainable Practices: Supporting farmers in transitioning to organic and sustainable farming practices not only benefits the environment but also taps into the growing demand for organic produce.

Capacity Building and Skill Development: Providing training and capacity-building programs for farmers, particularly in advanced cultivation techniques and entrepreneurship, empowers them to make strategic decisions.

Research and Development: Continued investment in research and development is essential for the introduction of new and climate-resilient varieties and the development of pest and disease management strategies.

Policy Reforms and Support: Regular review and reform of policies related to credit access, insurance, and market linkages can ensure that they remain relevant and effective in supporting the horticulture sector.

By implementing these strategies in a coordinated and sustained manner, India can overcome the challenges faced by the horticulture sector and foster sustainable growth. This, in turn, will contribute significantly to the nation's agricultural and economic prosperity, while ensuring the well-being of millions of stakeholders in the horticulture value chain.

Acknowledgment

We are thankful to the authors whomever we have mentioned in this chapter as their precise content made the whole process easier to collect specific information about the Economic Importance of Horticultural Crops in India's GDP.

References

Anonymous, (2020-21). National Horticulture Board, Indian Horticulture Database. Ministry of Agriculture & Farmers Welfare, Government of India. www.agricoop.nic.in

Chadha, K.L. (2019). Hand Book of Horticulture. Second Edition, Volume 1, Indian Council of Agricultural Research, New Delhi.

Gajanana, T. M., Murthy, D. S., & Sudha, M. (2011). Post-harvest losses in fruits and vegetables in South India–A review of concepts and quantification of losses. Indian Food Packer, 65(6), 178-187.

Hazra, C. R. (2001). Crop diversification in India. Crop diversification in the Asia-Pacific Region.(Minas K. Papademetriou and Frank J. DentEds.). Food and Agriculture Organization ofthe United Nations. Regional Office for Asia and the Pacific, Bangkok, Thailand, 32-50.

Jaiswal, D., Mathur, R., & Sonawane, S. (2023). Impact of MIDH on the Horticulture Sector In India. The Online Journal of Distance Education and e-Learning, 11(2).

Kaul, G. L. (1997). Horticulture in India-production, marketing and processing. Indian Journal of Agricultural Economics, 52(3), 561-573.

Keatinge, J. D. H., Yang, R. Y., Hughes, J. D. A., Easdown, W. J., & Holmer, R. (2011). The importance of vegetables in ensuring both food and nutritional security in the attainment of the Millennium Development Goals. Food Security, 3(4), 491-501.

Kondal, K. (2014). Trends in area and production of horticulture sector in India. ANVESAK Journal, 44(2), 1-11.

Malhotra, S. K. (2017). Horticultural crops and climate change: A review. Indian Journal of Agricultural Sciences, 87(1), 12-22.

Matthiessen, J. N., & Kirkegaard, J. A. (2006). Biofumigation and enhanced biodegradation: opportunity and challenge in soilborne pest and disease management. Critical Reviews in Plant Sciences, 25(3), 235-265.

Mitra, A., & Panda, S. (2020). Horticulture and economic growth in India: An econometric analysis. Journal of Applied Horticulture, 22(3).

Nabi, S. U., Raja, W. H., Dar, M. S., Kirmani, S. N., & Magray, M. M. (2017). New generation fungicides in disease management of horticultural crops. Indian Horticulture Journal, 7(1), 01-07.

Nath, P. (2011). Role of Horticulture in Livelihood Security. Horticulture and Livelihood Security, PNASF, Bangalore and Scientific Publishers, Jodhpur, India, 1-11.

Ravichandra, N.G., & Ravichandra, N. G. (2014). Horticulture and its role in the national economies. Horticultural Nematology, 1-3.

Ridolfi, C., Hoffmann, V., & Baral, S. (2018). Post-harvest losses in fruits and vegetables: The Kenyan context. Intl Food Policy Res Inst.

Sharangi, A. B., & Datta, S. (Eds.). (2015). Value addition of horticultural crops: recent trends and future directions. New Delhi: Springer India.

Singh, K. M., Ahmad, N., Pandey, V. L., & Sinha, D. K. (2022). Impact of national horticulture mission on Vegetable and Fruit sectors of India. Indian Journal of Economics and Development, 18(1), 66-75.

Singh, P., & Agrawal, G. (2020). Development, present status and performance analysis of agriculture insurance schemes in India: Review of evidence. International Journal of Social Economics, 47(4), 461-481.

Singh, R. P. (2013). Horticultural (high value agricultural) crop diversification in eastern India: II–employment opportunities and income generation strategies. International Journal of Innovative Horticulture, 2(1), 28-43.

Ugochukwu, A. I., & Phillips, P. W. (2018). Technology adoption by agricultural producers: A review of the literature. From agriscience to agribusiness: theories, policies and practices in technology transfer and commercialization, 361-377.

2

Diseases of Mango and Their Management

Shubham Yadav

Department of Fruit Science, Surdar Vallabhbhai Patel University of Agriculture and Technology, Meerut, Uttar Pradesh

Abstract

Mango (Mangifera indica) is one of the most economically important tropical fruit crops worldwide. However, mango trees are susceptible to various diseases caused by fungi, bacteria, and viruses, which can lead to significant yield losses and reduced fruit quality. Effective disease management strategies are essential to ensure a healthy mango orchard and maximize fruit production. This review provides an overview of common mango diseases, their symptoms, and management options. The diseases covered include anthracnose, powdery mildew, bacterial black spot, downy mildew, fruit rots, and viral infections. Management approaches encompass cultural practices, sanitation, chemical control, and the use of disease-resistant cultivars. Understanding and implementing these disease management strategies are crucial for sustainable mango production and ensuring a consistent supply of high-quality mangoes to global markets.

Keywords: *Mango, disease management, anthracnose, powdery mildew*

Mango (*Mangifera indica*) is a tropical fruit tree that is well-known and cherished for its delicious, sweet, and juicy fruit. It is one of the most widely grown and economically important fruit crops in the world, primarily cultivated in tropical and subtropical regions. Mangoes are not only enjoyed for their taste but also for their nutritional value, as they are rich in vitamins, minerals, and antioxidants.

However, like all agricultural crops, mangoes are susceptible to a range of diseases that can impact their growth, yield, and fruit quality. These diseases can be caused by various pathogens, including fungi, bacteria, viruses, and nematodes.

1. Powdery mildew (caused by *Oidium mangiferae*) is a fungal disease that affects mango trees, leading to significant losses in fruit production if left unmanaged.

DOI: 10.1201/9781042014460-2

Symptoms

Infection Site: Powdery mildew infections typically start from the inflorescence (the flowering part of the tree) and then spread downwards, covering the floral axis, tender leaves, and stems.

Leaf Damage: Infected leaves exhibit several symptoms, including twisting, curling, and distortion. These leaves may eventually fall off the tree.

Floral Damage: The disease can severely damage floral parts, causing them to drop off prematurely. This can lead to poor fruit set as the flowers are affected before fruit development.

Fruit Impact: If fruits manage to set despite the infection, they often do not grow to their full size and may drop before reaching pea-sized. This can result in a significant reduction in fruit yield.

Yield Loss: The overall loss due to powdery mildew can be substantial, with estimates ranging from 70% to 80% if the disease is not adequately managed.

Management

Sulfur Sprays: Two preventive sprays with wettable sulfur at a concentration of 0.3% are recommended. The first spray should be applied before the flowers open, and the second after fruit set.

Dusting with Sulfur: Dusting the mango trees with fine sulfur powder two or three times during the growing season can help suppress the disease.

Chemical Fungicides: The application of fungicides like Karathane at a concentration of 0.1% can be done before flowering and after fruit set to control the disease.

Resistant Varieties: Certain mango varieties have shown resistance to powdery mildew, reducing the need for intensive management. Some examples of resistant mango varieties include:Neclum, Zardalu, Banglora, Torapari-khurd and Janardhanpasand. These varieties are less susceptible to powdery mildew, and planting them can be an effective way to mitigate the disease's impact on mango production.

2. Anthracnose

Caused by –*Colletotrichum gloeosporioides*

Anthracnose is a common and destructive fungal disease of mango trees caused by the pathogen *Colletotrichum gloeosporioides.* This disease affects various parts of the mango tree, including leaves, flowers, twigs, and fruit, leading to significant economic losses if not managed effectively.

Symptoms

Leaf Symptoms: Anthracnose on mango leaves appears as circular, light yellow-brown spots initially, which later enlarge with dark brown to black margins. Infected leaves may also exhibit curling and distortion.

Blossom Blight: Infected inflorescence stalks and flowers show the development of minute spots, which increase in size. This can lead to blossom blight, where affected flowers may drop prematurely, reducing fruit set.

Twig Blight: Anthracnose can cause wilting and blight of twigs, leading to dieback in severe cases.

Fruit Symptoms: The most significant economic impact of anthracnose is on fruit. Sunken spots appear on fruits near the stem end, initially light brown and later turning black. The fruit pulp beneath these spots becomes hard, and the fruit can eventually crack and decay during ripening.

Management

Orchard Management: Avoid overcrowding in orchards by providing adequate spacing between mango trees. This can help improve air circulation and reduce humidity, which can favor disease development.

Fungicide Sprays: Regularly spray fungicides to protect mango trees from anthracnose. Commonly used fungicides include carbendazim, thiophanate methyl (at 0.1% concentration), or chlorothalonil (at 0.2% concentration). These sprays should be applied at 14-day intervals from fruit set until harvest.

Post-Harvest Treatment: To prevent the spread of anthracnose during storage and transportation, mangoes can be treated before storage. This can be done by dipping the fruits in hot water (50-55°C) for 15 minutes or immersing them in a solution of fungicides such as benomyl (at 500 ppm) or thiobendazole (at 1000 ppm). Alternatively, mangoes can be exposed to ammonia and sulfur dioxide gases to control fungal pathogens.

3. Downy Mildew

Downy mildew, caused by the oomycete pathogen *Pseudoperonospora cubensis*, is a significant disease affecting a variety of crops, including cucumbers, squash, and other cucurbits.

Symptoms

Leaf Symptoms: Downy mildew typically starts with yellow-green lesions on the upper side of leaves. These lesions often appear angular and limited by leaf veins. As the disease progresses, the lesions may turn brown and necrotic, leading to significant damage.

Leaf Curling: Infected leaves may curl upward, giving them a distorted appearance.

Reduced Fruit Set: Downy mildew can impact fruit development and set. Infected plants may produce fewer fruits, and the quality of the fruit can be compromised.

Management

Pruning and Sanitation: Remove and destroy infected plant parts to reduce the disease's spread. Prune affected leaves and vines, especially in the early stages of the infection.

Crop Rotation: Avoid planting cucurbits in the same location in consecutive growing seasons. Crop rotation can help break the disease cycle.

Spacing: Properly space plants to improve air circulation and reduce humidity, which can create favorable conditions for downy mildew development.

Chemical Management

Mefenoxam: Sold under various trade names, such as Ridomil Gold, this fungicide is effective against downy mildew when applied preventatively or at the first signs of the disease.

Copper-based Fungicides: Products like copper hydroxide and copper sulfate can help manage downy mildew. Copper-based fungicides work best when applied before the disease becomes severe.

Mancozeb: Fungicides containing mancozeb, such as Dithane and Manzate, are often used to control downy mildew.

Chlorothalonil: Fungicides like Bravo and Echo can also be effective against downy mildew in cucurbits.

4. Bacterial Black Spot

Bacterial Black Spot, caused by the bacterium *Xanthomonas axonopodis*, is a common disease that affects various fruit trees, including citrus trees like oranges and grapefruits.

Symptoms

Leaf and Stem Symptoms: The primary symptoms of bacterial black spot-on citrus trees include dark, raised lesions on leaves and stems. These lesions often have yellow halos around them, making them distinctive. The lesions can vary in size and shape but typically have a circular or irregular appearance.

Fruit Symptoms: Bacterial black spot can also affect citrus fruit, causing dark, raised lesions on the fruit's surface. These lesions can lead to reduced fruit quality and may result in fruit drop.

Defoliation: Severe infections can cause defoliation, where the tree sheds its leaves prematurely, leading to reduced photosynthesis and overall tree health.

Management

Sanitation: Remove and destroy infected plant material, including leaves, stems, and fruit, to reduce the disease's spread.

Pruning: Prune affected branches to improve air circulation and reduce the humidity that favors disease development.

Copper Sprays: Copper-based bactericides are commonly used to control bacterial black spot. Examples include Bordeaux mixture (a combination of copper sulfate and hydrated lime) and various copper-based fungicides and bactericides available under different trade names.

Chemical Management

Bordeaux Mixture: This traditional copper-based fungicide contains copper sulfate and hydrated lime. It has been used for many years to manage bacterial and fungal diseases in citrus and other crops.

Copper Hydroxide: Copper hydroxide-based products, such as Kocide and Champ DP, are effective against bacterial black spot when applied according to label instructions.

Copper Oxychloride: Products like Copper Count-N can also be used for controlling bacterial black spot in citrus.

5. Malformation (Mango Witches' Broom)

Malformation, also known as Mango Witches' Broom, is a complex physiological disorder that affects mango trees (*Mangifera indica*). This disorder is characterized by the development of abnormal, dense vegetative growth, often referred to as "witches' broom," on affected shoots. While the exact cause of malformation is not fully understood, it is believed to involve a combination of factors, including fungi, mites, and environmental conditions.

Symptoms

Abnormal Growth: The most striking symptom of malformation is the development of dense, bushy growth at the tips of affected branches. These abnormal growths often resemble broom-like structures, hence the name "witches' broom."

Reduced Flowering and Fruit Production: Malformation can lead to reduced flowering and fruit set on affected branches, resulting in lower fruit production and quality in those areas.

Management

Pruning: Pruning is one of the primary methods for managing malformation. Affected branches should be pruned and removed from the tree to prevent the disease's spread to healthy parts of the tree. Pruning should be done during the dormant season.

Sanitation: Proper sanitation practices, such as removing and disposing of infected plant material away from the orchard, can help reduce the disease's inoculum.

Environmental Control: Maintaining optimal growing conditions, including adequate irrigation, nutrition, and pest control, can help reduce stress on the mango tree and minimize susceptibility to malformation.

Chemical Management

Paclobutrazol: This plant growth regulator has been used experimentally to manage malformation. It can reduce the vegetative growth of affected branches and promote more balanced growth.

6. Fruit Rots

Fruit rots are a significant post-harvest problem that affects mangoes and various other fruits. Mangoes can be susceptible to various fruit rot pathogens, including Colletotrichum, Diplodia, and Rhizopus species. Effective management involves a combination of pre-harvest and post-harvest practices, including chemical control.

Symptoms

Surface Lesions: Infected mangoes develop soft, watery, and sunken lesions on the skin, often with fungal growth visible on the surface.

Internal Decay: The rot can extend into the fruit's flesh, causing a brown or black, mushy, and foul-smelling interior.

Fungal Growth: Fungal spores and mycelium may be visible on the fruit's surface or within the lesions.

Management

Harvesting: Mangoes should be harvested at the appropriate maturity stage, avoiding overripe or damaged fruit.

Handling: Gentle handling during harvest and subsequent sorting and packing can help prevent physical damage that can facilitate fungal infection.

Sanitation: Proper sanitation in the orchard and packing houses, including cleaning and disinfecting equipment and storage containers, can reduce the risk of fungal spore contamination.

Storage: Mangoes should be stored at the recommended temperature and humidity levels to minimize fungal growth. Cold storage can also slow down the development of fruit rots.

Chemical Management

Thiabendazole: Thiabendazole is an effective post-harvest fungicide used to control fruit rots in mangoes. It is often applied as a dip or spray treatment.

Imazalil: Imazalil is another fungicide used for post-harvest treatment of mangoes to control fungal pathogens that cause fruit rots.

Hot Water Treatment: Hot water treatment at specific temperatures and durations can also be used to control fruit rots. It involves immersing mangoes in hot water to kill fungal spores on the fruit surface.

Wax Coatings: The application of wax coatings can create a physical barrier that reduces the risk of fungal infection during storage.

7. Viruses

Viral infections in mango trees can have significant negative impacts on tree health and fruit quality. Two common mango viruses are Mango Malformation Virus and Mango Decline Virus. These viruses are typically spread by insect vectors and cannot be cured once a tree is infected. Prevention through the use of virus-free planting material and the management of insect vectors is essential.

Mango Malformation Virus

Symptoms: Mango Malformation Virus can cause various symptoms, including the development of abnormal growth, or "witches' broom," on branches. Leaves may exhibit mottling or yellowing, and affected trees often produce fewer and lower-quality fruit.

Management: There is no chemical cure for viral infections in mango trees. Management primarily involves preventing the introduction of the virus through the use of virus-free planting material and controlling insect vectors.

Mango Decline Virus

Symptoms: Mango Decline Virus can lead to stunted growth, leaf mottling, and a decline in overall tree health. Infected trees may also produce smaller and less flavorful fruit.

Management: Similar to Mango Malformation Virus, there is no chemical cure for Mango Decline Virus. Prevention through the use of virus-free planting material is crucial. Additionally, controlling insect vectors, such as aphids, can help reduce the spread of the virus.

Preventive Measures

Use Virus-Free Planting Material: Ensure that the mango saplings used for planting are free from viral infections. Selecting disease-free material is the most effective way to prevent virus introduction.

Regular Monitoring: Regularly inspect mango trees for symptoms of viral infections, such as leaf mottling, stunted growth, and abnormal growth patterns. Infected trees should be removed to prevent further spread.

Insect Vector Management: Manage and control insect vectors, such as aphids, that can transmit viruses. Insecticides can be used to reduce vector populations.

Orchard Sanitation: Maintain proper orchard sanitation practices to reduce the chances of virus transmission. Remove and destroy infected plant material.

8. Environmental Stress

Environmental stress factors can have a significant impact on the health and susceptibility of mango trees to diseases. Mango trees are sensitive to various environmental stressors, and exposure to adverse conditions can weaken their immune system, making them more vulnerable to diseases. Here's an elaboration on how environmental stressors affect mango trees and examples of diseases that can result from or be exacerbated by these stressors, along with references:

Effects of Environmental Stress

Waterlogged Soils and Poor Drainage: Mango trees require well-drained soil to thrive. Waterlogged soils and poor drainage can lead to root rot diseases such as

Phytophthora root rot. These conditions limit root oxygenation and can weaken the tree's overall health.

Extreme Temperatures: Mango trees are sensitive to both extreme cold and extreme heat. Frost or cold temperatures can damage leaves and flowers, while extreme heat can stress the tree, reducing its ability to resist diseases.

Drought Stress: Prolonged drought conditions can weaken mango trees, making them more susceptible to various diseases. Drought stress can lead to reduced vigor and compromised immune responses.

Phytophthora Root Rot: Waterlogged soils and poor drainage create favorable conditions for the development of Phytophthora root rot. This disease affects the tree's roots, leading to poor nutrient and water uptake and, ultimately, reduced overall tree health.

Anthracnose: Extended periods of rain and high humidity resulting from environmental stress factors can create ideal conditions for the development of anthracnose, a fungal disease that affects mango leaves, flowers, and fruit.

Dieback and Decline Diseases: Environmental stressors, including extreme temperatures and water stress, can weaken mango trees and make them more susceptible to dieback and decline diseases, where the tree exhibits a gradual loss of vigor and canopy dieback.

References

Control of Postharvest Diseases of Mango. Horticultural Reviews 26.1 (2000): 253-292.

Dhingra, O.D., and Parvin Kumar. "Diseases of Fruits and Vegetables: Volume II: Diagnosis and Management." Springer, 2004.

Diseases of Banana (Musa spp.). Food and Agriculture Organization (FAO). http://www.fao.org/3/X0670E/x0670e07.htm

Dissanayake, Muditha L., et al."Anthracnose of mango: A review of control methods."Crop Protection 41 (2012): 9-15.

Dutcher, James D. "Control of Bacterial Spot of Citrus with Copper Bactericides."Plant Disease 68.1 (1984): 41-44.

El-Shereif, A. M., et al."Studies on some mango malformation control methods."Zagazig Journal of Agricultural Research 35.6 (2008): 1349-1361.

Etiology and management of mango malformation: a review. Plant Pathology Journal 32.2 (2016): 91-101.

Gottwald, Tim R., et al."Disease progression of Asiatic citrus canker following simulated overhead irrigation and natural rainfall."Plant Disease 88.6 (2004): 643-649.

Jones, David R. "Diseases of Banana, Abacá and Enset." In: The Diseases of The South Pacific Agriculture. Edited by Waterhouse GM. South Pacific Commission. 1982.

Kumar, Satish, et al."Chemical control of post-harvest diseases of mango (Mangifera indica L.)."Journal of Food Science and Technology 48.5 (2011): 540-547.

Lingaraj, S., et al."Anthracnose (Colletotrichumgloeosporioides) of Mango: A Comprehensive Review."International Journal of Current Microbiology and Applied Sciences 7.4 (2018): 4497-4505.

Malformation of Mango: A Review. International Journal of Current Microbiology and Applied Sciences 6.7 (2017): 3633-3642.

Management of Mango Diseases and Disorders. Food and Agriculture Organization (FAO). http://www.fao.org/3/Y4757E/y4757e05.htm

Management of mango malformation. Indian Journal of Horticulture 63.1 (2006): 94-96.

Mango Anthracnose: Economic Importance and Current Knowledge. The Plant Pathology Journal 33.2 (2017): 115-129.

Mango Decline Disease. Plantwise Knowledge Bank. https://www.plantwise.org/knowledgebank/factsheet/190561002219

Mango Diseases. Indian Institute of Horticultural Research. https://www.iihr.res.in/mango-diseases

Mango Diseases and Their Management. Indian Institute of Horticultural Research. https://www.iihr.res.in/mango-diseases

Mango Malformation Virus. Plantwise Knowledge Bank. https://www.plantwise.org/knowledgebank/factsheet/190561012619

Plant Stress Physiology. Edited by Heribert Hirt and Karin Grossmann. Academic Press, 2010.

Plant Viruses: Diversity, Interaction, and Management. Edited by George Loebenstein and William O. Dawson. Academic Press, 2010.

Ploetz, Randy C. "Management of Banana and Plantain Diseases." In: Diseases of Banana, Abaca, and Enset. Edited by Jones DR. CABI Publishing. 2000.

Postharvest Management of Mango Fruit Rots: An Overview. International Journal of Current Microbiology and Applied Sciences 7.5 (2018): 1230-1241.

Raman, L. "Powdery Mildew of Mango."Crop Protection Compendium. https://www.cabi.org/cpc/datasheet/38017

Root Rot and Other Disease Problems of Mango in Florida. University of Florida IFAS Extension. https://edis.ifas.ufl.edu/publication/pp124

Saha, Sudip, et al."Management of postharvest diseases of fruits and vegetables by chemical control: a review."Food Research International 44.5 (2011): 1306-1313.

Sharma, D. R., et al."Efficacy of fungicides for the management of powdery mildew (Oidiummangiferae) of mango."Plant Disease Research 24.1 (2009): 103-104.

Timmer, Larry W., et al."Bacterial spot of citrus." University of Florida, IFAS Extension. https://edis.ifas.ufl.edu/publication/pp164

Viral Diseases of Mango. Plant Pathology Circular No. 438. Florida Department of Agriculture and Consumer Services. https://www.fdacs.gov/content/download/23742/file/Plant-Pathology-Circular-438.pdf

3

Diseases of Banana and Their Management

Harshita Yadav, Arvind Yadav, Shashank Shekhar and Shyam Lal

Department of Plant Pathology, Acharya Narendra Deva University of Agriculture and Technology, Kumarganj, Ayodhya-224229

Abstract

Banana (Musa spp.) is the most important fruit crop grown commercially in many nations around the world for use as a dessert and as a staple food in some parts of the world. Bananas can be grown in soil ranging from sandy loam to clay. It is a good source of carbohydrates, vitamins A, B6, C, minerals like potassium, calcium, and phosphorus, and fiber. Diseases caused by fungi, bacteria, and viruses are the major limiting factor in crop quality production, and nearly all commercial banana cultivars are highly susceptible to certain lethal diseases. In this chapter, we discussed several major diseases, including Panama disease, Sigatoka leaf spot disease, Moko disease, Bunchy top of banana disease, and Anthracnose disease, as well as their causal agent, symptomatology, pathogen characteristics, and treatment options.

Keywords: *Banana, disease, pathogen, management*

Banana

Botanical Name	*Musa parasidica*
Family	Musaceae
Chromosome Number(2n)	22

Introduction

The banana (*Musa parasidiaca*) belongs to the Monocotyledons class, the Zingiberales order, and the Musaceae family. It is indigenous to Southeast Asia. They thrive in tropical and subtropical climates and are widely grown in countries such as India, China, the Philippines, Ecuador, Brazil, Uganda, Colombia, and Indonesia. A banana's average length is 7 to 9 inches, and its diameter is 2 to 3 inches. Bananas are frequently planted during the rainy or monsoon season,

DOI: 10.1201/9781042014460-3

when the soil is moist and the young plants can absorb enough water. In tropical and subtropical climates where they are commonly grown, the planting season typically runs from April to June. Planting may take place in September or October in areas with a dry season, following the rainy season. Temperatures between 20°C and 30°C are ideal for growing bananas, with a preferred temperature range of 25°C to 28°C. High humidity levels are also necessary for healthy growth and fruit production. To support growth and fruit production, bananas require at least 1000 mm of rainfall or irrigation per year. Bananas thrive in well-drained, fertile soils rich in organic matter. They require adequate aeration and a pH range of 5.5 to 7.0. The crop is propagated vegetatively by "suckers" (shoots that develop from the mother plant base). Suckers are carefully removed from the mother plant and planted in a new location where they will develop into their own plant. Another method of propagation is through "rhizome" or rootstock." "Corm" cuttings are pieces of the banana plant's underground stem that can be used to start a new plant.

Common Diseases

1. Panama Wilt

Causal Organism: *Fusarium oxysporum f. spcubense*

Panama wilt is a fungal disease that enters the plant through the roots and wounds caused by nematodes. It is especially dangerous in poorly drained soil. Infected suckers are to blame for disease spread.

Symptoms

Wilting and yellowing of the leaves, extending from the margin to the midrib, is the most noticeable symptom of this disease. Splitting of the pseudo stem and brown or red streaks on the vascular vessel walls are characteristic symptoms.The fungus spreads through use of infected rhizomes or suckers, farm implements or vehicles, irrigation water. Acid alluvial soil is most conducive to disease.

Chlorosis of leaves due toFusarium wilt

Splitting of pseudostem

Wilting Symptoms

Pathogen

Mycelium is septate, hyaline and branched. Micro, macro conidia and chlamydospores are produced by fungus. Micro conidia - Single celled or rarely one septate, hyaline elliptical or oval. Macro conidia - Sickle shaped hyaline, 3-5 septate and tapered at both ends. Chlamydospores are thick walled, spherical to oval, hyaline to slightly yellowish in colour.

Disease Cycle

Spores of *Fusarium* germinate and grow towards banana plants & roots due to chemical compounds. Infection occurs on secondary and finer roots, leading to primary roots and rhizome. The xylem network consists of vessels with perforated end walls, blocking spore movement. Plants can prevent infection by producing gels and tyloses, but multiple infections can result in a total invasion. The virulence of tropical race 4 on 'Cavendish' bananas suggests that resistance mechanisms are less effective. (Daly and Waldeck, 2006).

Mode of spread and survival

The pathogen is soil borne. It can persist in soil for prolonged periods of time as chlamydospores. The primary spread of the disease is through infected rhizomes and secondary spread is through irrigation water. Continuous cultivation results in accumulation of inoculum.

Management

- Affected plants should be uprooted and burnt
- Avoid growing of Rasthali, Monthan, Red banana and Virupakshi. Grow resistant cultivar Poovan

- Remove and destroy infested plant material after harvest
- Apply*Bacillus subtilis* @ 2.5kg/ha bactericide along with farmyard manure and neem cake
- Soil drenching with Carbendazim 0.1 per cent solution around the pseudo stem at bimonthly intervals beginning five months after planting

2. Sigatoka disease

Causal organism: *Mycospharella musicola*

Sigatoka is the name of the valley where the disease first attracted attention. It is a fungal disease. During the rainy season, it causes huge economic damage and spreads quickly. Attacks primarily on leaves.

Symptoms

Small pale yellow or brownish green narrow streaks appear on the leaves. They grow to be linear, oblong, brown to black dots with a dark brown brand and a yellow halo. Photosynthetic area is reduced and in severe cases fruit does not mature at all. Rapid drying and defoliation of leaves are the characteristic feature of this disease. Poor fruits are produced by early-diseased plants.

Yellowish and green streaks

Elliptical brown streaks

Pathogen

Conidia are elongated, narrow and multi septate and measure 20-80 x 2–6-micron meter. Perithecia are dark brown to black, and asci are oblong, clavate measuring 28.8-36.8 x 8.0-10.8-micron meter. Ascospores are one septate, hyaline, obtuse with upper cell slightly broader.

Disease Cycle

The disease cycle begins when fungal spores (conidia) formed on infected banana leaves are carried by wind and rain to healthy banana plants. The conidia can also be transported by equipment, clothing, or other means.Conidia germinate and pierce the cuticle of a susceptible banana leaf after landing on it. The fungus subsequently colonises the leaf, forming lesions. As the fungus spreads, it produces black, necrotic (dead tissue) lesions on the banana leaves. These lesions are often encircled by a yellow halo, and they can coalesce, leading to extensive damage. Within the lesions, the fungi produce new conidia. These conidia are easily spread to other parts of the plant and to neighbouring banana plants through rain, splashes, wind, or other modes of transmission. Secondary infections caused by the discharged conidia might affect other leaves and even nearby banana plants. This process can result in the rapid spread of the disease. Sigatoka leaf spots can cause severe defoliation of the banana plant if left untreated over time. The loss of leaves can impair the plant's ability to produce energy through photosynthesis affecting fruit production. The fungus may enter a dormant phase during harsh conditions or when the host plant is less appropriate for infection, allowing it to live in the environment for a prolonged period of time. (Viljoin et. al., 2017).

Management

- Removal and destruction of affected leaves
- Keep the banana field as weed free and remove suckers on time
- Avoid planting in close quarters
- Provide good drainage facility and avoid water logging in the fields which promotes infection
- Spray three times every 10-15 days with Carbendazim 0.1%, Propiconazole 0.1%, or Mancozeb 0.25%, and teepol (sticking agent), as the disease starts with the formation of leaf specks.
- Some treatments for the conditions include spraying Thiophanate methyl 1 g/l, or 1% Bordeaux mixture + 2% linseed oil, or Captan 2 g/l.

3. Moko disease

Causal organism: *Pseudomonas solanacearum / Burkholderia solanacearum*

Symptoms

The leaves turn yellow and grow upwards. The petiole breaks and leaves hang. When cut open, there is discoloration in the vascular region ranging from pale yellow to dark brown. The discolouration is concentrated in the central portion of the corm. Internal rot of fruits with dark brown discoloration. When the pseudostem is cut transversely bacterial ooze can be seen.

Bacterial ooze symptoms

Pathogen

It is gram-negative, rod-shaped bacterium with single polar flagellum.

Infected plant material, vegetative propagative organs, wild host plants, and soil are medium for the survival of bacterium.

Disease Cycle

The moko bacteria are a year-round illness that can be spread via infected plant materials such as roots, stems, bunches, fruit, peel, suckers, or leaf debris. Infection from root to root is also conceivable.

Wherever plants are growing, moko bacteria can be found in the soil and can persist for more than 18 months. The disease can be transmitted via animals, water runoff, hands, equipment, machinery, shoes etc. Furthermore, pruning has the potential to spread the pest. Male buds that are infected might spread the infection to distant flowers. Contaminated fruit shipments can potentially transmit it (Plant Health, Queensland Government, Australia).

Mode of spread and survival

The pathogen is soil borne, it survives in susceptible hosts like banana and Heliconia spp.

Management

- Remove infected plants
- Use clean planting material
- Expose soil to direct sunlight

- Allow for fallowing and crop rotation
- Disinfection of pruning equipment's
- Providing adequate drainage

4. Banana bunchy top

Causal organism: Banana bunchy top virus

The disease is covered by domestic quarantine regulations. Every year losses were estimated to be Rs.4 crores with a 100% loss occurring if infected suckers are planted.

Symptoms

On the leaves, petioles and vein dark broken bands of green tissues are seen. Plants are extremely stunted. Leaves shows marginal chlorosis, curling, upright and become brittle. Top of the tree is densely crowded with leaves. Severely infected banana plants rarely fruit, but if fruit is produced, the banana hands and fingers are likely to be distorted and twisted. It is transmitted by infected suckers and banana aphid (Dongre et al., 2023).

BBTV symptoms on leaf sheath

Bunchy top symptoms

Mode of spread

Secondary spread is through the aphid vector *Pentalonia nigronervosa.*

Disease Cycle

The primary means of BBTV transmission is through banana aphids. These aphids acquire the virus while feeding on infected banana plants. When banana aphids feed on a BBTV-infected plant, they acquire the virus in their bodies. The virus can live in the aphids for their entire existence. This allows transmission of virus to healthy banana plants during subsequent feeding. Infected aphids then land on healthy banana plants and start feeding on the plant's sap. During the feeding process, the virus is transmitted from the aphid's mouthparts into the plant's vascular system. Once the virus is introduced into the plant, it spreads throughout thevascular system of banana plant. As a result, virus infects various parts of the plant, including the leaves, stem, and fruit. Infected banana plants begin to exhibit BBTV infection symptoms. These symptoms may include the characteristic "bunchy top" appearance of the leaves, stunting of the plant, and a reduction in fruit quality and yield. As infected plants grow and produce new shoots or suckers, the virus can also be found in these new plant parts. This permits BBTVto continue to spread within the plantation. Farmers sometimes propagate banana plants using infected plant material, including infected suckers or corms. If these materials are used for planting, they can introduce the virus to new areas.

Management

- Use virus free planting materials
- Remove and rouge infected banana plants
- Maintain clean, weed free field for early detection of infested suckers
- The aphid should be managed to prevent spread of the disease by spraying with Metasystox (0.1-0.5%)
- The plants should be injected with 4 ml of Fernoxone solution (50g in 400 ml of water)
- Infected plants should be destroyed with 4ml of 2, 4, D (50g in 400 ml of water)

5. Anthracnose

Causal organism:*Gloeosporium gloeosporioides*

Symptoms

At the initial stage, small, circular, black raised spots develop on the affected fruits. Then these spots enlarge in size and turn brown in colour.The fungus produces masses of conidia which form a pinkish coat. In severe cases, the entire fruit and bunch are affected. Sometimes main stalk of bunch diseased. The bunch becomes black and rotten. The disease is favoured by high atmospheric temperature and humidity, wounds and bruises caused in the fruit and variety's susceptibility **(Agnihotri et al., 1996).**

Dark brown-black patches on immature fruits

Pathogen

Acervuli are often rounded or sometimes elongated. Conidiophores are cylindrical, hyaline and septate tapering at ends. Conidia are hyaline, aseptate, oval to elliptical in shape.

Disease Cycle

Anthracnose is a fungal disease that occurs when fungal spores contact banana plants, usually through wounds or direct penetration. Infection thrives in conditions of high humidity and warm temperatures. The fungus enters plant tissues and reproduces, causing characteristic symptoms like small, water-soaked lesions on leaves, fruit, and pseudostems. The fungus produces spore-bearing structures called acervuli, which release pink or orange spores that can initiate new infections. Conidiospores, asexual spores, also play a crucial role in spreading the disease. Environmental conditions, such as temperature, humidity, and rainfall, significantly impact the disease cycle (Siddiqui and Ali, 2014).

Mode of spread and survival

The spread of the disease is by air borne conidia and numerous insects which frequently visit banana flowers also spread the disease.

Management

- Burn the infected materials
- Proper field sanitation
- Keep the field free of weeds and provide good drainage

- Fruit should be free of infection and as ripe as possible before it is transported, stored, and ripened
- Banana bunches should be harvested at correct stage of maturity
- Post harvest dipping of fruits in Carbendazim 400 ppm, or Benomyl 1000 ppm, or Aureofungin sol 100 ppm
- Minimising bruising; proper sanitation of handling and prompt cooling to 14°C are essential in minimising the disease in cold storage

6. Erwinia Rhizome Rot

Causal organism: *Erwinia caratovora* sub.sp. *caratovora* and *Erwinia chrysanthemii*

Symptoms

Yellowing of the leaves, necrotic drying with or without heart rot, and brown or yellow water-soaked areas in the rhizomes with dark brown borders were the signs observed. Tissue deterioration, cavity formation, a brown, stench-filled slime, and toppling were also noted (Loganathan et al., 2021).

Erwinia Rhizome rot

Pathogen

A number of the phytopathogenic organisms in the Gram-negative Enterobacteriaceae family, which includes the Erwinia genus, cause soft rots in fleshy fruits, vegetables, and ornamentals. These infections have established persistent cycles of plant-to-plant infection, frequently with the help of flies and bees as insect vectors.

Disease Cycle

The disease cycle typically begins with the introduction of the Erwinia pathogen into the banana plantation. This can happen through infected plant material, contaminated tools, or other means. Once the pathogen is present in the plantation, it can infect various parts of the banana plant, including the rhizome (underground stem), pseudostems (above-ground stem), and leaves. In the case of rhizome rot, the pathogen primarily targets the rhizome. Infected banana plants begin to show symptoms of rhizome rot. These symptoms may include wilting, yellowing of leaves, and soft, water-soaked lesions on the rhizome. As the disease progresses, the rhizome can become discoloured, mushy, and foul-smelling.

Mode of Spread and Survival

Bacterial cells in the ground and in plant suckers that are sick and used for planting. Sometimes it also survives in soil debris.

Favorable conditions

The crop's early stage is quite vulnerable. Early crop stages are characterized by a high frequency of 37°C temperatures; high temperatures and waterlogging facilitate *E. chrysanthemii* infection; and vulnerable cultivars, including as Cavendish and Tella chekkarakeli, are grown.

Management

- Planting susceptible varieties in areas with high rainfall or waterlogged soils should be avoided.
- Crop rotation with rice or sugarcane for three to four years should be followed.
- Light and frequent irrigation should be used to maintain optimal soil moisture.
- Cowpeas should be grown in interspaces to lower soil temperature.
- Infected plants and their roots should be removed.
- Infected suckers should be soaked in a bleaching powder solution (1.5%).
- Limit to one or two ratoons according to the frequency of the disease

Conclusion

The banana is a high-energy fruit crop. It contains vitamin B6, as well as small amounts of vitamin C, manganese, and fiber. This chapter discusses the diseases that threaten commercial cultivation as well as small-scale subsistence farming. Panama wilt is caused by a *Fusarium sp.* soil fungus, which turns the lower leaves yellow and causes petiole collapse around the pseudostem, which splits upwards from ground level. Sigatoka is a fungal leaf spot disease caused by *Mycospharella* spp. that was first identified in Fiji in 1963 or 1964. It causes large areas of the plant's leaf surface to die prematurely. *Pseudomonas solanacearum* causes moko disease, which causes discoloration of vascular tissues. Bunchy top of banana virus (BBTV) causes the leaves to be short and stunted, resulting in a bunched appearance at the apex of the plant. Another one is Anthracnose disease which is caused by *Gloeosporium gloeosporioides* fungus which results in black raised spots on affected fruits and severely in whole bunch. Erwinia rhizome rot caused by *Erwinia caratovora* sub.sp. *caratovora* results in yellowing and necrotic of leaves and tissue deterioration. Hence to reduce the spread of these pathogens and inoculum level, certain management strategies have been discussed in this chapter to enhance the fruit production.

References

Agnihotri, P., Prakash, O., Kishen, R., & Misra, A. (1996). Anthracnose diseases of tropical and subtropical fruits. Disease Scenario in Crop Plants, 1, 1-27.

Daly, A. M., &Walduck, G. (2006). Fusarium wilt of bananas (Panama disease) (*Fusarium oxysporum* f. sp. cubense).

Dongre, M. A., Ashraf, M. S., Koshariya, A. K., &Reddypriya, P. (2023). Plant Pathology and Plant Diseases. AG Publishing House (AGPH Books).

https://agritrop.cirad.fr/544321/1/544321.pdf

https://plantvillage.psu.edu/topics/banana/infos

https://vikaspedia.in/agriculture/crop-production/integrated-pestmanagment/ipm-for-fruit-crops/ipm-strategies-for-banana/diseases-andsymptoms

https://www.planthealthaustralia.com.au/wp-content/uploads/2013/01/Moko-FS.pdf

Loganathan, M., Thangavelu, R., Pushpakanth, P., Muthubharathi, K., Ramesh, R., Selvarajan, R., & Uma, S. (2021). First report of rhizome rot of banana caused by Klebsiella variicola in India. Plant Disease, 105(7), 2011.

Plant Health, Queensland Government, Australia,

Siddiqui, Y., & Ali, A. (2014). Colletotrichum gloeosporioides (Anthracnose). In Postharvest Decay (pp. 337-371). Academic Press.

Viljoen, A., Mahuku, G., Massawe, C., Ssali, R. T., Kimunye, J., Mostert, G., & Coyne, D. L. (2017). Banana pests and diseases: field guide for disease diagnostics and data collection. International Institute of Tropical Agriculture: Ibadan, Nigeria, 82.

4

Diseases of Citrus and Their Management

Avdhesh Kumar[1], Jagveer Singh[2], Prabhat Kumar[1] and Imran Ali[1]

[1]Department of Fruit Science, Acharya Narendra Dev University of Agriculture & technology, Kumarganj, Ayodhya, Uttar Pradesh
[2]Department of Fruit Science, Acharya Narendra Dev University of Agriculture & Technology, Kumarganj, Ayodhya, Uttar Pradesh

Abstract

Citrus fruits are among the most consumed fruits in the world. Citrus fruits are very common and the most important fruit crops in India, and are grown in 10.58 lakh hectares with a total production of 140.32 lakh tons Annually (Anonymous, 2020). Low productivity in citrus is caused by many factors such as, plant pathogens, insect-pests, nutritional imbalances, and physiological disorders. However, a number of obstacles prevent the global output of citrus fruits from growing. Citrus quality and productivity have also suffered due to the spread of pests such Mediterranean fruit flies, mites, and mealybugs. Thus, this review's primary goals were to list the major species that are detrimental to citrus crops and to emphasize the best ways to manage them.

Keywords: *Citrus, disease, etiology and management*

Introduction

The innermost layer of the pericarp is the endocarp. The segments are also called "liths", and the space inside each lith is a locule filled with juice vesicles, or "pulp". From the endocarp, stringlike "hairs" extend into the locules, which provide nourishment to the fruit as it develops. Citrus fruits are notable for their fragrance, partly due to flavonoids and limonoids (which in turn are terpenes) contained in the rind, and most are juice-laden. The juice contains a high quantity of citric acid giving them their characteristic sharp flavour. The genus is commercially important as many species are cultivated for their fruit which is eaten fresh, pressed for juiceor preserved in marmalades and pickles. Many citrus fruits, such

DOI: 10.1201/9781042014460-4

as oranges, tangerines, grapefruits, and clementines, are generally eaten fresh. They are typically peeled and can be easily split into segments. Grapefruit is more commonly halved and eaten out of the skin with a spoon. There are special spoons (grapefruit spoons) with serrated tips designed for this purpose. Orange and grapefruit juices are also very popular breakfast beverages. More acidic citrus, such as lemons and limes, are generally not eaten on their own. Meyer Lemons can be eaten 'out of hand' with the fragant skin; they are both sweet and sour. Lemonade or limeade are popular beverages prepared by diluting the juices of these fruits and adding sugar. Lemons and limes are also used as garnishes or in cooked dishes. Their juice is used as an ingredient in a variety of dishes; it can commonly be found in salad dressings and squeezed over cooked meat or vegetables.

The most significant fruit crop in Central India is citrus. Because of the severe harm that insect pests cause, it suffers greatly. They often attack this crop, causing noticeable harm from the nursery stage until harvest. There are around a dozen of them. However, there are very few that take this seriously enough that timely and effective control is necessary to guarantee both the amount and quality of the produce.

Diseases

Virus and Virus-Like Diseases

1. Tristeza (CTV): It is caused by a phloem limited, 2000 nm long filamentous virus.

Symptoms: Depending on the citrus cultivars, the surrounding conditions, and the viral strain, the tristeza virus can cause a variety of field symptoms. It is recognized that there are five distinct strains of tristeza: mild, seedling yellows, decline on sour orange, stem pitting on grapefruit, and stem pitting on sweet orange types. A tristeza infection is indicated by sudden collapse or rapid wilting, which gives rise to the term "quick decline," followed by defoliation in sweet orange, grapefruit, and mandarin on sour orange rootstocks.

Compared to other cultivars, CTV causes greater harm to delicious oranges. The leaves turn dull, lose their deep green sheen, and eventually get chlorosis along the midrib and lateral veins. Soon after discoloration, the leaves start to fall, exposing the limb and eventually causing die-back symptoms. Several of the twigs also get defoliated at this point. A specific CTV isolate experiences symptoms of vein corking (Fig 4). Occasionally, abnormally large crops of prematurely coloured fruits serve as the first obvious sign of a disease. Vein flecking on leaves, leaf cupping, and stem pitting are typical symptoms on kagzi lime. Mandarins are generally tolerant, yet certain strains of CTV can cause sickly appearance on Nagpur mandarin, dropping leaves, upright branches with yellowing foliage, and even stem pitting on Coorg mandarin. Additionally, grape fruit, pummelo, tangelo, and sweet orange trunk and branches develop stem pitting due to tristeza strains.

CTV induced quick decline of sweet orange

CTV induced mosambi leaves showing cholorosis of mid veins and lateral veins and leaf yellowing.

CTV induced declined mosambi plant.

Vein corking in mosambi

Leaf cupping in acidlime

Stem pitting

Black and brown

Aphid - Vector of CTV

Host range: CTV infected nearly all species, cultivars, and intergeneric hybrids of citrus and some citrus relatives. The only known *nonrutceous* host is *Passiflora.*

Transmission: Tristeza can be transmitted by grafting, budding, dodder and by aphid vectors. *Toxoptera citricidus, T. aurantii, Myzus persicae* and *Aphis craccivora* are the common aphid vectors and the most efficient among these is *T. citricidus*(Fig 7).

Detection/ Identification

- The greatest plant to detect CTV when vein clearing, vein darkening, leaf cupping, stem pitting, or vein corking symptoms appear is Mexican lime, also known as kagzi lime.
- To identify the stem pitting strain of CTV grapefruit seedlings, Madame Vinous or Pera sweet oranges ought to be utilized.
- The yellow strain of CTV seedlings, which have small yellow leaves and exhibit compact and stunted growth, should be utilized to identify grapefruit or sour orange seedlings.
- After allowing all shoots to grow rapidly for the first three growth flushes, they should be pruned and trained as a single shoot. Test sample graft inoculation is prepared for eight to ten months old plants.

- For each test, a minimum of three to five indicator plants are required.
- Requirement temperature: 24-27 ^{0}C maximum days /18-21^{0}C min. night.

2. Ring Spot (CRSV)

Most kinnow mandarin orchards in North India have up to 100% incidence of ring spot disease. Mature leaves exhibit the field symptoms, which include a loss of pigmentation characterized by discrete rings ranging in diameter from 2.1 to 24.2 mm, with green tissue centered. Each leaf can have one or more of them. The Kinnow mandarin, Malta orange, and Kagzi lime exhibit these symptoms prominently. Infected plants have mottling and epinasty, chlorotic specks on their leaf lamina. Later, the diseased leaves begin to drop early. Plants that are severely impacted exhibit signs of decline and dieback. The disease is also linked to blisters on twigs and leaves. Fruit quality in Kinnow Mandarin is negatively impacted by decreased fruit size, juice, total soluble solids, and vitamin C levels.

Ring spot on kinnow

Malta orange

Kagzilime

Transmission: There is currently no known natural method of disease transmission besides bud transfer. There have been reports of viral transmission through pollen.

Detection/ Identification

- The mechanism of transmission to *Chenopodium quinoa* results in the development of evenly spaced chlorotic lesions in 4–8 days.
- When employed as marker plants, sweet orange and grapefruits exhibit shock and a juvenile leaf pattern four to six weeks following graft inoculation with positive samples.
- For indicator plants to display symptoms, a cool screen house temperature of 24-27^{0}C during the day and 18-21^{0}C at night is required.

3. Mosaic

The disease known as citrus yellow mosaic is common, especially on sweet oranges. The culprit responsible for the disease has been determined to be the citrus yellow mosaic badna virus group. The disease has an incidence of 10–70% in the acid lime, Sathgudi, and Chinni sweet orange districts of South India.

Symptoms

Under field conditions, bright yellow mottling or mosaics on sweet orange leaves are typical indicators of infection. Pummelo likewise displayed signs of a similar nature. Acid lime leaves with citrus mosaic virus infection exhibit sporadic patches of light green color. In the leaf lamina of severely affected leaves, there were yellow patches mixed with green ones. Some of the fruits have patches of variegation in light green and yellow colors. The size of the fruit and leaves has decreased somewhat.

Mosaic on *Mosambi*

Yellow mottling on sathgudi orange

Mosaic/golden yellowing

Transmission

In addition to being transmissible through dodder, mosaic can be spread by graft and mechanical inoculation from citrus species. The virus's potential vector is thought to be the mealy bug (*Planococcus sp.*).

Detection/ Identification

The best indicator plants are sweet orange and rangpur lime, which exhibit different degrees of bright yellow mosaic and chlorotic patches.

4. Exocortis

Five different families of viroids are known to spontaneously infect citrus trees. Of these, citrus exocortis viroid (CEVd), which is prevalent in practically all citrus-growing countries, is the most significant viroid disease commercially.

Symptoms: On trifoliate and Rangpur lime rootstock, field symptoms range from minor bark cracking to very severe bark scaling/splitting, accompanied by varying degrees of stunting and mild to moderate tree decline. Yellow streak-like blotches may appear on the twigs and branches of CEVd-infected trifoliate or Rangpur lime, particularly on the younger, greener twigs. Although rough lemons are rather robust, they can also have slight bark cracking without showing any signs of loss. Despite the fact that many commercial citrus cultivars have no symptoms, trees grown on rootstocks that are typically thought to be resistant may exhibit some degree of stunting.

Exocortis induced severe bank splitting
on Rangpur lime rootstock

Transmission

The main way citrus viroids spread is by introducing and spreading diseased budwood. They can also spread mechanically through contaminated tools, knives, and other hedging equipment. It is not known that CEVd or any other citrus viroids are vector-transmitted.

5. Witches' broom

This is one of the most recent acid lime illnesses, occurring in small areas and being brought on by Candidatus *Phytoplasma aurantifolia*. Blastomania, also known as numerous sprouting, is a phytoplasma disease that is similar to *C. aurantifolia, C. limonia*, and *C. reticulate*.

Symptoms: Shorter internodes and tiny chlorotic leaves with numerous proliferating shoots are symptoms. On the canopy of the infected lime or mandarin tree, twisted twigs resembling typical witches' broom symptoms are displayed on infected branches, causing leaves to drop early. The diseased plant displayed dieback symptoms in its advanced stages. Not every limb on the afflicted trees displayed the telltale signs of a witches broom. There were no fruits or blossoms on infected twigs.

Symptom on canopy of acidlime

Dried broom like symptom on mandarin

Transmission: Graft and dodder (*Cuscuta reflexa*) transmitted. Natural transmission takes place probably by leaf hopper vector *Hishimonus phycitis.*

Detection/ Identification

- Mexican lime indicator plant.
- Immunofluorescence, electron microscopy, DAPI staining.

6. Other diseases

- **Yellow Corky Vein Disease (CYCV) in Kagzi Lime:** The etiology of Yellow Corky Vein Disease (CYCV) in Kagzi Lime is unknown, although it has long been recognized to occur in India. The primary vein's yellowing and the development of corky tissue on the leaf's underside are the disease's hallmarks. The illness has a viroid etiology and is mechanically transmissible.
- **Rubery wood**: is a phytoplasmal disease that is more prevalent in Mandarin. It produces thin, linear shoots with branches that are remarkably flexible or elastic. The afflicted tree does not have blooms or fruits, and the plant becomes stunted with little chlorotic leaves.
- **Woody gall/vein enation:** It is especially responsible for galls on rough or volkamer lemons. On the other hand, damage from insects or other causes. Presence of typical enations or protuberance on veins on underside of leaves is diagnostic of the disease.

Yellow corky vein symptom

Woody gall on rough lemon rootstock on Kagzi lime

Diagnosis

A brief summary of the various diagnostic approaches is provided below. Sensitive and trustworthy diagnostic procedures are crucial for formulating strategies to avoid infections in propagating material.

Guidelines for Biodiagnosis

A. Seed collection and raising seedlings of indicator plants

- Gather fruit from plants that appear healthy and develop branches that are higher than one meter above the ground.
- To release the seeds from the pulp, remove the seeds and give them a good wash in soapy water.
- Surfaces sterilize seed by combining 0.1% wetting agent and 0.5% sodium hypochlorite for ten minutes, after which the seed is properly rinsed with water.
- Apply dried seed on a shaded surface. After drying, the seed can be treated with a powdered fungicide like thiram or dipped for three minutes in a 1% solution of 8-hydroxyquinoline.
- Seeds should germinate, especially in sterile soil mixtures.
- Once seeds have sprouted, they can be moved into sterile soil mixtures in facilities free of insects.
- The seedling should be utilized for biodiagnosis once it has grown to a suitable size.

B. Seedling indexing

- Gathering implements (e.g., secateurs, knives) to be sterilised by dipping in a solution of sodium hypochlorite (1%–2%).
- Samples of bub wood should be taken from each tree's four quadrants at the very least.
- Take out the thorns and leaves, but don't cut the petioles. Label, put the plant in a plastic bag, and keep it at 25 ^{0}C degrees Celsius or lower—ideally in an ice chest.
- After gathering the bud wood, give it a thorough wash with water and a light detergent, rubbing it dry with a towel or air drying it.
- Typically, indicator seedlings are inoculated with bud, chip, blind bud, side graft, or wedge graft. The recommended norm for inoculation is two buds, with the exception of recalcitrant cases where many side grafts or leaf grafts are favoured for detection of greening. Sufficient positive and negative controls must be included for each test.
- After inoculation, plants require careful attention. For example, Mexican lime seedlings' developing young side shoots should not be clipped during the first three growth flushes in order to collect the maximum number of leaves to check for vein clearing, which is crucial for CTV detection. The side shoot should be cut after the third growth flush, and the most strong terminal shoots should be connected to a stack and trained to grow as a single shoot so that stem pitting may be examined later.
- On corresponding indicator plants, a cooler screen house temperature is crucial for the expression of pathogen symptoms such as CTV, ringspot, greening, mosaic, etc., whereas a higher temperature is necessary for the formation of viroids and phytoplasma symptoms.

Integrated Management of Viral Diseases

The best way to handle agricultural plant virus diseases, especially those affecting perennial horticultural plants, is to utilize an integrated approach that includes vector management, sanitation, cultural practices, host resistance, planting material free of viruses, and regulatory measures.

a) Use of virus-free certified planting material/Bud wood certification: Using planting material free of viruses is essential for crops grown vegetatively, such as citrus. Testing must be done carefully since even bud wood taken from plants that seem healthy may contain latent viruses. The best way to accomplish this is with a robust certification scheme for healthy bud wood that is equipped with sophisticated diagnostic tools. Producing virus-free plants using in vitro "shoot-tip grafting" is also highly beneficial, but these plants must also be certified.

b) Sanitation: Since infected plants can serve as foci for secondary virus spread, employing virus-free plants won't be very helpful unless the orchards using them are also free from other sources of infection, such as old diseased plants and collateral hosts. The citrus relatives are the most evident alternate host for citrus diseases. Citrus infections, however, can also be found in completely unrelated genera. Apple stem grooving virus, which is primarily comparable to citrus tatter leaf virus in terms of particle, biological characteristics, and host range, is hosted by pome and lily fruits. Citrus-infecting viroids also infect a broad range of hosts outside the Rutaceae. The discovery of CEVd variations in vegetable crops in a region free of citrus suggests that viroids may be more common than previously thought. According to reports, *Murraya koenigii, Ruta graveolens*, and *Evodia hupelensis* are asymptomatic carriers of tristeza; periwinkle is another organism that is thought to be a host for greening. For this reason, it is imperative to implement a specific program for getting rid of unhealthy plants and other infection sources and replacing them with approved healthy planting material.

c) Vector control

- Gathering instruments, such as secateurs and knives, to be sterilised by immersing them in a solution of sodium hypochlorite 1%–2%.
- Different insect vectors transmit different viruses or virus-like diseases of citrus. To stop the spread of viral infections, biological and non-biological approaches should be used to control these insect vectors. Among the non-biological techniques include the application of appropriate insecticidal sprays, insect traps, reflective mulches, and so on.
- Aphids can be managed by spraying water containing acephate (1.5 g/litre), monocrotophos (1.5 ml/litre), or dimethoate (2 ml/litre) to control psylla, mealybugs, and leafhoppers.
- The timing of these pesticides' application is also crucial. Spraying should be done in January and February to get rid of aphids. In the months of March through April, spraying should be done for psylla as well as leaf hoppers and mealybugs as soon as these insects are observed.
- Various biocontrol agents consist of microorganisms, predators, and parasitoids. Insects are the specific hosts of parasitoids. *Tamarixia radiata* is an efficient parasitoid of citrus psylla, while the genera *Aphelinus, Mesidia*, and *Mesidiopsis* of Aphelinidiae (superfamily: Chalcidoidae) are parasitoids of aphids.
- Aphids are effectively hunted down by coccinellids, which vary in similar ways. Commercial applications of entopathogenic fungi, such as *Verticillium lecanii* and *Paecilomyces farinosus*, are made against *T. citricidus* and *Aphis gossypii*.

d) Cultural Practices

- Neglected crop hygiene is a major contributor to mechanically transmitted diseases such as ring spot, mosaic, and exocortis, which are spread by field tools used in intercultural operations, harvesting, and other activities.

- Before collecting fruit, grafting, or cutting any tree or nursery plant, all pruning and grafting instruments should be thoroughly cleaned with 1-2% sodium hypochlorite to prevent the spread of illnesses that are carried by mechanical means.

e) Cross Protection

- When a particular plant is previously inoculated with a mild strain of the same virus, a process known as "cross protection" occurs where resistance develops against the severe form of the virus.
- The citrus industry has been the one to apply cross protection the most extensively.
- Aims may be made to find a stable, moderate strain virus that is successful in protecting trees in the field.

f) Quarantine / Regulation

- The most crucial method for avoiding the introduction of novel, unusual, and potentially harmful viral infections is still isolation.
- One of the primary issues with quarantine is the entry of diseases for which there are no indexing techniques or for which there are inadequate infrastructures.
- Bud wood imports ought to be limited to the most necessary items, and several stages of treatment should be completed prior to introduction.
- Viral infections can be prevented from spreading by enforcing tight quarantine regulations and limiting the transportation of citrus bud wood.

Bacterial Diseases

7. Greening

Symptoms: Season, the kind and degree of infection, the age, and the nutritional state of the trees all affect the range and intensity of symptoms. Citrus leaves with greening infections are often tiny, erect, and often exhibit chlorotic interveinal regions and green veins. Additionally, diseased leaves exhibit several kinds of chlorotic mottling. One of the best indicators of greening is leaf mottling. In extreme situations, leaves nearly turn chlorotic and have sporadic dark green patches. Plants become severely stunted, their internodes get shorter, and many of them never bear fruit. Often, simply a branch or twig will exhibit symptoms, or even only a portion of the canopy. Trees that are continuously infested have scant foliage, significant twig die-back, and a rapid decline to an unproductive state. The term "greening" originated from the undersized, misshapen, and often prematurely falling infected fruits. The surviving fruits on the tree do not properly color, staying green on the shaded side.

Greening symptoms in Nagpur Mandarin

Leaf chlorosis

Citrus psylla, vector of greening

Causal Agent: It is currently understood to be caused by an uncultured, phloem-limited bacteria with a Gram-negative cell wall.

Identifying and Sending: Citrus psylla in orchards spreads the greening disease by infected budwoods. As an adult, Psylla carries on the pathogen that it first encountered as a nymph. Additionally, greening can spread through importation and replication in nurseries, particularly across long distances and globally.

Identification: Citrus greening diagnosis is critical to managing the illness. It can be very challenging to correctly identify the illness in the field since it might be mistaken for other stress-related leaf symptoms, root rot, or mineral deficiencies. Additionally, both visual detection and bioassays are challenging due to the disease's uneven distribution inside the tree and its sluggish progress. The most straightforward and affordable method of identifying greening infection at the moment is biological indexing with indicator plants.

Management

- The management of greening disease entails the elimination of infected, unproductive trees and their substitution with disease-free budded plants grown on enhanced rootstock.
- Bud wood should be picked from parent trees that do not exhibit greening through an appropriate indexing method.
- Strengthening regulatory procedures, such as quarantine, is necessary to restrict the transit, sale, and use of contaminated bud wood or nursery stock.
- To stop the spread of disease, nurseries must be strictly regulated by a licensed disease-free certification process.
- Given that the illness is also transmitted by the citrus psylla vector, it is recommended to restrict its spread by spraying appropriate insecticides, such as monocrotophos at 0.7 ml/litre or quinalphos at 1.0 ml/litre of water.
- When an infestation is observed on sprouts in June-July, January-February, and October-November, spraying should be done at the bud burst stage.
- Complete removal of the pathogen cannot be achieved by trunk injection of tetracyclines, despite reports that they decrease greening signs.

- To put it briefly, the most effective technique to combating greening illness is an extensive and comprehensive management strategy. This covers quarantine protocols, the use of planting material free of disease, and the reduction of inoculums by the removal of strongly affected trees, pruning of infected branches, and psylla management.

8. Canker- *Xanthomonas campestris pv. citri* (Hasse) Dye

Symptoms: Canker lesions begin as tiny dots and grow to be 2–10 mm in diameter. The age of the host tissue during infection and the cultivar have the most effects on their final size. Initially round, the lesions can later develop irregularly, frequently aggregating at the leaf tips, leaf border, or within a specific area of the leaf. About 7 days after infection, the lesion becomes obvious on the underside of the leaves, and it appears on the upper surface shortly after. The yellow halo surrounding the lesion is a telltale sign of the illness on leaves. Especially on the underside of the leaf, young lessons are elevated or pustular. Eventually, the pustules developed into corky, crateriform lesions with a raised border and a sunken center. Damage on on fruit and stems extend 1-3 mm in depth and are superficially similar to those on leaves.

Canker on leaves

Canker on twig

Canker on fruit

Canker spread through leaf minor

Disease Cycle

Management

- The only effective way to lessen the canker illness is through prevention.
- Prevent damage to leaves from wind-blown soil.

- Installing wind barriers can help lower wind speeds.
- Affected orchards should not be worked in when the trees are damp from rain or dew.
- Control harm from leaf miners makes infection easier.
- Careful examination of quarantines in nurseries and orchards, along with on-site burning of diseased or clipped trash before to the monsoon.
- Two to four monthly applications of 0.3% copper oxychloride (COC) mixed with 100 ppm streptomycin at the beginning of the monsoon season.

Fungal Diseases

9. Phytophthora Diseases (*P. citrophthora, P.nicotianae (= P. parasitica),* and *P. palmivora*)

Symptoms: In well-established orchards, phytopthora is the culprit behind foot rot, root rot, crown rot, gummosis, leaf fall, and brown rot disease. Foot rot lesions can appear on the trunk up to 60 centimeters above ground, and they can also spread to the crown roots below the surface of the earth as crown rot. When the lesion's dead bark is scraped, a discolored, slick, brown patch is revealed. Gummosis, or brownish-black oozing gum, is the result of such active lesions and appears on the trunk.

Dead bark stiffens, separates from healthy bark, coils, and splits during the dry season. Typically, growers are unaware of the illness since it only affects feeder roots. The initial sign of such damaged plants is dull, chlorotic foliage, with the remaining portion of the leaf remaining normal in color while the midrib, main lateral veins, and bands of leaf tissue surrounding them turn yellow.

In severe circumstances, the damaged plant will exhibit signs of famine, reduced canopy volume with bare branches, die back, and steady decline if the feeder root regeneration is unable to keep up with the rate of destruction. When there is persistent rain for at least 24 hours, Phytophthora splashes down with raindrops and causes low-hanging fruits to decay and leaves to fall. Due to severe moisture stress brought on by foot and root rot, plants either bear erratically and lose yield, or they bear excessively and collapse while the fruits are still on the plant.

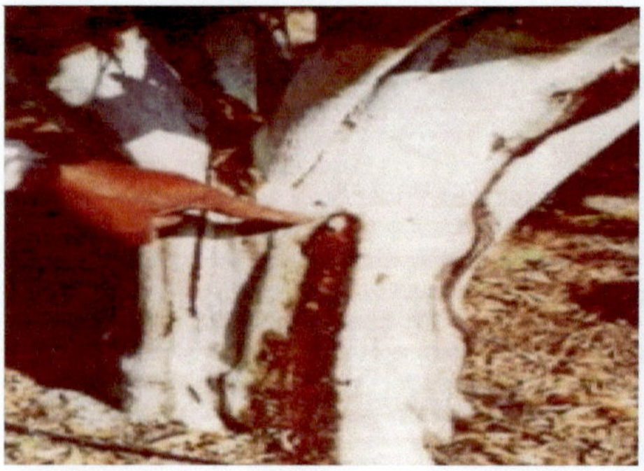

Gummosis

Foot rot

Root & Collar Rot

Brown Rot

Decline due to foot & root rot

Reason for Severity

- Because *P. nicotianae* and *P. palmivora* can withstand temperatures of up to 35.°C, they are most active in the summer, when temperatures range from 28 to 32 °C. As temperatures rise, their activity decreases.
- In nurseries and irrigated orchards, phytophthora is a year-round pest.
- Diseases and inoculum accumulation are made worse by heavy, poorly drained soil, over-irrigation, using susceptible rootstock, and extended water contact with tree trunks.

Mode of disease spread

- Use of vulnerable root stock.
- Flood irrigation, flatbed system.
- Prolonged water retention in beds.
- Extended periods of precipitation.
- low budding.
- Recurrent use of the same site for nursery growing.
- Placement of nurseries close to historic orchards
- Frequent contamination of soil and irrigation water is a risk factor for P*hytophthora* infections.

- Through a cut or damage on the tips of hanging foliage or roots.
- Through organic material that contains oospores or Chlamydospores.

Management

(i) **Use of resistant rootstocks:** Rough lemon is sensitive to *Phytophthora* despite growing quite quickly. There is a great deal of variation between the strains of Rangpur and rough lemon. Trifoliate oranges and sour oranges are quite tolerant. Nonetheless, sour oranges can be employed in situations when tristeza is not an issue despite being sensitive to the virus.

(ii) **Customs and traditions:** Since the pathogen is soil-borne, field nurseries may find it difficult to completely eradicate it; nevertheless, by implementing the following modified cultural techniques, the disease can be controlled:

- To reduce the likelihood of pathogens, the nursery soil beds should be fumigated using soil fumigant such as Dazomet or solarized during the summer.
- It is important to maintain bud union as high as possible during planting to prevent irrigation water from coming into contact with the scion.
- Maintaining adequate drainage of the soil is crucial, as is avoiding flood irrigation and prolonged water stagnation in the basin.
- It's also important to prevent farm operations from injuring the trunk and root system.
- It is important to arrange irrigation emitters such that the trunk is not left moist for an extended amount of time.
- When *Phytophthora* destroys a significant amount of the root and collar of old budlines or elite mother plants in an orchard, the plants can be efficiently saved by inarching or bridge grafting with resistant rootstock. It is possible to revive these diseased plants in the orchard to extend their productive lives.

(iii) Chemical control

- To control *Phytophthora* diseases, copper fungicides are applied as trunk paste, foliar spray, and basin drenching. Two sprays with soaking, applied 40 days apart, covering the entire plant canopy and basin of the afflicted plant, using either Aliette (2.5 g/l) or Ridomil MZ-72 (2.75 g/l).
- When applied appropriately, copper fungicides can effectively prevent gummosis, foot and root rot, and other citrus plant diseases.
- As a preventative step, bordeaux paste should always be applied to tree trunks prior to the start of the monsoon. Before applying Bordeaux paste, areas damaged by gummosis or foot rot should be carefully scraped off with a sharp knife so as not to harm the wood.

- Use fosetyl-A1 and matalaxyl (or its isomer, mefenoxam) as
- Alternatives both fungicides should be practiced to minimize the risk of the development of fungicide resistance.

Precautions

- Planting material free of diseases ought to be chosen for planting.
- Flood irrigation should be avoided in order to prevent water logging and stagnation for extended periods of time.
- Irrigation water should never come into contact with a tree trunk.
- To control the disease at its early stages, regular monitoring for the symptoms as stated should be carried out.
- If there is any uncertainty about the illness, soil and affected plant samples should be tested for *Phytophthora* in a laboratory.

References

Anonymous, (2020). Horticultural Statistics at a Glance 2020. Govt. of India, Ministry of Agriculture and Farmers Welfare. Department of Agriculture, Cooperation and welfare horticulture statistics division.

Boina, D. R. and Bloomquist, J. R. (2015). Chemical control of the Asian citrus psyllid and of huanglongbing disease in citrus. Pest Management Science, 71(6), 808-823.

Gade, R. and Lad, R. (2018). Biological management of major citrus diseases in Central India-a review. International Journal Current Microbiology and Applied Science, 6, 296-308.

Gohel, N. M., Prajapati, B. K. and Srivastava, J. N. (2022). Major diseases of citrus and their management. Diseases of Horticultural Crops; Apple Academic Press: New York, NY, USA, 155-167.

ICAR (2020) .https://www.rvskvv.net/images/Disease_Hort_Crops_20.04.2020.pdf

NHM (2012). https://farmer.gov.in/imagedefault/ipm/citrus.pdf

Ploetz, R. C. (Ed.). (2003). Diseases of tropical fruit crops. Cabi Publishing.

T.K. Bose (2021). Citrus. Fruits: Tropical and Subtropical. Daya Publishing House New Delhi, 2, 1-164

TNAU (2002) https://agritech.tnau.ac.in/crop protection/crop/citrus.pdf

5

Diseases of Papaya & Their Management

Ankita, Prince and Priya Devi Maurya

Department of Plant Pathology, Sardar Vallabhbhai Patel University of Agriculture & Technology, Meerut, Uttar Pradesh

Abstract

The papaya, or Carica papaya, is a fruit belonging to the Caricaceae family that is valued for its high nutritional and therapeutic content and its commercial importance. Papaya fruit have several benefits, it is rich in nutritional value, help in digestion, immune support, antioxidant properties, provide health to the skin, eye & heart. Ripe fruits are consumed fresh or used in the preparation of jam, jelly, smoothies, desserts, nectar, soft drink, ice cream, flavouring crystallized fruits & canned syrup. Unripe fruits (rich in papain enzyme) are used as vegetable. Productivity of papaya are reducing because of several diseases. In this chapter we discuss about major diseases of papaya like the foot rot of papaya, powdery mildews, papaya mosaic disease, papaya ring spot disease, anthracnose, papaya leaf curl, nematodes etc. & their management.

Keywords: *Carica papaya, immune, antioxidant, papain, management*

Introduction

Carica papaya is a tree-like herb that grows 2 to 10 meters tall and is often unbranched, cylindrical & hollow stem Fruits are large, cylindrical, with fleshy orange pulp. Numerous, small, black, round seed with gelatinous aril. Papaya leaves have historically been used to treat thrombocytopenia brought on by a dengue infection. (Chinnappan et. al. 2016; Sharma et. al. 2019). Papaya fruits have healing capabilities against cancer, antimicrobial anti-malarial, anti-tumor, anti-inflammatory, antioxidant, immune-boosting, blood sugar-lowering, wound-healing, gastro-protective, and capable of curing fever, asthma, jaundice, irregular menstruation, and beriberi. The pharmaceutical and wellness industries have seen a steady rise in demand for papaya as a result of researchers' recent discovery of their therapeutic potential.

DOI: 10.1201/9781042014460-5

Papaya originates from Central America and southern Mexico but now it is grown in many other parts of the world. It is grown widely under tropical and sub-tropical climates. In 2020, 13.9 million tonnes of papayas were produced worldwide, India led 43% of the total world production (FAOSTAT) 2022. India is among the leading papaya producing nation worldwide. Cultivated papaya is susceptible to a wide range of disease pathogens. Determining the economics of disease incidence or severity assessment and fruit loss are crucial for papaya production.

Major disease of papaya	Name of Disease	Causal organism
1	Anthracnose	*Colletotrichum gloeosporoide*
2	Foot rot of Papaya	*Pythium* spp., *Phytophihora* spp. and *Rhizoctonia solani*
3	Powdery mildews	*Oidium indicum/Oidium caricae*
4	Papaya leaf curl disease	*Papaya leaf curl virus,* PaLCuV
5	Papaya mosaic disease	*Papaya mosaic virus,* PapMV
6	Papaya ring spot disease	*Papaya ringspot virus*, PRSV-p, *Papaya ringspot virus*- PRSV-w
7	Root Knot Nematodes	*Meloidogyne* spp

Diseases of papaya

A. Fungal Diseases

1. Powdery Mildew: Odium indicum Kamat, Odium caricae Noack, Leveillula taurica (Lev.) Arnauld
2. Leaf- Blight: Corynespora cassiicola (Burk and Curt).
3. Damping-Off : Pythium aphanidermatum Nirvan; Rhizoctonia solani Subramaniam.
4. Foot Rot : Pythium aphanidermatum Nirvan.
5. Anthracnose: Colletotrichum gloeosporioides Penz.

1. Anthracnose

Causal Organism

Colletotrichum gloeosporoide perfect stage *Glomerella cingulata* (Karunakaran and .: Nair, 1980) especially in hot, humid climates, both stages are quite common. It belongs to class ascomycetes. The mycelium is branched, septate, hyaline. Conidiophore are unbranched and hyaline. The conidia are subhyaline having variable shapes. Setae are also found in acervuli. The sexual spores are ascospores which formed in perithecia.

Symptoms

According to Misra and Prakash (1986), there were two different kinds of pre-harvest symptoms that were brought by *C. gloeosporioides*. Very young fruits (only 5 to 10 days old) were discovered to be infected with a white fungal growth covering while slightly larger fruits displayed symptoms,the entire fruit surface. Disease occurs both in green leaves and fruit also. Fruit spots begin as superficial skin discolorations that range from brown to black and eventually form into 1 to 3 cm diameter, round, somewhat depressed regions. On the fruit's surface, brown sunken patches appear first, these spots grow to become a water-soaked lesion, and the affected area becomes soft and begins to decay. Infection at early stages results in mummification of fruit, there is reduction in yield quantity or quality.

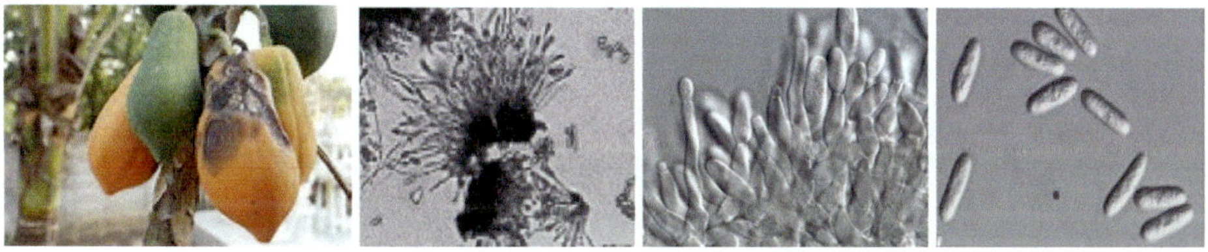

Symptoms on fruits Acervuli Conidia with Conidia Conidiophore

Fig. 1: Symptom & Microscopic structure of anthracnose in papaya

- The mycelium is branched, sparsely septate and hyaline hyphae. Setae are 1-4 septate and swollen at the base.
- Conidiophores are hyaline and unbranched. The conidia are sub hyaline and variable in shape.
- The sexual spores are the ascospores born on perithecia.
- The perithecial stage develops on stromatoid cushions in which the perithecia are immersed.
- Anthracnose causes considerable losses and is very commont belong to the Kingdom-Fungi, Class-Filamentous, Ascomycetes; Order-Phyllachorales.
- The mycelium is branched, sparsely septate and hyaline hyphae. Setae are 1-4 septate and swollen at the base.
- Conidiophores are hyaline and unbranched. The conidia are sub hyaline and variable in shape.
- The sexual spores are the ascospores born on perithecia.
- The perithecial stage develops on stromatoid cushions in which the perithecia are immersed.
- Anthracnose causes considerable losses and is very common

Epidemiology: The pathogen is more severe at 25°c-30°c & 85-90% relative humidity. Conidia are spread by blowing wind & rain splashes. Pathogen survive saprophytically on fallen leaves, petioles & fruits.

Disease Cycle: Primary source of inoculum: the fungus survive as dormant mycelia on fallen fruits and leaves from the field. Secondary source of inoculum: Conidia spread by wind and rain splash.

Management

Cultural practices

- By managing irrigation and fertilizing in a balanced manner. of
- Elimination of fully ripe fruits, particularly those with infections, and leaves, both on the plant and on the ground.
- Cleaning of equipment and installations to eliminate sources of inoculum.
- Export purpose fruit should be treated with hot water & fungicidal wax.

Chemical control

- Apply a foliar spray twice a week with thiophanate methyl@0.1% and copper oxychloride or chlorothalonil@0.2%, spaced 15 days apart.
- Ten days prior to harvest, the final spray should be applied.

2. Foot rot of Papaya

Causal organism

Phytophthora spp, *Rhizoctonia solani*, and *Pythium* spp. are some of the causative organisms, although *Pythium aphanidermatum* is the main culprit. According to Bhargava (1941), during the 1940 wet season, an epidemic of *Pythium aphanidermaium* caused stem and foot rot devastated the papaya in Naini, Allahabad. Coenocytic mycelium, 500 x 20 mm lobulate sporangia, vesicles generated during germination, smooth, spherical oogonia, and widely clavate antheridia and oospores are thick-walled.

Symptoms

Foot rot of papaya is distinguished by the development of wet areas on the stem at ground level, which enlarges and ultimately destroyed the root. There are reports of disease in every area where papayas are grown, particularlly in rainy periods, in heavy wet and poorly drained soils. The pathogen attack as damping off in the nursery with seedlings. The disease may be carried by seedlings grown on contaminated soil. Under favorable circumstances, these seedlings eventually get stem rot. Commonly, trees that are two to three years old exhibit stem rot.

Symptom at ground level

Lodging of tree in severe attack

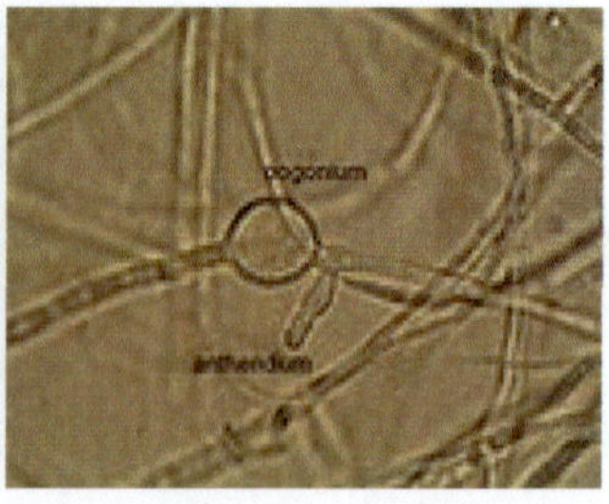

Oogonium & Anthredium of *Pythium* spp.

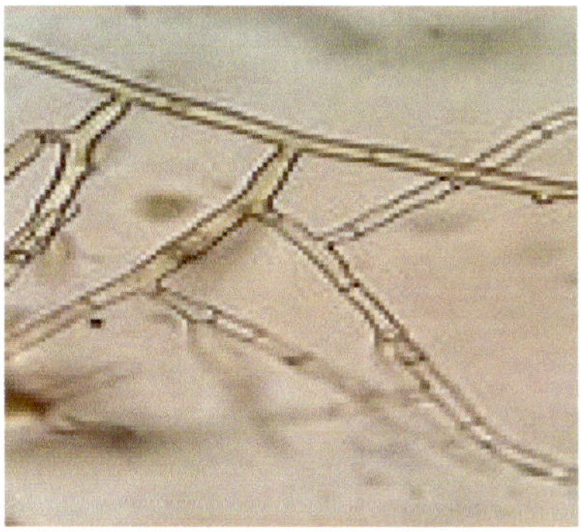

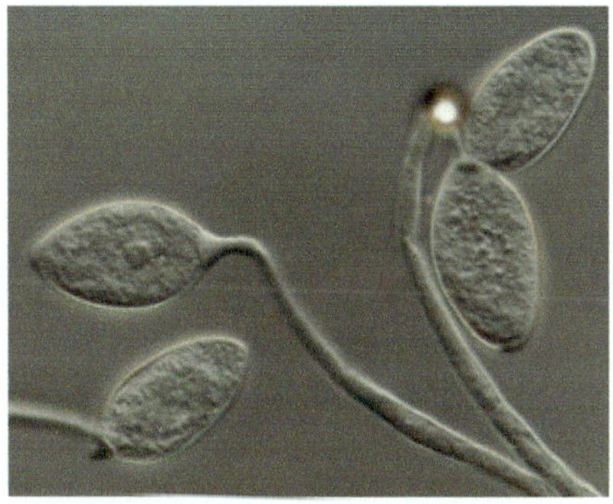

Acute angle branching of *Rhizoctonia* spp. *Sporangium of Phytophthora* spp.

Symptoms & Microscopic structure of foot rot of papaya

Epidemiology

Injuries, poorly drained soils, high humidity, and temperatures between 28-36°C are sufficient triggers for the disease to start. Zoospores and sporangia can survive for short periods while resting spore oospores (*Pythium*) or sclerotia (*Rhizoctonia*), can survive for long periods, in the soil and in fallen fruits. It germinates and release zoospores along with irrigation water spread throughout the field.

Disease Cycle: The pathogen is soil inhabitant & survive on the residue left in the soil where it produce abundant oospore.

Management

- For controlling foot rot, principle of escape, exclusion and eradication for management should be followed.
- Plantation avoid in more clay soils with inadequate drainage. For seedlings, use sterile soil and healthy, fungicide-treated seeds.
- Use metalaxyl or chlorothalonil-based fungicides to apply preventative sprays to the plant collar region; additionally, perform stem surgery to remove damaged tissue and treat the surrounding area with cupric paste (1 kg of copper sulfate + 2 kg of lime + 10 liters of water).
- Soak the stem's base with metalaxyl@0.1% or Bordeaux mixture@1.2%

3. Powdery mildew

Causal Organism: *Oidium indicum/Oidium caricae*

According to Chidderwar (1955), *Oidium indicum* is the cause of papaya powdery mildew. The illness brought on by *Oidium caricae* Noack was first documented by Chona and Lal (1959) in Delhi. Pathogen belongs to class ascomycetes. It is an obligate parasite. Mycelium are hyaline, septate, ectophytic the internal development of haustoria inside epidermal cells. Asexual hyaline spore are conidia born in chain at tip of conidiophore. Sexual spores are ascospore which are globose & each ascocarp contains one ascus.

Symptoms: White, powdery spots are visible on the disease leaves undersides. The infection site causes patches of pale green or yellow colour on the upper surfaces of the leaves. In addition to slowing down the growth of the plants, the illness may cause damage to the leaves, which will impact photosynthesis and, in turn, the commercial quality of the fruits.

Powdery substance at upper surface, lower surface of leaf, fruit & severe infection in fruits respectively

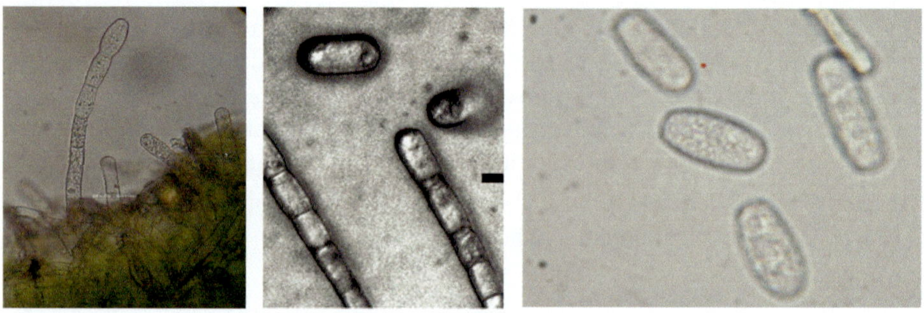

Conidia with conidiophore Arrangement of conidia Conidia

Symptoms & Microscopic structure of powdery mildew of papaya

Epidemiology

The disease occurs when the weather is gloomy, the average temperature fluctuates between 21 and 24ºC, and the air's relative humidity is less than 70% Strong relative humidity is necessary for the conidia to germinate. In orchards with drip (or micro) irrigation systems, the illness is more severe because sprinkler watering is not conducive to the fungus.

Disease Cycle: Ascospore in cleistothecium act as resting spore present on debris. Primary infection initiated by conidia or ascospore. Spores falling on healthy leaves germinate & produce germ tube and spread new diseases. If spores are created in these new infections, the disease may spread secondary.

Management

- Field sanitation are most important factor in management.
- Sulfur dust or wettable sulfur, applied every two weeks starting when the first symptoms develop.

- When there are more than 25% symptoms of the fungus on a leaf chemical management should be adopted.
- Spray triademefon @0.1% repeated at every 15 days after symptom appearance.
- Spray calixin 75EC (5ml/10litre of water) repeated at every 15 days also help to reduce disease incidence (https://www.studocu.com/in/document/).

4. Papaya leaf curl disease

Causal Organism: *Papaya leaf curl virus,* PaLCuV belongs to family *Geminiviridae* & genus *Begomovirus* The nucleic acid have single stranded DNA.

Symptoms: Leaves become curled, wrinkled, twisted with yellow veins. The petioles become short in size & thick. The plants either have no fruits or having small fruits with irregular shape. Infected aged plant growth are checked. Papaya leaf morphology, physiology, and bioactive characteristics were significantly changed by PaLCuV infection. During the later stages of infection, infected plants exhibit defoliation, are unable to produce flowers or fruits, and have restricted growth (Summanwar), 1993

Curling of leaves

Severe infection

Whiteflies

Symptoms & vector of papaya leaf curl disease

Epidemiology: Leaf curl virus have wide host range tobacco, tomato, sunhemp, chilli, petunia, zinnia, datura. Transmission occurs through the grafting & whiteflies vector.

Management

- Rouging and destroying of diseased plants.
- Planting of non-host crop& trap crop to distract whiteflies.
- Use four to five foliar sprays of dimethoate (0.05%) and metasystox (0.02%) spaced ten days apart to manage the population of whiteflies.

5. Papaya mosaic disease

Causal Organism: *Papaya mosaic virus,* PapMV belonging to genus *Potexvirus.*

Symptoms: The disease symptom appears first on young leaves of the plant. The leaves are reduced their size. Vein clearing in mosaic pattern (irregular distribution of dark green & yellow patches) as well as the leaves are cupping downward.. Virus infected plants lead retarted growth of plant and reduces the size of fruits.

Mosaic & vein clearing symptoms on leaves Aphid

Symptoms & vector of papaya leaf curl disease

Epidemiology

Spread of disease occurs through aphid.

Management

- Some cultural practices must be followed e.g. growing of non-host crop, rouging of diseased plant & sanitation of field.
- Apply nuvacron 0.05%,@ metasystox @0.02% at 10 days interval.
- Use tolerant & certified seeds.

6. Papaya ring spot disease

Causal Organism

Papaya ringspot virus belongs to the genus *Potyvirus* and family *Potyviridae*. The flexible rod-shaped particles of the virus have a diameter of 12 nm and a length of 760–800 nm.

The papaya ringspot virus has two main biotypes: the papaya biotype (PRSV-p) identified by Dahal et al. (1997) and the cucurbit or watermelon biotype (PRSV-w) identified by Baker et al. (1991) The virus can spread through grafting, inoculation with sap, and the use of aphid vectors such as *Myzus persicae* (aphid of green peaches), *Aphis gossypii* (aphid of cotton), *Aphid fabae* (aphid of black beans), *Aphid coreopsidis*, and *Toxoptera citricidus* (aphid of black citrus).

Symptoms

Blisters, severe deformations, and mottling could happen. The initial ringspots have a diameter of around 1 mm and are dark green to tan. Later on, the rings may become necrotic and a few of the spots may collapse to form irregularly shaped spots as large as 8 mm. On ripe fruits dark orange-brown ring like symptom appear. Canopy of the tree becomes smaller than healthy ones. Market value of fruit will be reduce.

Symptom on leaf   Ring spot on fruit   Enlarging of ring spot   Aphids

Symptoms & vector of papaya ring spot disease

Epidemiology: Spread of disease occurs through aphid. The virus transmitted plant to plant by aphid in non- persistant manner. It doesn't survive on plant residues. The spread of virus (PRSV-p) occurs between papaya to papaya plant.

Management

- Once this virus established in a papaya orchard, it is disseminated rapidly and challenging to manage that's why quick rouging is needed to reduce secondary infection control & affeccted plant of bigger ones should be cutted.
- Measures for quarantine should be followed to stop the entrance of virus in to new area.
- Cross protection (use of a mild virus strain to shield plants from infection by a severe strain of the same virus) is a good management technique .
- For controlling aphid apply carbofuran (1kg a.i./ha) in nursery bed during sowing of seed.
- Do two to three foliar spray applications of phosphamidon (0.05%) every ten days.

7. Nematodes

Causal Organism

Meloidogyne spp. Body of nematode are pear-shaped, length of stylet is knob shaped measures about 2-3cm.

Symptoms

In many tropical and sub-tropical climates, papaya is considered to be an excellent host for *Meloidogyne* spp. It results in reduced leaf production, fading of the leaves, and early falling of the papaya fruits (Inserra and Cartia, 1977; Roy and Das, 1980 and Khan, 1989). The growth and gall formation caused by root-knot nematodes inhibits the roots, ability to absorb water and nutrients. Strong nematode infestations can limit a papaya tree's productive life and induce wilting, stunting, decreased plant vigor, lower yields, and other symptoms (Anon., 2008).

Infection in roots

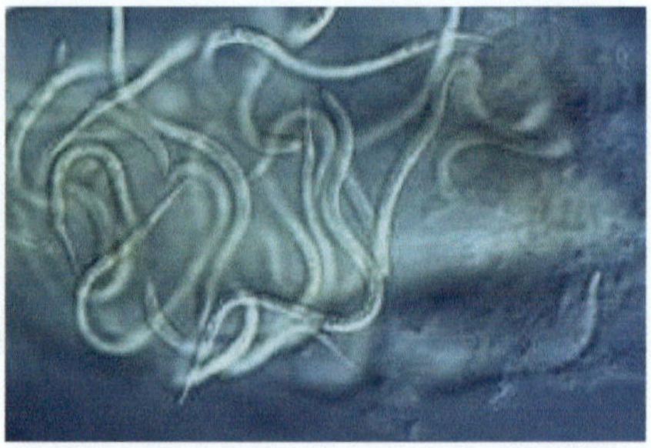

Meloidogyne nematode

\Symptoms & vector of nematode disease in papaya

Epidemiology

They have three life stages egg, juvenile & adult. Huge number of (350)eggs are surviving in soil, which can produce 14–17 generations annually if they are maintained in oothecas. Juvenile (J_2) comes from the egg causes primary infection to the plant. Second-stage juveniles penetrate the root, usually close to the root tip. When the female starts feeding in the central cylinder region of the root, enormous cells are generated. José Aires Ventura, (2007).

Management

- Growing of biochemical having antagonistic plants e.g. marigold, mustard, garlic etc. can kill the nematode & reduces the population.
- Sterilize the field before planting with vapam.
- Use of bioagent e.g. *Pseudomonas putida* significantly reduces the population of nematode, Nesal et.al. (2019).

Conclusion

Numerous pathogen cause diseases in papaya (*Carica papaya*). *Colletotrichum gloeosporioides* infections are the cause of anthracnose perfect stage is *Glomerella cingulata*. *Pythium* spp., *Rhizoctonia spp.* and *Phytophthora* spp. cause root, foot and trunk rots. Aphids & whiteflies transmit the viral diseases like leaf curl virus, mosaic virus, ring spot virus *Meloidogyne* nematode. This chapter help to the student & expert to detect the disease of papaya. So that we can follow different cultural & chemical management practise.

References

Anonymous (2008). ikisan, Agriculture Informatics and Services. Retrieved from (http://www.ikisan.com/tnPapaya-nematode-management.html). Access: 30 dez. 2002.

Baker, C.A.; Lecoq, H.; Purcifull, D.E. (1991). "Serological and Biological Variability Among Papaya Ringspot Virus Type-W Isolated in Florida"(PDF). Phytopathology. 81 (7): 722–728.

Bhargava, K.S., 1941. Pythium aphanidermatum on Carica papaya. Curr. Sci., 10 : 212-213.

Chidderwar, P.P., 1955. A powdery mildew on Carica papaya L. Curro Sci., 24 : 239-240.

Chinnappan, S.; Tamilarasu, V.S.R.K.; Krishnan, U.M.; Pillai, A.K.B.; Rajendiran, S. Inhibition of platelet during Dengue infection: An in vitro study. Viral Immunol. 2016, 29, 164–168.

Chona, B.L. and Girdhari Lal, 1959. A new powdery mildew disease of Carica papaya in India. Indian Phytopath . 12 : 186.

Dahal, G.; Lecoq, H.; Albrechtsen, S. E. (1997). "Occurrence of papaya ringspot potyvirus and cucurbit viruses in Nepal". Annals of Applied Biology. 130 (3): 491.

https://www.studocu.com/in/document/amrita-vishwa-vidyapeetham/diseases-of-fruit-plantation-medicinal-and-aromatic-crops/disease-of-papaya/30426392

Inserra, R. M. and Cartia, G. (1977). Meloidogyne javanica su Papaya in Sicilia. Nematologia Mediterranea, 5, 137-139.

José Aires Ventura, Hélcio Costaand, Joseli da Silva Tatagiba (2007). Papaya Diseases and Integrated Control. In: S.A.M.H. Naqvi (eds), Diseases of Fruits and Vegetables Diagnosis and Management Volume: 201-268.

Karunakaran, P. and Nair, M.C., 1980. Anthracnose disease of papaya. Agr. Res. J. Kerala, 18 : 137-138.

Khan, T. A. (1989). Studies on a disease complex of Papaya caused by Meloidogyne incognita and Fusarium solani, Annual progress report, D. S. T. Project, Department of Botany, Aligarh Muslim University, Aligarh, pp. 55

Misra, A.K. and Prakash, Om, 1986. Studies on diseases of fruit crops. Annual Report. CIHNP, Lucknow. pp. 67-68.

Nesal A. Patel and Ashok D. Patel. (2019). Management of Root-Knot Nematodes (Meloidogyne spp.) Using Different Bio-Agents in Papaya Nursery. Int.J.Curr.Microbiol.App.Sci. 8(08): 1934-1940. doi: https://doi.org/10.20546/ijcmas.2019.808.227

Papaya production in 2020; Crops/Regions/World list/Production Quantity (pick lists)". UN Food and Agriculture Organization, Corporate Statistical Database (FAOSTAT). 2022.

Roy, S. and Das, S. N. (1980). Plant parasitic nematodes associated with field and vegetable crops in Orissa. Journal of Research, QUATXL, pp.71-76

Sharma, N.; Mishra, K.P.; Chanda, S.; Bhardwaj, V.; Tanwar, H.; Ganju, L.; Kumar, B.; Singh, S.B. Evaluation of anti-dengue activity of Carica papaya aqueous leaf extract and its role in platelet augmentation. Arch. Virol. 2019,1 64, 1095–1110.

Summanwar, A.S.; Ram, R.D. Virus disease of papaya. In Advance in Horticulture Fruit Crops; Chaddha, K.L., Pareek, O.P., Eds.; Malhotra Publishing House: New Delhi, India, 1993; pp. 1439–144.

6

Diseases of Guava and Their Management

***Suraj Kumar Patel*[1]*, Reecha J*[1] *and Shashank Shekhar*[1]**

[1]Department of Plant Pathology, Sardar Vallabhbhai Patel University of Agriculture and Technology, Meerut, Uttar Pradesh

Abstract

The Guava (Psidium guajava L.), also referred to as the Poor Man's Apple, is a fruit crop of great importance in India, falling under the Myrtaceae family. Around the world, guava is a commercial fruit crop in tropical and subtropical regions. Vitamins, minerals, fibre, and dietary antioxidants are abundant in the fruit. It is rich in high amount of Vit-C. India accounts 44.4% of global production, which is the biggest and most important guava producer in the world followed by China. The production of guava in India is 40.54 lakh MT, with an average yield of 15.3 MT/ha, on an area of 2.65 lakh hectares. Uttar Pradesh comes in first in production, but Andhra Pradesh has the best productivity at 24.12 MT/ha. The productivity of guava was decreased by various diseases caused by fungal, bacterial and nematode infection i.e., rust, wilt, canker, anthracnose, root knot nematode. In this chapter, major diseases of guava is discussed.

Keywords: *symptoms, epidemiology, management*

Guava

Scientific name	:	*Psidium guajava* L.
Family	:	Myrtaceae
Chromosome number (2n)	:	22

Introduction

The Myrtaceae family includes the guava (*Psidium guajava* L.), a significant fruit crop in India (Nakasone and Paull, 1998). It was originated from Mexico or Central America and later spread to Asia, Africa and Europe. Some of the guava cultivars, which are used to produce seeds-free fruits, are triploid (2n=3x=33) while some are diploid (2n=22). It is primarily grown in tropical and subtropical nations

DOI: 10.1201/9781042014460-6

including India, China, Thailand, Pakistan, Bangladesh, Mexico, Indonesia, and Thailand. The nutritionally rich, vitamin and mineral-rich guava is referred to as the "poor man's apple".

India accounts 44.4% of global production, which is the biggest and most important guava producer in the world followed by China. The production of guava in India is 40.54 lakh MT, with an average yield of 15.3 MT/ha, on an area of 2.65 lakh hectares. The largest guava-producing states in the country include Uttar Pradesh, Madhya Pradesh, Bihar, West Bengal, Punjab, Haryana, Maharashtra, Gujarat, and Karnataka. In terms of both area (0.49 lakh ha) and production (9.28 lakh MT), Uttar Pradesh comes in first, but Andhra Pradesh has the best productivity at 24.12 MT/ha. **(Anonymous, 2018).**The productivity of guava was decreased by various diseases caused by fungal, bacterial and nematode infection.

1. Rust- *Puccinia psidii*

In Brazil in 1884, the disease was first identified on guavas (Anonymous, 1985). The Caribbean, Central America, and South America are the only regions in which it is now found. Melaleuca in Florida has been reported recently (Rayachhetry *et al.*, 1997).

Symptoms

- Initial orange coloration of leaves is followed by crimson coloration.
- Reddish-brown coloration also develops in young shoots, flowers, and fruits.
- Circular rust pustules appearing on the leaves.
- Plant becomes weak and stunted
- Leaves that fall too soon
- Deformed leaves

Guava rust symptoms

Causal agent

Clusters of pale yellow, amphigenous uredia, measuring 0.1 to 0.5 mm in diameter, are produced by *Puccinia psidii* as either dark or brownish patches, having a diameter of 5 mm. (Laundon and Waterson, 1965).

Epidemiology

Between 18 to 22°C and 8 hours of darkness, uredospore germination takes place (de Toledo Piza and Ribeiro, 1988). More than 90% relative humidity and temperature between 15 to 25°C were associated to severe disease on eucalyptus leaves and twigs in Bahia (Carvalho *et al.*, 1994). Cross-infection was possible between guava, eucalyptus, and Syzygium jambo isolates (de Castro *et al.*, 1983).

Management

- Use of chlorothalonil, copper oxychloride, mancozeb, oxycarboxin and triforine as protective control.
- Triforine, oxycarboxin, and tridimenol showed therapeutic efficacy (Ruiz *et al.* 1991).
- Strict quarantine regulations must be followed in places where the disease is not present in order to prevent its introduction. Only under extremely strict regulations are plants of susceptible species from guava rust zones allowed to be brought into Australia. that for a minimum of a year after arrival, involve a post-entry quarantine and disease testing. (Anonymous, 1985).

2. Dieback and Anthracnose- *Colletotrichum gloeosporioides*

In every country that grows guavas, anthracnose is the most common fruit disease both before and after harvest. It may result in significant postharvest losses.

Symptoms

Die back phase

- The plant starts to dry out from the top of a branch downward.
- The growing tip's greenish colour changes to a dark brown and then to black necrotic patches that expand backwards.
- The fungus grows on the contaminated twigs, dropping petioles and immature leaves that may fall or drop, leaving the dry twigs leafless.

Fruit and leaf infection phase

- Infection of fruit and leaves is common during the rainy season.
- On mature fruits, pin-head dots appear and get bigger over time.
- Dark brown in colour, the spots are sunken, round, and contain tiny black stromata in the centre of the lesion, which when wet, create creamy spore masses.
- A number of spots combine to form larger lesions.

- When an unripe fruit has a severe infection, the affected area becomes tough and resilient, and frequently develops cracks.
- The dropping of unopened buds and flowers is also caused by this.

Anthracnose symptoms in guava

Causal agent

C. gloeosporioides (teleomorph: *G. cingulata*) is the most prevalent anthracnose pathogen, however *C. acutatum* (teleomorph: *Glomerella acutata*) (Cooke, 1982).

In guavas, anthracnose is caused by Colletrotrichum gloesporiodies, which multiplies over the course of storage. One of the main alterations is to the cell wall, which has an impact on the fruit's appearance and increases its susceptibility to the pathogen's attack. This has a big impact on marketing and storage. (Botelho *et. al*., 2016). Guava anthracnose, a fungal disease that affects guava trees and their fruit, is primarily caused by the fungus *Colletotrichum gloeosporioides*. This pathogen is responsible for the characteristic symptoms of anthracnose, which include dark lesions or spots on the fruit, often with a sunken appearance, as well as the rotting of affected fruit.

Epidemiology

The primary method by which the infection and disease are dispersed in the field is by conidia. Their generation occurs on decaying twigs, leaves, and other host tissues; on necrotic fruit lesions, and they are dispersed through rain splash. Latent infections—which stay dormant for months—are frequent, even though symptoms could appear soon after infection. The greatest time for disease propagation is in moist conditions. Lesions can appear on fruit at any stage of growth, while they grow fastest at 30°C (Pandey *et al*., 1997).

Management

- The most effective way to manage this disease is through resistant cultivars. "Allahabad Safeda" and "Banarsri Surkha" provided resistant hybrids (Naresh *et al*.,). Some hybrids produced by crossings with "Apple Colour" exhibited resistance (Sharma, 1981).

- Using hot water in combination with benomyl and carbendazim as postharvest treatments and in the field helps lower postharvest anthracnose.
- Thiophanate methyl, thiabendazole and benomyl at 5–50 g /l all significantly prevented the formation of acervulus (Butt *et al.*, 1995).
- Vitigran blue prevented the developement of acervulus at 50 µg in 1 litre water, while boric acid decreased colony expansion more than bleaching powder or borax.

3. Fruit canker- *Pestalotia psidii*

Reported of Fruit Canker from Bombay indicated that it was caused by *Pestalotia psidii*. (Chibber, 1911) and later from Dharwar, Thane, Poona (Patel *et al.*, 1950), Mysore (Narsirnhan, 1938, Venkatakrishniah, 1952), Himachal Pradesh, Ponta Valley (Verma and Sharma, 1976) and Lucknow(Mishra and Prakash, 1986).

Symptoms

- Usually, symptoms appear on green fruits, although they are infrequent on leaves.
- Fruits first develop tiny, unbroken, round, necrotic patches that are brown or rust in colour.
- The lesion exhibits a raised edge and a visible internal depression. Fruits have a more pronounced crater-like look than leaves do.
- White mycelium with many spores is seen in older cankers.
- If the condition is severe, there are more raised cankerous areas and the fruits split open to reveal seeds.
- Affected fruits stay immature, becoming tough, distorted, and eventually fall off.
- Occasionally, the leaves develop tiny, sharp brown dots.

Fruit canker symptoms

Causal organism

- Fruit canker is caused by *Pestalotia psidii* Pat. (Chibber, 1911). Narsirnhan (1940) and Venkatakrishniah (1952) found *C. psidii* Curzi, *G. psidii* (Del.) Sheld. and *Pestalotia psidii* Pat. related with canker. Venkatakrishniah (1952) secured that *Colletotrichum psidii* is a general parasite and *Pestalotia psidii* is a specialized to guava.
- Both species are found on immature green and mature fruits and foliage, however the true source of canker is thought to be *Pestalotia psidii* (Patel *et al.*, 1950).
- Pathogen produces round, dark-black pycnidia with conidia and conidiophore. Usually measuring 13–31 x 5–10 urn, the conidia are oblong, clavate or elliptic–fusoid, erect, hardly constricted at septa, and have three medium-sized, highly brownish cells. The central cell is the thickest and most enlarged, while the other cells are relatively hyaline. The apical conidia are slender, hyaline, and have elongated appendages, while the basal cell is obtuse, erect, and has a small pedicel.
- Young cultures have thin, septate, cottony white to pinkish mycelium that is branched sporadically. The hyphae are somewhat thicker in older cultures (Patel *et al.*, 1950).

Epidemiology

- Maximum diseases occurs at 25-30°C and high RH. (Kaushik *et al.*, 1972).
- *Pestalotia psidii* spores germinate best at 30°C, that spore do not germinate at temperatures below 15°C or above 40°C (Ramaswamy *et al.*, 1984).
- Guava fruit extract is the most effective germination medium, high relative humidity (98%) is required for germination.

Management

- During the early stages of infection, the disease can be controlled by spraying 1% Bordeaux mixture and lime sulphur in a ratio of 1:25 three to four times at 15-day intervals (Venkatakrishniah, 1952).
- Using arsenic oxide with potassium iodide prevents the formation of pathogens (Khanna and Chandra, 1977).
- Spore germination is inhibited by leaf extract of *Ocimum sanctum* and *Azadirachata indica*. Either before or after inoculation, dipping guava fruits in these extracts works well. It is advised to use *Ocimum sanctum* extract because it has no effect on the flavour of the fruit (Pandey *et al.*, 1983).

4. *Phytophthora* fruit rot

In Cuba, there have been reports of phytopthora fruit rot cases. (Sohi and Sridhar, 1971; Singh *et al.*, 1976, 1978; Gupta *et al.*, 1977; Sharma *et al.*, 1978; Gupta, 1988; Mathur *et al.*, 1992), Malaysia (Lim and Chin, 1987), Hawaii (Ko *et al.*,

1982), and India (Ariosa, 1982). Chin and Lim (1987) also mentioned guava seedlings and plants suffering from Phytophthora foliar blight.

Symptoms

The fruit developed a wet lesion as the first symptom. The lesions become grey in colour and, in humid environments, develop fine, fluffy, cottony masses of white mycelia. Ripe yellow fruit has a lower susceptibility to the illness than young, mature green fruit. (Ko *et al.*, 1982). Affected leaves have a damp, withered appearance, and seedling young shoots may wither back. (Lim and Chin, 1987).

Phytophthora fruit rot in guava

Causal agent

There are at least two taxa involved: *P. nicotianae* (Cuba, India, and Malaysia) and *P. citricola* (Hawaii) A2 mating type (Ariosa, 1982; Ko *et al.*, 1982; Singh *et al.*, 1978; Lim and Chin, 1987). There have only been reports of *P. nicotianae* mating type A1 in Malaysia. Using guava fruit or specific media as bait, it is feasible to isolate both species from the soil (Singh *et al.*, 1978). *P. nicotianae* generates thick-walled chlamydospores.

Epidemiology

P. nicotianae is the predominant source of inoculum, producing thick-walled chlamydospores that remain in the soil. In the soil and on tissue of dead hosts, the fungus can endure for 12 to 16 months. (Prasad and Mehrotra, 1985). Fruit infection secondary sources of inoculum are infected leaves and twigs.

Relative humidity more than 90% is favorable for this disease. Mathur *et al.* (1992) found that the ideal conditions for the development and transmission of the disease are at least three to four days of rain each week, with minimum temperatures of 21 to 22°C and maximum temperatures of 28 to 30°C.

Management

- The disease was 5-20% susceptible in red, pink flesh, Apple Guava, and Allahabad Safeda, variety (Singh *et al.*, 1976).

- Chitidar and Apple Guava exhibited comparatively high resistance. Highly vulnerable were Allahabad safeda, Lucknow 49 and Banasri Surkha (Gupta *et al.*, 1977).
- Using sterile soil and preventing soil splashing during irrigation will help lower the disease's incidence on guava seedlings in nurseries.
- Foliar sprays of fosetyl-Al, metalaxyl, or their combinations with mancozeb and etridiazole can be used to control the disease (Lim and Khoo, 1990).
- Additionally, control was achieved using aureofungin and zineb (Sohi and Sridhar, 1971) and zineb and captafol (Gupta, 1988).
- It has been reported that copper fungicides are phytotoxic. (Sohi and Sridhar, 1971; Lim and Khoo, 1990).

5. Guava wilt- *Fusarium oxysporum* f. sp. *psidii*

Kurosawa was the first to report the guava wilt disease in Taiwan in 1926. Guava wilt with nearly similar symptoms emerged for the first time in South Africa in 1980 and was later discovered in Malaysia's Johor Province in 1995 (Schoeman, 1997). That same fungus, or one closely linked to it, is probably the source of the ailments in these countries.

A 270 ha commercial planting of the clonally propagated pink variety "Beaumont" in Malaysia was affected by the disease in 42 percent of cases (Schoeman, 1997). The disease initially struck South Africa's Mpumalanga Province in the southeast. It quickly spread throughout the guava-growing regions in ten years. All guava-producing regions in Taiwan are affected by the disease, which lowers orchard life expectancy from 25 to 10–12 years. In Taiwan, infection seems to mostly come from pruning wounds, but in South Africa, it's more of a soil-borne illness (Grech, 1986)

Symptoms

- When the monsoon season begins, pale yellow leaves appears along with a decrease in turgidity and epinasty.
- In later stages, premature shedding and defoliation are observed.
- A portion of the plant twigs get dry and don't sprout new leaves or blossoms.
- Every branch that is impacted still has undeveloped, stony, and hard fruits. The plant finally dies and loses all of its leaves later.
- Additionally, the basal portion of the roots exhibits rotting, and the bark can be simply separated from the cortex. Vascular tissues also exhibit a little brown colour.
- Both young and old fruit-bearing trees are attacked by the pathogen, but older trees are more vulnerable to the infection.

Causal agent

There is confusion in the classification of the fungus causing guava wilt. The fungus that caused the disease in Taiwan was found to be *Myxosporium psidii* (Kurosawa, 1926). It is, however, *Fusarium oxysporum* f. sp. *psidii* that causes the illness in India (Pandey and Dwivedi, 1985). *Macrophomina phaseolina* is the cause of a less prevalent wilt in this nation. (Dwivedi, 1990).

Epidemiology

30°C is the ideal temperature for the guava wilt pathogen. Guava is especially vulnerable in the summer when it's hot (Schoeman, 1995). Injuries accelerated the spread of the disease; trees exhibited serious symptoms in three months and died six months later. When square plants were utilised, the disease expanded radially in fields that received sprinkler irrigation and in rows where furrow irrigation was employed. If not, the distance from infected trees determined the spread.

Within 60 days, seedlings put in soil infected with *F. oxysporum* f. sp. *psidii* in India withered and perished (Pandey and Dwivedi, 1985).

Management

- One way to slow down spread is to disturb up afflicted plants (Leu *et al.*, 1979). The fungus can live on root fragments for more than a year, but it can only endure a few months in soil. As such, it is imperative to clear the soil of as much root debris as possible prior to replanting.
- Using copper oxychloride and benomyl to treat pruning wounds.

Before planting, add an arbuscular mycorrhizal inoculum to the soil to reduce disease.

6. Nematode

On guava, nematode species from sixteen genera have been documented. The most troublesome and pervasive, though, are those in Meloidogyne.

Symptoms

Chlorosis is one of the aboveground signs, along with decreased leaf yield, size, and growth. These infections cause significant root distortion and a reduction in fine root densities below ground; *Meloidogyne spp*. develop small to large numerous galls.

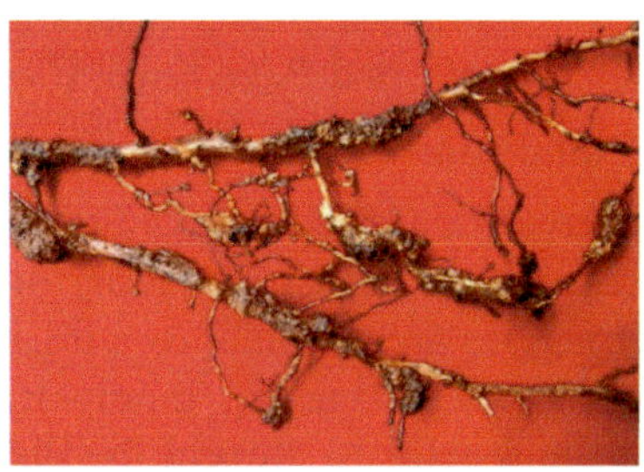

Root knot nematode

Causal agent

The root knot nematodes, *Meloidogyne spp*., have been studied the most in relation to guavas. There have been reports of *M. incognita* in Cuba, Brazil, Malaysia and Venezuela (Fernandez and Silveira, 1975; Razak and Lim, 1987; Crozzoli *et al*., 1991). Using distinct host differentials, races 1 and 2 have been identified (de Moura and de Moura, 1989; Crozzoli and Casassa, 1998). There have been reports of *M. arenaria* in Venezuela and Cuba (Fernandez and Silveira, 1975; Crozzoli *et al*., 1991), *M. javanica* in Mexico (Carillo *et al*., 1990) and *M. mayaguensis* in South Africa (Willers, 1997).

Epidemiology

The guava-related biology of *M. incognita* and *M. arenaria* was described by Fernandez et al. (1986). After hatching, the larvae pierce roots in 24 hours, and their life cycle spans 26–30 days. The ideal pH for hatching is 6-7, and the ideal temperature ranges for *M. arenaria* and *M. incognita* are respectively 27–30°C and 25–30°C. Plant waste and contaminated soil spread the larvae of root knot nematodes, which are found in light-textured soils and thrive there (Razak and Lim, 1987).

Management

- The most effective control approach is exclusion. By utilizing de-infested soil and non-infested planting supplies, this can be accomplished at the nursery stage (Lim and Khoo, 1990).
- According to Fernandez et al. (1987), Meloidogyne spp. were controlled with 50 g of 98% methyl bromide m-2 and 100 g of dazomet m-2, producing 100% healthy seedlings.
- Fenamiphos worked well; fruit samples taken one week to two months following soil treatment showed no observable chemical residues. (Lim and Khoo, 1990). Carbofuran is also effective.
- After a second treatment, no residues were found in the fruit, and nematode numbers were decreased by ethoprophos. At comparable rates, etoprophos outperformed fenamiphos and carbofuran (Casassa *et al*., 1996).
- It was suggested to use *Psidium friedrichsthalianum* as a rootstock since it exhibited greater resistance to *M. incognita* and *M. arenaria* than *Psidium molle* and *P. guajava*. (Fernandez and Silveira, 1975).

Conclusion

Guava diseases pose a significant challenge to guava growers worldwide, affecting both the quality and quantity of fruit production. Effective disease management strategies are essential to mitigate these issues and ensure a healthy guava crop. Use resistant variety, timely irrigation, maintaining proper spacing, sanitation of field helps to prevent the disease incidence. Use of chemical as the curative measures. Integrated disease management provide effective management of the diseases in the guava orchards.

References

Adisa, V.A. (1985) Fruit rot diseases of guava in Nigeria. Indian Phytopathology 38, 427–430.

Anonymous (1985) Guava rust. Australia Commonwealth Department Primary Industry Plant Quarantine Leaflet No. 45.

Anonymous (2018). Horticultural Statistics at a Glance, Department of Agriculture Cooperation and Farmers Welfare, Horticulture Statistics Division, Ministry of Agriculture and Farmers" Welfare, Government of India. 458 pp.

Ariosa, T.I.M.D. (1982) Una neuva enfermedad de la guayabe (Psidiumguajava L.) en la provincia Sancti Spiritus. Centro-agricola 9, 3–7 (in Spanish).

Benade, E., Kemp, G.H.J., Wingfield, M.J. and Kock, J.F.L. (1991) Comparison of Acremonium diospyri with the guava wilt pathogen in South Africa. Phytophylactica 23, 98.

Botelho, L. N. S., Rocha, D. A., Braga, M. A., Silva, A., & de Abreu, C. M. P. (2016). Quality of guava cv.'Pedro Sato'treated with cassava starch and cinnamon essential oil. Scientia horticulturae, 209, 214-220.

Butt, A.A., Nasir, M.A. and Bajwa, M.N. (1995)Invitro evaluation of different chemicals against Gloeosporiumpsidii, the cause of anthracnose of guava. Pakistan Journal of Phytopathology 7, 92–93.

Carillo, R.J., Carrillo, F.C. and Dominguez-Alvarez, J.L. (1990) Nematods asociados al cultivo de la guayabe (Psidium guajava L.) y control quimico en el Canon de Juchipila, Zac., Mexico. Revista Chapingo 15, 67–68, 94–97 (in Spanish).

Carvalho, A. de O. de, Alfenas, A.C., Maffia, L.A. and Carmo, M.G.F. do (1994) Avaliacao do progresso de ferrugem (Pucciniapsidii) em brotacoes de Eucalpytus cloeziana no sudeste da Bahia. Revista Arvore 18, 265–274 (in Portuguese).

Casassa, P.A.M., Matheus, C.J.M., Crozzoli, P.R. and Casanova, A. (1996) Control quimico de Meloidogyne spp. em el cultivo del guayabo (Psidium guajava L.) en el Minicipio Mara del Estado Zulia, Venezuela. Revista de la Facultad de Agronomia, Universidad del Zulia 13, 303–312 (in Spanish).

Chibber, H. M. (1911) A working list of diseases of vegitable pests of some of the economic plants, occuring in the Bombay Presidency. Poona Agric. Coll. Mag., 2: 180- 198.

Cooke, A. (1982) Guava. In: Persley, D. (ed.) Diseases of Fruit Crops. Department of Primary Industries, Queensland, p. 59.

Crozzoli, P.R., Cassa, P.M., Rivas, G.D. and Matheus, C.J. (1991) Nematodos fitoparasitos asociados al cultivo del guayaba en el estado Zulia, Venezuela. Fitopatologia Venezolana 4, 2–6 (in Spanish).

Crozzoli, R. and Casassa, A.M. (1998) Especies y razas de Meloidogyne en el cultivo del guayabo en Venezuela. Revista de la Facultad de Agronomia, Universidad del Zulia 15, 107–108 (in Spanish).

De Castro, H.A., Krugner, T.L., Ideriha, C.H.F., Capello, M.S.C. and Marchi, A.B. (1983) Inoculacao cruzada de Eucalyptus, goiaba (Psidiumguajava) e Jambeiro (Syzygiumjambos) com Pucciniapsidii. Fitopatalogia Brasileira 8, 491–497 (in Portuguese).

De Moura, R.M. and de Moura, A.M. (1989) Meloidogyninose da goiabeira: doenca de alta severidade no estado de Pernambuco, Brasil. Nematologia Brasileira 13, 13–19 (in Portuguese).

De Toledo Piza, S.M. and Ribeiro, I.J.A. (1988) Influencia da luz e da temperatura na germinacao de uredosporos de Puccinia psidii. Bragantia 47, 75–78 (in Portuguese).

Dwivedi, S.K. (1990) Guava wilt incited by Macrophomina phaseolina. National Academy of Science Letters 13, 301–303.

Fernandez, D. and Silveira, M. (1975) El Psidium friedrichsthalianum como patron para guayabo, resistente a los nematodos del genero Meloidogyne. Revista de Agricultura Cuba 3, 80–85 (in Spanish).

Fernandez, E., Miralles, L., Navarro, A. and Sanchez, Y.M. (1987) Efecto de diferentes dosis de bromuro de metilo sobre Meloidogyne spp. en suelo para viveros de guayabo (Psidium guajava). Cienca y Tecnica en la Agricultura, Proteccion de Plantas 10, 59–71 (in Spanish).

Fernandez, E., Valdes, S. and Carrasco, J. (1986) Estudio del la biologia de especies de Meloidogyne que atacan el cultivo de la guayabe (Psidium guajava). Cienca y Tecnica en la Agricultura, Proteccion de Plantas 9, 27–41 (in Spanish).

Grech, N.M. (1985) First report of guava rapid death syndrome caused by Septofusidium sp. in South Africa. Plant Disease 69, 726.

Guava: Diseases and symptoms. https://vikaspedia.in/agriculture/crop-production/integrated-pest-managment/ipm-for-fruit-crops/ipm-strategies-for-guava/guava-diseases-and-symptoms retrieve on 20/05/2020

Gupta, J.H. (1988) Fungicidal control of fruit rot of guava caused by Phytophthora parasitica Dastur. Progressive Horticulture 20, 331–332.

Gupta, P.C., Madaan, R.L. and Suhag, L.S. (1977) Varietal reaction of guava fruits to Phytophthora nicotianae var. parasitica. Indian Journal of Mycology and Plant Pathology 7, 177.

https://plantvillage.psu.edu/topics/guava/infos

Khanna, K. K. and Chandra, S. (1977) Control of guava fruit rot caused by Pestalotia psidii with homeopathic drugs. Pl. Dis. Rep., 61 (5) : 362- 366.

Ko, W.H., Kinimoto, R.K. and Nishijima, W.T. (1982) Fruit rot of guava caused by Phytophthora citricola. Plant Disease 66, 854–855.

Kurosawa, E. (1926) Guava Tachigare Byd. Report of the Taiwan Museum 83, 47–61 (in Japanese).

Laundon, G.F. and Waterson, J.M. (1965) Puccinia psidii. CMI Descriptions of Pathogenic Fungi and Bacteria No. 56. Common wealth Mycological Institute, Kew, UK.

Leu, L.S., Kao, C.W., Wang, C.C., Liang, W.J. and Hsieh, S.P.Y. (1979) Myxosporium wilt of guava and its control. Plant Disease Reporter 63, 1075–1077.

Leu, L.S. and Kao, C.W. (1979) Artificial inoculation of guava with Myxosporium psidii. Plant Disease Reporter 63, 1077–1079.

Lim, T.K. and Chin, C.L. (1987) Foliar blight of guava seedlings caused by Phytophthora nicotianae var. nicotianae. Fitopatologia Brasileira 12, 251–254.

Lim, T.K. and Khoo, K.C. (1990) Guava in Malaysia Production, Pests and Diseases. Tropical Press, Kuala Lumpur, Malaysia.

Mathur, S., Bhatnagar, M.K. and Mathur, K. (1992) Occurrence and epidemiology of fruit rot of guava caused by Phytophthora nicotianae var. parasitica. Indian Phytopathology 45, 217–220.

Misra, A. K. and Prakash, Om (1986) Studies on diseases of fruit crops. Annual Report, Central Institute of Horticulture for Northern Plains, Lucknow. pp. 67-68.

Nakasone, H. Y., and Paull, R. E. (1998). Tropical fruits. Cab International.

Naresh, M., Madaan, R.L. and Daulta, B.S. (1987) Evaluation of guava hybrids to anthracnose fruit rot under field conditions. Haryana Journal of Horticultural Sciences 16, 82–85.

Narsimhan, M. J. (1938) Pestalotia psidii on guava in India. Annual Administrative Report, Agriculture Department, Mysore, 1936 - 37. pp.169-173.

Narsimhan, M. J. (1940) Annual Report, Mysore Agriculture Department for the year 1938-39. pp. 96 - 97.

Pandey, R.R., Arora, D.K. and Dubey, R.C. (1997) Effect of environmental conditions and inoculum density on infection of guava fruits by Colletotrichum gloeosporioides. Mycopathologia 137, 165–172.

Pandey, R.R. and Dwivedi, R.S. (1985) Fusarium oxysporum f. sp. psidii as a pathogen causing wilt of guava in Varanasi district, India. Phytopathologiese Zeitschrift 114, 243–248.

Pandey, R. S., Bhargava S. N., Shukla, D. N. and Dwivedi, D. K. (1983) Control of Pestalotia fruit rot of guava by leaf extracts of two medicinal plants. Revista Maxicana de Fitopathologia, 2 (1): 15-16.

Patel, M. K., Kamat, M. N. and Hingorani, G. M. (1950) Pestalotia psidii Pat. on guava. Indian Phytopath., 3 (2): 165-176

Prasad, G. and Mehrotra, R.S. (1985) Persistence of Phytophthora nicotianae var. parasitica in artificially infested soil at different inoculum levels in dead host tissues. Indian Journal of Mycology and Plant Pathology 15, 279–282.

Rayachhetry, M.B., Elliot, M.L. and Van, T.K. (1997) Natural epiphytotic of the rust Puccinia psidii on Melaleuca quinquenervia in Florida. Plant Diseases 81, 831.

Razak, A.R. and Lim, T.K.(1987) Occurrence of the root knot nematode Meloidogyne incognita on guava in Malaysia. Pertanika 10, 265–270.

Ruiz, R.A.R., Alfenas, A.C., Ferreira, F.A. and Zambolim, L. (1987) Fungicidas protetoes e sistemicos para o controle da ferrugem do eucalipto causada por Puccinia psidii. Revista-Avore 11, 56–59 (in Portuguese).

Ruiz, R.A.R., Alfenas, A.C. and Demuner, N.L. (1991) Efficiencia de fungicidas para o controle da ferrugem (Pucciniapsidii) em goiabeira (Psidiumguajava). Summa-Phytopathologica 17, 147–153 (in Portuguese).

Schoeman, M.H. (1995) Effect of temperature on guava disease development. ITSC Information Bulletin, August, 5–6.

Sharma, B.S., Champawat, R.S. and Bhanagar, G.C. (1978) Effect of chemicals and hot water treatments on Phytophthora fruit rot of guava. Indian Journal of Mycology and Plant Pathology 8, 40.

Sharma, S.K. (1981) Control of anthracnose of guava fruits caused by Glomerella cingulata (Stonem.) Spauld. and Schrenk. Thesis/Abstract 7, 157.

Singh, G., Chohan, J.S. and Mann, S.K. (1976) Fruit rot of guava – a new disease problem in the Punjab. Indian Journal of Mycology and Plant Pathology 6, 77

Singh, K.B., Prasad, G., Bhargava, K.S. and Mehrotra, R.S. (1978) Studies on the occurrence of fruit rot of guava due to Phytophthora nicotianae var. parasitica. Indian Phytopathology 31, 263.

Sohi, H.S. and Sridhar, T.S. (1971) Controlling fruit rot of guava. Indian Horticulture 16, 9–10.

Venkatakrishniah, N. S. (1952, September).Glomerella psidii (Del.) Sheld. and Pestalotia psidii Pat. associated with a cankerous disease of guava. In Proceedings of the Indian Academy of Sciences-Section B (Vol. 36, pp. 129-134). Springer India.

Verma, B.R. and Sharma, S.L. (1976) Seasonal variation in symptoms caused by Pestalotia psidii on guava fruits. Indian Journal of Mycology and Plant Pathology 6, 97–98.

Vos, J.E., Schoeman, M.H., Berjak, P., Watt, M.P., Toerien, A.J. and Verhoyen, M.N.J. (2000). In: In vitro selection and commercial release of Guava wilt resistant rootstock. XXV International Congress, Part 3: Culture techniques with special emphasis on environmental implications, Brussels, Belgium, August 2–7, 1998. Acta Horticulturae 513, 69–79.

Willers, P. (1997) First record of Meloidogyne mayaguensis Rammah and Hirschmann, 1988. Heteroderidae on commercial crops in the Mpumalanga province, South Africa. In ligtings bulletin Insituut vir Tropiese en Subtropiese - Gewasse no. 294, 19–20.

7

Diseases of Grapevine and Their Management

Shyam Lal [1], Priya Devi Maurya [2], Harshita Yadav [3] and Shivani Chaudhary [4]

[1]Department of Plant Pathology, Chandra Shekhar Azad University of Agriculture and Technology, Kanpur- 208002, Uttar Pradesh
[2,4] Department of Plant Pathology, Sardar Vallabhbhai Patel University of Agriculture and Technology, Meerut, Uttar Pradesh
[3]Department of Plant Pathology, Acharya Narendra Deva University of Agriculture and Technology, Kumarganj, Ajodhya, Uttar Pradesh

Abstract

One of the fruits that is grown worldwide most commonly, grapes are said to have a longer history of human interaction than either rice or wheat. Apart from their nutritional significance, grapes are today regarded as one of the most lucrative crops. The grapes are farmed for a variety of uses, including the manufacture of raisins, wine, juice, canning, and table grapes. More grapes are produced worldwide than any other fruit crop. The grapesare high in calcium and phosphorus and has about 20% sugar in an easily digested form. Around the world, its primary uses are in the production of wine (82%), raisins (10%), and table grapes (8%). However, it is primarily eaten as fresh fruit in India, where a small amount is also used to make liquor and dry fruits like raisins. Approximately 11 million hectares are devoted to growing grapes, which yield 75 million tonnes yearly. More than half of the world's supply of grapes was produced in Spain, Italy, and France. Grapes are cultivated in India in both tropical and subtropical climates. In Maharashtra, Karnataka, Andhra Pradesh, and Tamil Nadu, commercial grape production only started in the last 40 years. In this lesson we have explain the major several disease of grapevine viz. Downy mildew, Powdery mildew, Anthracnose, Black rot, Rust, Bacterial canker with their causes, symptoms and management.

Keyword: *Grapevine, Causes, Symptoms, Management etc.*

DOI: 10.1201/9781042014460-7

Grapevine

One of the fruits that is grown worldwide most commonly, grapes are said to have a longer history of human interaction than either rice or wheat. Apart from their nutritional significance, grapes are today regarded as one of the most lucrative crops. The grapes are farmed for a variety of uses, including the manufacture of raisins, wine, juice, canning, and table grapes. More grapes are produced worldwide than any other fruit crop. Approximately 11 million hectares are devoted to growing grapes, which yield 75 million tonnes yearly. More than half of the world's supply of grapes was produced in Spain, Italy, and France. Grapes are cultivated in India in both tropical and subtropical climates. In Maharashtra, Karnataka, Andhra Pradesh, and Tamil Nadu, commercial grape production only started in the last 40 years. Grapevine production affect many sever diseases like downy mildew, powdery mildew, Anthracnose, Black rot, Rust, Bacterial canker and abiotic causes.

Major disease of grapevine given under given following

1. Downy mildew : *Plasmopara viticola*
2. Powdery mildew : *Unicnual necator*
3. Bird's eye spot/Anthracnose : *Gloeosporium ampelogum*
4. Black rot : *Guignardia bidwelli*
5. Rust : *Phakospora euvitis*
6. Bacterial canker - : *Xanthomonas campestris pv. viticola*

1. Downy mildew

Grape downy mildew is a serious problem in grapevines grown area. The pathogen infected all green plant tissues, such as leaves, new stems, flowers, and developing berries etc.In an attempt to address the threat posed by the Phylloxera aphid, which feeds on plant roots, Europe imported grape rootstocks that were resistant to the pest from North America in the year 1878.At the same time, the downy mildew infection expanded rapidly over Europe after being introduced to France, most likely as oospores on the imported rootstocks. Bordeaux mixture was initially employed to combat downy mildew in 1885by P.M.A. Millardet . According to legend, a farmer used this copper sulphate and lime mixture to coat the grapes growing next to the road with a chemical coating to deter people from stealing them. Bordeaux combination, the first commonly used chemicalsubstance to shield plants from fungus diseases, was created as a result of lengthy experimentation by Millardet.

Causal organism: *Plasmopara viticola* ((Berk. & M.A. Curtis) Berl. & De Toni, (1888)

Systemic position

Kingdom: *Chromista*

Phylum: Oomycota

Order: Peronosporales

Family: Peronosporaceae

Genus : Plasmopara

Species : ***P. viticola***

Symptoms

The first signs of grape downy mildew, which is induced by ***Plasmopara viticola,*** often appear onthe leaves 5 to 7 days following infection. Foliar symptoms (also called oilspots) are round, yellow spots with an oily look. A brownish-yellow halo surrounds newly formed oilspots on newly formed leaves. The oilspot ages and the halo disappears.The majority of the leaf surface may become covered in oil spots if the weather is favourable. The lower side of the leaves and other diseased plant parts will develop a white downy fungal growth (sporangia) after a series of suitably warm, humid night. This downy development is what gives the illness its nickname "downy mildew.

(a) **(b)** **(c)**

(a) An oily mark observed on the upper side of recently developed grape leaves, **(b)** An oily mark present on the upper side of mature grape leaves, **(c)** white downy fungal growth (sporangia) lower surface of grapes leaf

Etiology

The pathogen, producesexual oospore and asexual biflagellate zoospores, aseptate (coenocytic) mycelium. It belongs to the Peronosporales suborder of oomycete. The genera Bremia, Peronospora, and Sclerospora each contain species that are significant downy mildew infections. The majority of taxonomists now place oomycetes in the kingdom Chromista rather than include them in phylogenetic groups that also include ascomycetes or basidiomycetes.

The obligatory parasite ***Plasmopara viticola*** uses globose haustoria to absorb foodfrom the living host. Most of the hyphae are located inside the host. In the host tissue, antheridia and oogonia fuse to produce sexual reproduction. It was only recently discovered that *Plasmopara viticola* is heterothallic. An oospore, the pathogen's stage of survival and rest, is the sexual spore that results from this process.

The main source of infection is an oospore, which has the ability to survive the winter in leaf litter, get released into the soil during leaf decomposition, or be buried by decomposing organisms. Oospore populations may continue to sprout during the entire growing season in certain regions. In many cases, they begin germinating in significant numbers right after the emergence of grape buds. The germination process involves the formation of a single germ tube that culminates in a sporangium.

Zoospores are created inside the sporangia and subsequently discharged. Only through active stomata, or only on green host tissue, zoospores germinate and enter the plant. Desiccation is a simple process for sporangia and zoospores. Most infections occur shortly after their release, as they perish within 2 to 3 hours of being exposed to low humidity and sunlight. However, under cool, humid environments, they may remain on leaf surfaces for longer than 24 hours.

The infection can also spread secondarily through sporangiaArising from a mycelial mat embedded in the host tissue, treelike sporangiophores with white, lemon-shaped sporangia emerge through stomata. Wind and splashing rain spread sporangia. Zoospores released from the sporangia encyst close to a stoma by swimming in open water on the grapevine surface. After germinating, zoospores pass through a stoma using a germ tube.

Cycle of the disease

The pathogen overwinters in the form of oospores within deceased leaves and other host tissues found on the vineyard floor. Oospores may be released onto the soil's surface as decaying plant material breaks down. Typically, oospores give rise to sporangia, which, in turn, produce zoospores. Sporangia and zoospores are dispersed onto the lower leaves and tissues of grapevines through rain and wind splatter.

Oospore germination necessitates damp soils with temperatures at least 10°C (50°F) above freezing. Some regions use a guideline of at least 10 mm (0.04 in) of rain within a 24-hour period to ensure adequate soil moisture. Once there is a minimum of 45 degree-hours of leaf wetness, infection occurs following the formation and distribution of zoospores to host tissues.Zoospores encysting then germinate and go through stomates. Most infections happen in the morning because sporangia normally become inactive After being exposed to sunlight for a period of 5-7 hours (Kennelly et al., 2007). Following infection, the pathogen multiplies intracellularly and creates haustoria.

Sporangia for secondary infections are produced on sporangiophores that extend through the stomata of affected leaves and other vegetative parts of grapevines, including bunches. It need at least 4 hours of darkness, 95 to 100% relative humidity, and temperatures that are initially higher than 13°C (55°F) for the production of sporangiophores and sporangia. How frequently foliar lesions were made to sporulate affects their lifetime and productivity. Lesions often maintained their maximal capacity to generateSporangia persist for a minimum of 22–24 days, potentially extending up to 2-3 months in the absence of sporulation opportunities or exceptionally high temperatures (Kennelly et al., 2007).

According to Rossi et al. (2013), the incubation period—the amount of time between infection and the onset of new symptoms—Manifest within a range of 5 to 21 days, influenced by temperature, the specific tissue targeted, and ontogenic resistance of the host.At typical temperatures of 20–25 °C (68–77 °F), the incubation period is the shortest (5 days). The incubation duration extends to a minimum of 14 days when average temperatures are 12°C (54°F) or lower.Sexual reproduction takes place in the latter part of the season. The resultant oospores have strong walls and function as spores for survival.

Disease Management

Cultural practices

When made the grape orchard, the place, drainage, methods of irrigation, and training and pruning system should all be chosen to minimize the attack of disease. Grapevines ought to be planted in effectively drainedareas with sufficient air flow since moisture encourages the growth of downy mildew. To help reduce leaf wetness, grapevines ought to be arranged in rows that take advantage of the design of natural wind circulation. In mature vineyards, there are few cultural management measures available for dealing with downy mildew.To reduce leaf wetness, the leaf canopy can be managed via trellising systems and trimming. To reduce secondary infection, avoid nighttime increases in humidity and leaf moisture.Whenever feasible, it is advisable to steer clear of overhead irrigation or schedule it in a way that facilitates rapid drying of the leaves. Pruning and training of grapevine plants also diminish the quantity of overwintering inoculum.

Use of genetic resistance

All varieties of the Eurasian species *Vitis vinifera* are believed to be susceptible to downy mildew, with certain types such as Chardonnay, Pinot Noir, and Sultana considered more prone than Cabernet Sauvignon and Semillon. While *V. labrusca* cultivars like Niagara and Catawba are highly susceptible, several North American species like *V. rotundifolia* and *V. labrusca* display resistance to downy mildew. Crossbreeding between North American species and *V. vinifera* has resulted in cultivars exhibiting enhanced wine-grape characteristics and increased resistance to downy mildew. However, popular varietal wines are preferred by wine consumers over less-known and occasionally fruitier hybrids. Ongoing research projects are focused on genetically modifying *V. vinifera* cultivars to enhance disease resistance.

Chemical control

To combat downy mildew, fungicides are commonly utilized, both as pre-infection (protective) and post-infection (systemic or penetrant) treatments. Pre-infection chemicals are applied before the onset of an infection, with their optimal use occurring regularly between shoot lengths of 10 cm (4 in.) and when berries reach the size of peas, a period when the host is most vulnerable. Alternatively, pre-infection agents may be intermittently applied through misting just before anticipated weather conditions favoring P. viticola infection. Dithiocarbamates and copper-based fungicides like Bordeaux mixture (refer to the section on significance) are examples of pre-infection fungicides, and there are no reported instances of resistance to these fungicides in the pathogen. The frequency of sprays depends on the prevailing weather conditions.As post-infection fungicides are more costly than their pre-infection counterparts, it is advisable to use them judiciously. Their application is recommended as soon as possible after an infection incident and before the emergence of oil spots for optimal efficacy. Currently used post-infection fungicides include phosphonate (e.g., fosetyl-aluminum), phenyl amides (e.g., metalaxyl), QoI (e.g., azoxystrobin), and carboxylic acid amides (CAA; e.g., mandipropamid). The use of some of these chemicals for grape downy mildew management in the US depends on state regulations. However, fungicide resistance to certain substances like QoI or CAA fungicides has been documented in Australia, Japan, France, and the United States. Despite the availability of new chemistries, efforts should be made to prevent the development of fungicide resistance.

2. Powdery mildew of Grapevine

Powdery mildew of grape cause sever loss in drought condition. Fruit infections can cause serious disease losses, which occasionally cause the entire crop to fail. The powdery mildew fungus, which affects sensitive cultivars, also has an adverse effect on vine growth and winter hardiness in addition to fruit yield and quality.

Causal organism: *Unicnual necator*

Systemic Position

Kingdom : Fungi

Division : Ascomycota

Class : Leotiomycetes

Order : Erysiphales

Family : Erysiphaceae

Genus : Uncinula

Species : ***U. necator***

Symptoms

All of the grapevine's green components are susceptible to powdery mildew infection. The typical symptoms of the disease manifest as a dusty or white powdery growth appearing in patches on fruits (see Figure 1), leaves (see Figure 2), and vines.Recently emerging and growing leavesthat have an infection may develop distortions and stunting. In times of extreme heat and aridity, severely damaged leaves may curl upward. On sick vines, dark brown to black blotchy lesions develop (Figure 3).When berries from red or purple cultivars get infected during ripening, they don't color as well and seem blotchy when harvested. Most fruit is prone to infection from the time it blooms until a few weeks later. Contrarily, two to three weeks after bloom, concord grape berries show remarkable resistance. Berries that are affected often exhibit an irregular shape or display rusty spots on the surface. Severely diseased fruit may experience splitting.

Fig. 1: Fungal growth in the form of white substance on the surface of affected fruits

Powdery growth on leaf surface

Etiology

The ectophytic mycelium creates bilobate or multilobate appressoria. The tall, simple, multiseptate conidiophores contain a chain of three to four conidia. The fungus is seldom found in its ideal stage in India. Cleistothecia can develop later in the growth season on all infected tissues if the mating types are present.

Life Cycle

The fungus *Uncinula necator* is responsible for grape powdery mildew. The pathogen overwinters as cleistothecia, which are fungus fruiting structures that have settled into cracks in the grapevine's bark.

In the spring, cleistothecia release airborne spores (ascospores), which are then transferred by the wind to tissues that are vulnerable. On any green area of the growing vine, the spores germinate and cause initial infections. Following infection, the fungus establishes itself within the plant tissue and releases an alternative type of spore (conidia) on the surface. The conidia, along with the fungal mycelium, impart a powdery or dusty appearance to the affected parts of the plant. Throughout the rest of the growth season, conidia act as secondary inoculum for new infections. On diseased plant surfaces in the late autumn, cleistothecia form. Many of these fruiting bodies wash into cracks in the bark, where they overwinter, starting the disease's another cycle. The fungus can be active at temperatures ranging from 59 ^{O}F to 90°F, but 68°F to 77°F is ideal for infection and disease development. The powdery mildew fungus is suppressed by higher temperatures. The ideal range for spore germination and infection is 40% to 100% relative humidity. Powdery mildew just needs high humidity, unlike the majority of other grape diseases (such black rot and downy mildew), which also need free wetness for spore germination and infection. Rainfall, in particular, is harmful to the survival of conidia. This implies that powdery mildew can persist even under conditions that are too dry for other illnesses to manifest. Powdery mildew development is also aided by dim, diffuse lighting.

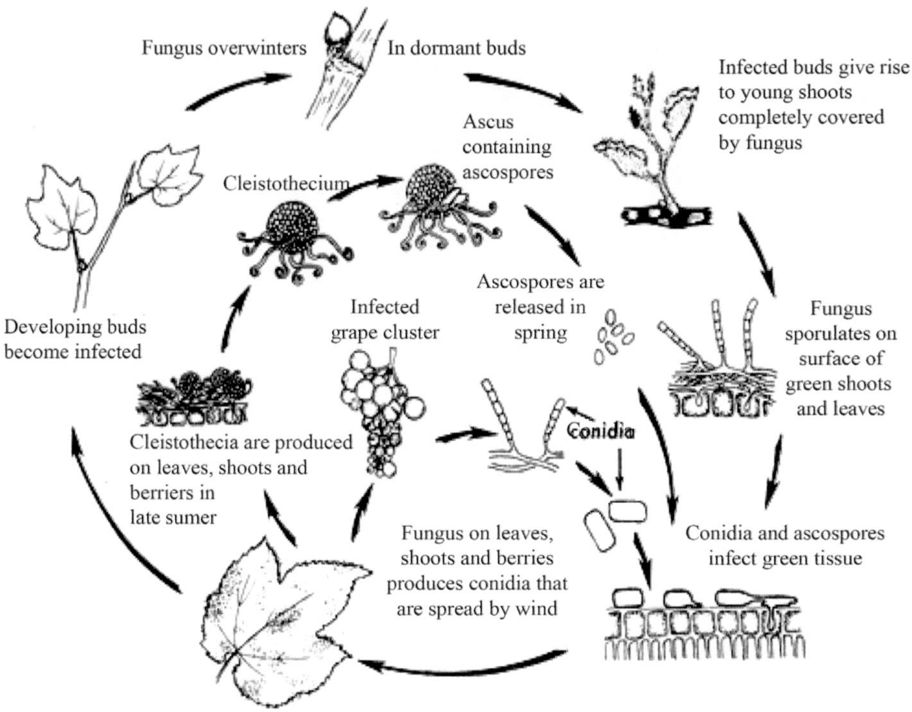

Disease Management

To manage the powdery mildew of grape use under given following practices

Cultural

To reduce the chance of powdery mildew attack to the proper air circulation and sunshine penetration, a healthy plant population should be kept and application of too much fertilizer must be avoided. long-term overhead Sprinklers are recommended. The diseased part of plant should be removed and destroyed. Grow resistant varieties like Chholth Red, Chholth White, Skibba Red,Skibba White etc.

Chemical

Spraying Wettable Sulphur (1.5 kg/200 l) will effectively control powdery mildew. It is important to use caution when spraying because they can burn the berry's skin and leave tiny black flecks. The disease is efficiently controlled by sulphur dusting

(20 kg/ha) in the morning. The dust should completely cover the afflicted plant portions. Systemic fungicides like Benomyl (5g/10 litres of water), Bayleton (1g/litre of water), Calaxin (3-4 ml/10 litres of water), and Bayleton (1g/litre of water) provide better and more durable disease control. From November to February, this should be treated with fungicidal spray. A chemical should never be sprayed more than twice. To prevent the emergence of disease resistance, a spray of wettable sulphur, a non-systemic fungicide, is applied between two subsequent systemic fungicide applications.

Biological method

Use active bacterium *Bacillus subtilis* in preventing the powdery mildew from invading the plant and Plant-based oils like jojoba oil or neem oil can also be utilize to prevent of powdery mildew infections. Plant oil spraying should be postponed for two weeks after Sulphur spraying.

3. Bird's eye spot/Anthracnose

Most of the world's humid and warm regions are affected by the anthracnose disease of grapes. Due to the beneficial environmental factors for this disease, it is a common occurrence over most of India. During the rainy season, the sickness manifests itself severely. Anthracnose significantly lowers the quality and yield of berries and severely weakens vines, particularly on young sprouts.

Causal Organism: *Elsinoe ampelina*

Systemic Position

Kingdom : Fungi

Phylum : Ascomycota

Class : Dothideomycetes

Subclass : Dothideomycetidae

Order : Myriangiales

Family : Elsinoaceae

Genus : Elsinoe

Species : ***E. ampelina***

Symptoms

The symptoms of anthracnose on shoots and berries are familiar to most winegrowers, but it can also affect fruit (Figure 1), leaves (Figure 2),stems (Figure 3), tendrils, and petioles. New leaves are more vulnerable to disease, making it more harmful. The spots on infected leaves typically have rounded or angular edges, a grey centre, and a brown to black perimeter. The spots' grey centres ultimately dry out and fall out, giving the appearance that shots were fired through the leaves. If spots impair the leaf veins or cover the entire leaf, the entire leaf gradually dries out. Small, round, crimson dots are the initial symptoms on shoots. The dots eventually get larger and have grey centres with rounded or angular borders, reddish brown to black rims, and so on.Shoots on lesions with infection may crack, become brittle, or wither away. The pre-flowering to veraison stages of berry development are all susceptible to infection. Infection symptoms first show up as small, rounded, reddish dots on berries, much like they do on shoots. After some time, spots expand and take on the shape of a "bird's eye" with grey centres and reddish brown to black edges. After that, infected areas may penetrate the pulp and result in berry shattering. In the lesions, fungus fruiting bodies, which are visible as pinkish spores, develop in favourable climatic circumstances.

Black spots on fruit spots on infected leaves

Bird eye lesion on older stem

Etiology

The anthracnose disease of grape is casued by *Elsinoe ampelina* (Shear 1829). It is the sexual stage of the fungus, and the asexual stage is *Sphaceloma ampelinum* (*Gloeorporium ampelophagum*). The fungus creates tiny, hyaline, single-celled conidia in acervuli. Ascostromata are where asci develop. An ascus, a 4-celled structure, produces ascospores. On the impacted plant sections, fungus also creates sclerotia. These sclerotia grew and produced numerous conidia.

Life Cycle

Infected woody tissue has dense mycelial masses of the anthracnose fungus, known as sclerotia. The next spring as well as throughout the year, stem cankers can emerge as a significant infection source. Microscopic spores (conidia) are released during rain and are transported to young, developing leaves and shoots as well as fruit via rain-splash. When free water is provided on the plants for 12 hours and the temperature ranges from 36 ^{O}F to 90°F, the conidia will germinate and infect. An infection will spread more quickly the higher the temperature. When the drizzly day, spores can be released throughout the spring and summer, causing an ongoing infection. The fungus grows more slowly in cool conditions.

Disease Management

To manage the anthracnose disease of grape use under given following practices

Cultural Method: Cultural method of disease management include sanitation, pruning, grow resistant variety. Removal and destruction of contaminated plant part as well as the eradication of infected wild grapes close to the vineyard. Planting resistant types and improving air circulation practices, such as using the right training system and removing leaves, are also beneficial. Fungicide sprays containing Sulphur or copper can be used if anthracnose does occur in the vineyard.

Chemical Method

To effectively control an infestation, preventative actions must be taken.While pruning, it is necessary to remove all twigs or canes exhibiting cankers.The pruned twigs and leaves need to be burned or buried far into the earth. In October and November, this illness causes more problems. During this time, it is important to spray protection on the young branches and new shoots. Spraying the grapevines with fungicides such Bordeaux combination @ 0.8%, Copper Oxychloride @ 0.25%, or Carbendazim @ 0.1% at the 3–4 leaf stage is beneficial against this disease.

4. Black rot of Grapes

The fungus *Guignardia bidwellii* (Ell.) Viala and Ravaz, which has as its imperfect stage *Phyllosticta labruscae Thum,* is responsible for grape black rot. The illness has been documented since 1802 and is currently seen in Canada and the United States east of the Rocky Mountains. Around 1880, this nation exported guinardia to Europe, where it quickly spread throughout the grape-growing areas. Where it occurs, black rot can significantly reduce grape production.

Causal Organism: *Guignardia bidwelli*

Systemic Position

Domain: Eukaryota

Kingdom: Fungi

Division: Ascomycota

Class: Dothideomycetes

Order: Botryosphaeriales

Family: Botryosphaeriaceae

Genus: Guignardia

Species: *G. bidwellii*

Symptoms

On leaves: The first indication of a disease onset is a small concentric region. The spots change from a grayish-brown appearance with a black line border to their characteristic brown color with a darker brown edge. As the leaf spot expands to a diameter of 3 to 8 mm, the black pycnidia of Phyllosticta appear, scattered across the entire surface or confined to a more or less concentric ring.

On Fruits: Small, round blanched spots that quickly turn whitish are the first symptoms. When the whitish patches are about 2 mm in diameter, they enlarge and develop reddish-brown edges.The spots grow quickly until they are about 8 and 9 mm in size and then divide into two or three dark rings. In most cases, this happens in a 24-hour period. The entire fruit shrinks into a distinctive dried, blackened mummy after 7–10 days, with the pycnidia of the Phyllosticta stage evident as minute black dots on the surface. Mummified grapes left in the vineyard from the previous autumn are what the fungus uses to evolve into its apothecial stage in the spring. To start a new disease cycle, the ascospores produced serve as the primary inoculum. Warm temperatures and wet spells lasting 1-2 days are favourable to the sickness.

Black rot of grape showing leaf spots,

Black rot of grape showing mummified caused by *Guignardia bidwellii. fruit* and *pycnidia (Phyllosticta)*

Etiology: The mycelium of *G. bidwellii* exhibits both intracellular and intercellular development and does not create haustoria. Mycelium that is older starts to turn brown. Mycelium that is produced fruiting bodies first becomes yellow, then brown, and finally darkens (Reddick, 1911). Ascospores are produced by ascocarps and conidia are produced by pycnidia, respectively, by *G. bidwellii*. G. *bidwellii* further produces pycnosclerotia and spermatia. Both pseudothecia and ascocarps are the fruiting bodies that produce ascospores.Pycnidia are rounded, ranging from 59 to 196 μm in diameter, appearing black, solitary, erumpent, and with an ostiole at the apex.The perithecia of *G. bidwellii* are spherical, black, distinct, and have a flat or papillate ostiole at the tip. Their diameters range from 61-199 cm. Numerous asci form in the ascal locule from the lower portion, where the fertile cells are located.Finally, asci create ascospores when they reach maturity (Janex-Favre et al., 1996) and Bitunicate asci can form ascospores. The perithecia of *G. bidwellii* lack paraphyses. Unicellular, hyaline, ascospores have hyaline mucilaginous apical caps and measure 10.6 to 18.4 cm in length and 4.8-9.0 cm in breadth (Northover, 2008).

Life Cycle

On the grapevine and on the ground, the black rot fungus overwinters in canes, tendrils, and leaves. The main source of infection the following spring is mummified berries on the ground or those that are still hanging to the plants. When it rains, many, dark fruiting structures (perithecia) release microscopic spores (ascospores), which are then transferred to developing, young leaves by air currents. These spores germinate in 36 to 48 hours in the presence of moisture and gradually infiltrate the leaves and fruit stalks. After 8 to 25 days, the infection is apparent. When the drizzly day, spores can be released throughout the spring and summer, causing an ongoing infection. The fungus grows more slowly in cool conditions.It needs warm weather for best development and 2 to 3 days of rain, mist, or fog.During extended periods of warm, rainy weather, new black rot infections persist until late spring and summer. The disease cycle is completed in August when the pycnidia change into the overwintering stage (pycnosclerotia), which produces perithecia inside which the spring ascospores are created.

Disease Cycle

Disease Management

For a very long time, it was believed that the best method for protecting vineyards from *G. bidwellii* was a mix of cultural techniques and the annual administration of fungicides to prevent downy and powdery mildews.

Cultural practices such as employing training methods that facilitate adequate air circulation within the canopy, preventing excessive shading, and planting vines in areas with favorable air movement and sunlight exposure have been recommended to minimize the period of canopy wetness in vineyards.These practises also improve spray penetration and coverage.Other cultural practises, notably the distraction of mummified clusters during winter pruning, were reduce the primary inoculum inoculum. The incineration of infected mummies in the soil has been advised, particularly in vineyards facing persistent black-rot epidemics or those adhering to organic management practices.The infection is avoided by applying the Bordeaux mixture (4:4:100) once or twice to young bunches. Spraying copper fungicides on bunches is preferable because they don't leave any observable deposits on the fruit surface.

5. Rust

Grapevine leaf rust (*Phakopsora euvitis*), a fungal disease, can lead to premature leaf shedding, compromised plant vitality, and reduced fruit yield and quality. Regions cultivating grapes in Eastern and Southern Asia, along with portions of North, Central, and South America and Europe, have reported occurrences of this disease. In 2001, grapevine leaf rust was identified in the Northern Territory but was promptly eradicated

Causal Organism: *Phakospora euvitis*

Systemic Position

Domain : Eukaryota

Kingdom : Fungi

Division : Basidiomycota

Class : Pucciniomycetes

Order : Pucciniales

Family : Phakopsoraceae

Genus : Phakopsora

Species : *P. euvitis*

Symptoms: Small, angular, yellow or brown lesions are the first symptoms that are visible on the upper leaf surface. On the lower leaf surface, tiny yellow-orange pustules known as uredinia, which contain spores, are visible to the unaided eye (Figures 2 and 3). Yellow lesions may turn necrotic (dead), turn black, and possibly spread to cover the entire leaf surface as the disease worsens (Figure 4). Leaf drop due to severe diseases might lower fruit yield and quality.

Grapevine leaf rust symptoms on the uppersurface of a grape leaf, which initially appear as small,angular, yellow-brown spots.

Characteristic grapevine leaf rust symptoms on the underside of a grape leaf. symptoms on the underside of a grape leaf. Yellow orange pustules containing rust spores (uredinia) are visibleto the naked eye.

Etiology

Uredinia are hypophyllous, tiny, subepidermical, and erumpent, encircled by cylindrical paraphyses that range in length from 32 to 55 µm. Obovoid, obovoid-ellipsoid, or oblong-ellipsoid urediniospores measure 17–28 x 12–18 µm. Hypophyllous and subepidermical Telia. Teliospores are oblong, oblong-ellipsoid, or cylindrical and measure 14–30 x 8–12 µm, usually grouped irregularly in 3–4 layers.

Life Cycle

Phakopsora euvitis has two life cycles and can reproduce asexually or sexually (also known as vegetatively).*Phakopsora euviti*s has a complex sexual life cycle that involves two separate host plant species (grapevines and *Melisoma myriantha*) and three different spore types (summer, overwintering, and sexual). Only in regions with the two host plants, grapevines (*Vitis species*) and *Meliosma myriantha* (an Asian tree), can the sexual life cycle be fully completed.The wind-borne aeciospores (also known as "summer" spores) produced by *Phakopsora euvitis* typically infect *Meliosma myriantha* in late spring to early summer. Afterward, grapevines (*Vitis specie*s) might become infected by these aeciospores, usually in the summer to late autumn. In the autumn, grapevines develop teliospores, which are referred to as "overwintering" spores because they can dormantly survive the winter in dead Vitis plant leaves. When the dormant teliospores germinate in the spring, the fungus begins to reproduce sexually by employing wind-borne basidiospores, or "sexual" spores. After that, *Meliosma myriantha*'s basidiospores restart the infection cycle.

Disease Management

To manage the rust of grapes, use cultural practices like cutting and pruning of infected plant part and removal and distraction of wild plant (alternate host), plant debris. Due to cutting and pruning provide proper air circulation and sun light reduce the chance of infection. Grow resistant variety if available.

Use fungicides that contain Bordeaux mixture, difolatan, propiconazole, tebuconazole, or azoxystrobin, which are believed to dramatically lower the pathogen incidence. Rust can be controlled in vineyards by administering 3–4 sprays of Baycor (0.1%) at fortnightly intervals during the following growing seasons.

6. Bacterial canker of Grapes

The grapes *(Vitis vinifera.)* is an important fruit grow throughout the world, temperate, subtropical, and tropical climates. It is affected by bacterial canker due to *Xanthomonas campestris pv. viticola* (Nayudu) Dye.In Maharashtra, Karnataka, and Andhra Pradesh, the bacterial bacterial canker disease has become a significant problem during the rainy season and is responsible for 60 to 80% of losses in vineyards that are badly infested.

Causal Organism: *Xanthomonas campestris pv. viticola*

Symptoms: The grapes plant infected by bacterial canker disease (*Xanthomonas campestris pv. Viticola*) symptoms show on leaves, tiny, irregular water spots with a necrotic pinhead-sized centre emerge. They are translucent to light. Large patches are formed when a number of spots combine. Sever infected plant sow defoliation of leaves.The disease begins on the stem as brown, black patches near nodes. Node girdling and cracking take place in the early stages. Finally, the plant eventually breaks down.

Bacterial canker on grapevine, **A.** Leaf spots; **B.** Marginal necrosis; **C. and D.** Cankers on canes; **E.** Dried inflorescences on a cluster; **F.** Grape cluster showing necrosis on the rachis and on a berry.

Etiology

The bacteria *Xanthomonas campestris pv. viticola* is Gram negative, rod shaped with rounded ends, motile by single polar flagellum and measure o.4-1.2 × 2.3 µm. At 41°C, the bacteria do not proliferate. On nutritional agar, yellow pigment is insoluble. Bacteria don't use carbon sources like L-arabitol and D-glucose.

Life Cycle

Pruning tools and irrigation water can spread the *X. campestris pv. viticoloa* virus. Warm, humid environments promote infection. The bacterium grows as an epiphyte, persists in plant-producing wood or planting material, and can be transmitted by cuttings. Mango, *Phyllanthus maderaspatensis*, and neem are also infected by the pathogen; these plants may act as an alternative host. The disease is more likely to occur in a temperature range of 25–30°C.The disease frequency increases at maximum rainy day.

Disease Management

To reduce the chance of bacterial canker of grapes disease caused by *X. campestis* pv. *Viticola* remove and destroy the infected plant part of vines. Cutting and pruning should be done in late October and use of healthy cutting for propagation reduce the chance of infection. Avoid excessive use of irrigation because due to irrigation pathogen transfer infected field to healthy field and Before the monsoon, use preventative sprays of copper oxychloride or Bordeaux combination.Antibiotics often have poor effectiveness, however in locations with little rainfall, streptomycin sulphate (500 ppm) may be sprayed at intervals of 5-7 days.

8

Diseases of Pomegranate and Their Management

Imran Ali[1], Prabhat Kumar[1], Avdhesh Kumar[1] and Upendra Maurya[2]

[1]Department of Fruit Science, Acharya Narendra Dev University of Agriculture & Technology, Kumarganj, Ayodhya, Uttar Pradesh
[2]Department of Fruit Science, Sardar Vallabhbhai Patel University of Agriculture & Technology, Modipuram, Meerut, Uttar Pradesh

Abstract

The pomegranate, or Punica granatum L., is a deciduous shrub that grows in dry and semi-arid regions of the world. It is a member of the Punnicaceae family and has chambers that hold globose fruits. Pomegranate, a word from Latin, means "apple with many seeds." One of the earliest known edible fruits, it is said to have come from present-day Iran, where it was initially grown around 2000 B.C. and then spread throughout the Mediterranean region. This fruit crop is attacked by a variety of nematodes, bacteria, and fungi, which causes large losses in yield both in terms of quantity and quality. A major physiological issue that causes greater financial losses for the pomegranate fruit crop not only in India but globally is fruit breaking in ripe fruits. Fruit rot, bacterial blight, wilt, and cracking are the main obstacles that need to be overcome. On the other hand, cercospora and anthracnose diseases are less common but can occasionally result in crop failure in favorable environmental circumstances. A range of bio-agents, plant products, bacterial inoculants, bacteriophages, and plant growth promoters and inducers have been investigated and used in disease control and to meet the growing demand for low-input agriculture due to the increasing restrictions on the use of chemical fungicides as well as concerns about the environment and human health. This chapter aims to provide an overview of the most recent advancements concerning significant pomegranate illnesses and their sustainable management.

Keywords: *Pomegranate, diseases, losses, management strategies.*

DOI: 10.1201/9781042014460-8

Introduction

Pomegranates are a tasty, juicy, sweet fruit that are rich in nutrients. *Punica granatum* is its scientific name, and Iran is where it was originally found. Because pomegranates may be cultivated effectively in regions of the nation with dry climates, they hold a distinctive place among the fruits that are grown there. Other names for pomegranates include Dadim in Sanskrit, Bedana in Bengali, Anar in Hindi, and Madulai in Tamil. The states of Maharashtra, Gujarat, Rajasthan, Karnataka, Andhra Pradesh, Tamil Nadu, and Uttar Pradesh are the primary locations for its cultivation. From the perspective of export, this fruit is also highly significant.

Pomegranate is referred to as a "Super Food" because of its many health and nutritional advantages. It is one of the greatest sources of antioxidants and has significant medicinal and nutritional value. Ripe fruit has a pleasant and nutritious juice. Juice from pomegranates in the summer, sherbet is incredibly refreshing. Pomegranates are regarded as a medication for long life and excellent health in the Ayurvedic medical system. In addition to treating sore throats, coughs, urinary tract infections, digestive issues, and skin conditions, it may also help prevent heart attacks and heart disease. Its fruits are used to make a variety of processed goods, including mouth freshener, juice, squash, jelly, and anardana. Patients with stomach issues are advised to drink the very nutritious juice. The stunning red flowers and attractive fruits of pomegranate beautify the garden.

Important Diseases of Pomegranate

The pomegranate plant is far more susceptible to a wide range of diseases brought on by pests and microorganisms. In India, these diseases result in 60–80% yield loss and are difficult to control after they have spread throughout the entire field (Mondal and Mani, 2009). Disease symptoms that are easily seen include leaves with spots of different sizes and colors, stems with blighting and dieback, and fruits with cutaneal to deep down fruit lesions that cause the fruit to rot (Xavier *et al.,* 2019).

1. Bacterial Blight Disease (BBD)/Oily spot disease (*Xanthomonas axonopodis* pv. punicae)

Symptoms

- A bacterial blight infestation on leaves is confirmed by the emergence of water-soaked patches on the underside of the leaf, which eventually turn into irregular blackish brown dots of varying sizes and a consistent yellow or chlorotic halo against light.
- If a fruit fractures or splits open due to BLB infection, similar water-soaked or brownish black spots also form on lemon-sized fruits.
- The bacterial slime that appears on the surface of blight lesions after rain, spray, or dew feels sticky to the fingertips and dries into a white, shiny encrustation.

- Water-soaked grey/black lesions form on stems at nodes, which then grow into various-sized cankers.
- Planting material derived from orchards afflicted by bacterial blight causes stem/nodal blight infection slightly above the soil's surface.
- Symptoms do not appear on blossoms or roots. At the roots Symptoms of Nodal blight in planting material and bacterial blight on stems (right)

Water soaked lesions on under surface of leaves

Necrotic spots with yellow halo

Water soaked spots on fruit

Necrotic spots with small cracks on fruit

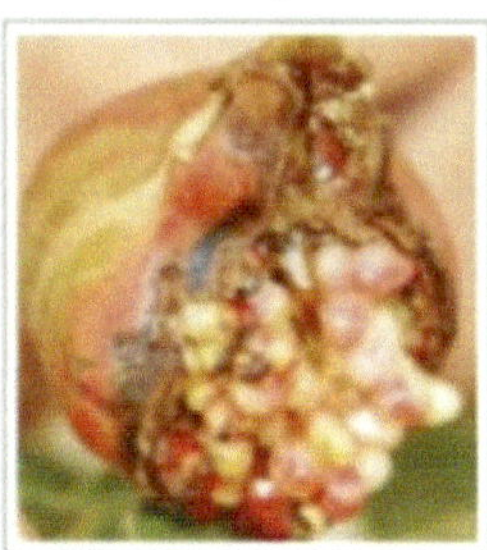

Splitting of fruit at necrotic spot

White shiny encrustation of bacterial ooze on infected fruit

Water soaked lesions on stem nodes tuming brown black

Various symptoms of pomegranate bacterial blight on different plant parts

Predisposing factors: Bacterial blight develops quickly under conditions that are between 25 and 30 ^{0}C + RH above 50%, rain, wind, larger nitrogen dosages than advised, deficits in micronutrients, and an excessive amount of undesired sprays.

Survival and spread: Infected cuttings are the main source of inoculum. Rains splattered by the wind disperse the secondary source of the inoculum.

Management: Orchards that exhibit severe cases of blight or those that are disease-free but close to an orchard that has been afflicted by blight should adhere to the timetable as prescribed (Annexure 3). It has been discovered that the six simple and affordable measures listed below are highly successful in reducing bacterial blight. Aside from that, the following lists the other strategies:

Six easy and economical steps to manage bacterial blight

The "Six Step" program for managing Bacterial Blight Disease (BBD) is intended for pomegranate producers who are experiencing losses from BBD in their mid- and late-season harvests. It must be handled in a community-based manner, with all orchards in an area affected by bacterial blight adhering to the timetable at the same time. .

I. Main Pruning: The primary pruning should be completed soon after fruit harvesting, which occurs between the end of December and the middle of February. This involves eliminating crowded, damaged, and dry branches to allow for adequate light penetration and aeration.

II. Rest period fertilizer dose and disease/insect pest protection: Apply rest period dose of fertilizer according to the age of the plant. For plants of 2 years and above apply:

a) (20 kilogram of thoroughly broken-down farm yard manure, or 15 kg of manure plus 2 kg of vermicompost plus 1–2 kg of neem cake per plant, or 7 kg of well-broken poultry dung plus 2 kg of neem cake each plant).

b) The amount of nitrogen given to each plant is 205 grams (446 grams of neem coated urea) + 50 grams of phosphorus (315 grams of single super phosphate) + 152 grams of potash (254 g of murate or 304 g of sulfate of potash).

III. Put crop on stress during hottest months for natural defoliation: In order to stress the crop and induce complete natural defoliation, stop watering it by the middle or end of March. Following full defoliation, use secateurs to cut the branch 2 to 4 inches below the canker and burn it outside the orchard in order to eradicate any visible blight cankers.

IV. Expose defoliated stems to solar radiation: The most recent adjustment and crucial step to remove the bacterium blight pathogen is to expose defoliated naked stems to solar radiation for 15 to 20 days prior to crop commencement in order to kill bacteria in the nodes. Keep a close eye on this time. Farmers should not wait 20 days to administer their first irrigation; they should do so as soon as they notice the branch or stem's 1 to 2 cm tip drying.

V. Light pruning and fertilizer application: Try lightly trimming the upper 8 to 10 inches of the branches. Eliminate any cankers as indicated previously. Follow the instructions in step I and paste using 10% Bordeaux paste.

VI. Follow crop season fertilizer and IDIPM spray schedule: Applying crop season fertilizer dose and irrigation.

Pictorial Six Steps in the management bacterial blight of pomegranate

Step I: Main Pruning: The secondary and tertiary branches with BBD cankers should be removed following fruit harvesting in December and February. All stems and fruits infected by BBD should be destroyed or buried to allow the soil to break down.

Step II: Rest period: Following the primary pruning, a recommended dosage of fertilizers for the rest period should be applied, and frequent application of insect/ disease sprays is advised.

Step III:Stress Induction: In order to stress the crop and induce complete natural defoliation, stop watering it by the middle or end of March. Eliminate any BBD cankers that are apparent on bare branches.

Step IV: Exposed Defoliated Stems to Sunlight: To eradicate bacteria in the nodes, expose defoliated naked stems to sunlight for fifteen to twenty days prior to crop start. Keep a close eye on this time. Farmers should not wait 20 days to apply their first irrigation; they should do so as soon as they notice the stems' 1-2 cm tips drying off.

Step V: Light pruning and fertilizer application: As previously said, remove any cankers by lightly clipping the top 8 to 10 inches of the branches. Follow the instructions in step I and paste using 10% Bordeaux paste.

Step VI: Adhere to the ICAR-NRCP's suggested IDIPM spray schedule for a successful pomegranate harvest.

(b) Integrated Management

- Skip the mrig bahar (rainy season) crop and switch to hasta bahar for a minimum of two to three years.
- Provide a well-rounded diet for plants by adding humic acid and sulfur (80%) at a rate of 10–30 g/plant, depending on the pH of the soil. Apply vermicompost (two to three kg per plant), poultry manure (0.5 to 1.0 kg per plant), FYM (30 to 40 kg per plant), and neem cake (3 kg per plant).
- Apply two soil treatments (one at blossom regulation and the other shortly after harvest) using bioformulations (*Aspergillus niger* AN 27 @ 1 kg/acre and *Mycorrhiza* [*Rhizophagusirregularis*Syn. *Glomus irregularis* @ 1 Kg/ acre).
- During the interval of rest and crop regulation, administer the prescribed dosages of phosphorus and potash. After the fruit reaches the lemon size, apply nitrogen in divided dosages.

- Give each micronutrient mixture four sprays at a concentration of two grams per liter (must contain manganese and zinc sulfate) and salicylic acid @ 300 ppm (30 g in 100 litre) at 1 month interval starting at pre-flowering stage.

2. Wilt

Symptoms

- Plants exhibiting symptoms of a nutrient shortage, such as partial or whole branch yellowing, drooping, or drying out of the leaves.
- Look at the roots and break open the lower part of the stems if you see anything

Yell owing /drooping /drying of leaves in 1-2 branches of a ptant

Plants with yellowing/drooping/drying of leaves and dried fruits attached

Drying and death of several neighbouring plants in a row due to *Ceratocystis* wilt

Brown/gray/black discolouration of xylem and surrounding wood due to *Ceratocystis* wilt

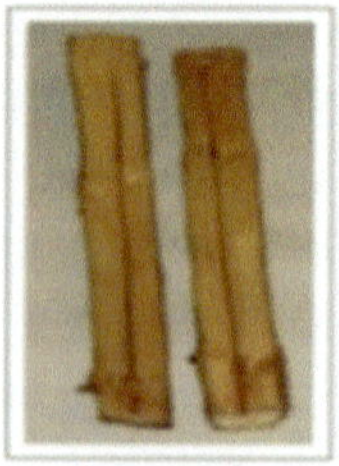

Browning of xylem due to *Fusarium* wilt

Pin holes in lower part of stems due to shot hole borer attack along with C. fimbriata

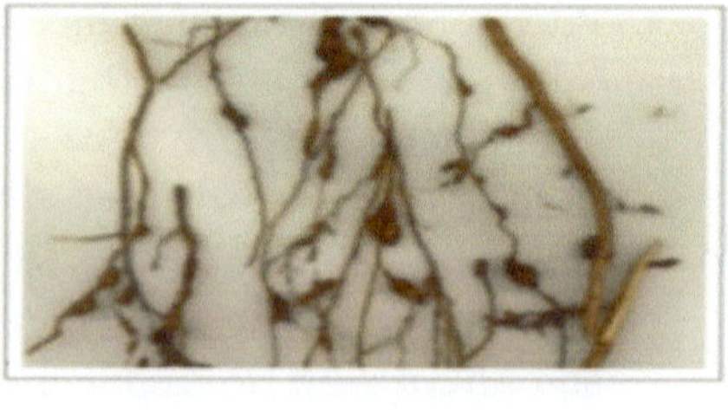

Knots on fine roots of wilted plant due to nematode infestation

Wilted plants due to *Ceratocystis fimbriata* fungus.

- The wilting plant's roots exhibit a brown, grey, or black discoloration on their wood, an indication of a *Ceratocystis fimbriata* fungal infection (Figure 2). In India, this is the most common fungus linked to pomegranate wilt. Early infection stages are characterized by yellow wood that smells like alcohol in the inner root. Plants that gradually turn a row of adjacent plants purple are typically caused by the spread of C. fimbriata by irrigation, weeding, fertilizer application, farm equipment, and farm animals like chickens and birds.
- Grey/black root rot is typically detected in areas with high levels of moisture surrounding the root zone and is caused by fungal species such as *Sclerotium* and *Macrophomina*.
- Caused by an infestation of either *Rhizoctonia sp. or Phytophthora nicotianae var. nicotianae*, collar rot/canker on stem at soil level (may also spread below and above soil level). Root rot may also result from symptoms that reach the root.
- *Rhizoctonia sp.* is the cause of the white wash/mesh-like fungal growth on the root surface. This is typically seen when soil is piled around the stem of plants and beds are raised after they have grown.
- If the center pith or xylem has a brown or red color, *Fusarium sp.* may be the cause of these symptoms. However, in many pomegranate locations, it is not a prominent cause of wilt.
- The shot hole borer is the cause of the pin holes on the outer and inner wood surfaces. In addition to attacking weak plants, these shot hole borers are drawn to the metabolites that *Ceratocystis fimbriata* produces.
- The presence of knots or galls on pomegranate roots indicates the presence of root-knot nematode infestation.

Survival and Spread

- The wilt fungus *C. fimbriata* can endure for several years (5–7 years) in the soil and in infected plant waste. Root-knot nematode-induced wilt has been documented in all types of soils; however, it is more severe in sandy loam soils with higher aeration levels.
- The majority of wilt pathogens, which include fungi and nematodes, propagate to other locations by contaminated planting material. Intercultural activities such as weeding, applying manure, using agricultural equipment and pruning tools, grafting roots, flooding/runoff irrigation water, and insects like shoot hole borer can spread the pathogens within and nearby orchards.

Management

1. Wilt pathogens such as fungi, nematodes, and insects should not be present in the planting material, which is saplings, or the potting mixture in which they are planted. Most soil-borne infections are eliminated when saplings are planted in soil that has been solarized or sterilized (potting mixture). Pomegranate orchard wilt incidence is decreased and aeration is increased when sampling plants are planted on raised beds.
2. The soil-borne pathogen can be eliminated by applying 50–75 LLDPE (Linear Low-Density Polyethylene) sheets to the soil and solarizing it for six weeks during the hottest months (April–May). For best results, prepare the planting plot, irrigate it, then cover it with an LLDPE sheet (Figure 2).
3. It is recommended to apply bioformulations that show promise, such as Aspergillus niger AN27 (IRG 07) and *Mycorrhiza [Rhizophagusirregularis Syn. Glomus irregularis*], at a rate of 1 kg/acre, as well as *Trichoderma harzianum, T. viride, Pseudomonas spp., or Paecilomyces spp*., at the time of planting and every six months thereafter. These are the most effective defense against any kind of wilt pathogen.
4. During the rainy season, grow green manure crops such as sunhemp (*Crotalaria juncea*) and dhaincha (*Sesbania aculeata*), and mix them into the soil prior to flowering.
5. Apply boron based on the results of the soil test.
6. Drenching should be done as soon as possible after harvest during the rest period to prevent residues in fruits. A day before treatment, give the plants a thorough irrigation. Two days should pass before irrigation is resumed following chemical drenching.
7. Starting in the second year, 10 liters of a preparation (paste) containing red soil (4 kg) + chlorpyriphos 20 EC (20 ml) + copper oxychloride 50 WP (25 g) must be applied on the stem surface from plant base up to 2 ft. in order to control shot hole borer (*Xyleborus spp*.), which is linked to wilt disease. Pasting ought to be carried out twice a year, right after harvest and right before crop regulation during defoliation..
8. Dig a trench that is three to four feet long between the healthy and diseased/wilted plants as soon as the wilt disease is identified in the orchard. Plants infected by partial wilt that exhibit early signs should be treated right away with the appropriate agrochemicals. Plants with more than 25% dry branches or those that are entirely dead should be pulled out and burned; they shouldn't be left to be piled up as wood.
9. Using the recommended agrochemicals, soaking should be applied to the root areas surrounding the main stem and 2-3 plants on each of the four sides as soon as the first signs of wilt appear in the orchard.

11. Plant cut ends should be coated with 10% Bordeaux paste, and pruning instruments should be cleaned after every usage. During the rainy season, 50 milliliters per liter of *Azadirachta indica* (neem) oil can be added to the paint. It is best to avoid pruning on monsoon or wet days.

3. Fungal Spots and Rots

Symptoms

The fungal pathogens cause leaf and fruit spots and rots. The major fungal diseases are discussed here.

Cercospora spots: The reddish brown, 1–4 mm *Cercospora punicae* dots on leaves can have a diffused halo that is either absent or barely noticeable. On colored variants, dark brown/black, discrete or merged, round to angular dots of varying sizes encircled by a diffused yellow halo. The absence of stickiness and fissures in the patches set it apart from bacterial blight. A long-term humidity of 60% or higher and temperatures between 20 and 30°C favor the quick development of illness. Conidia are hyaline to light olivaceous cylindric and septate, whereas conidiophores are olivaceous brown, short, fasciculate, and sparingly septate. Through conodia born of the wind, the pathogen spread.

Symptoms of Cercospora spots on leaves (left and on fruits (right)

Survival and spread

- Primary source: Diseased plant debris
- Secondary Source: Wind borne conidia

Fruit Scab: The fungus *Elsinoe (Sphaceloma) punicae* is the cause of fruit scabs. Fruit skin has a russet appearance due to rough, raised brown patches of different sizes. When impacted, fruits and flowers become warped. A climate of 22 to 28°C, 45 to 65% humidity, wind, and rain encourage the growth of disease.

Scab symptoms on pomegranate fruits

Anthracnose (on Fruit): Hard, various-sized, dark-black to brownish-black dots of varying shapes and sizes cover the fruit's surface. These signs and symptoms are linked to *Colletotrichum sp*. Infected leaves lose their color and turn yellow.

Symptoms of anthracnose on fruits caused by *Colletotrichum*

Survival and Spread

- Primary Source: Infected leaves
- Secondary Source of Inoculum: Windborne conidia

Favourable conditions

August through September, when there is significant humidity and temperatures between 20 and 27 ^{0}C, are the worst months for the condition.

4. Colletotrichum rot: It is *Colletotrichum gloeosporioides* that causes the disease. Fruit discoloration often begins at the calyx end, though occasionally spherical, irregular spots can be seen on the fruit's surface. The discolored spot hardens and turns from reddish brown to black. Under ideal circumstances, the infected region covers half to full fruit in a week. The rot goes past the rind and into the disintegrating, dark grey/brown-black arils, which are not wet. High humidity and temperatures between 25 and 30°C encourage the growth of illness.

Pomegranate fruit rot caused by Colletotrichum

5. Alternaria heart rot: *Alternaria alternata* is the culprit behind heart rot, which begins at flowering and progresses later. A temperature range of 25 to 30°C is ideal for heart rot development. The fruit does not exhibit any outward signs of heart rot, but in severe cases or when secondary rot fungi are present, the fruits may appear softer when slight pressure is applied and eventually become light weight.

Rot is only visible when the fruit is opened, and the inside arils turn brown-black. Rotten fruits may also display greyish black fungal growth/sporulation. The rind is usually unaffected, though minor discoloration may be noticed in severe cases.

6. Aspergillus rot: The pathogen causes brownish soft rots that have a sunken center. Later on, blackish fungus sporulation can be seen both inside and occasionally on the outside of the rots. Later on, the damaged skin's surface becomes lighter in color. Usually, the fungus can develop inside for a while without showing any symptoms outside of the body. Unless there is external injury, the fruits maintain their shape even while the rind and pulp underneath soften.

7. Penicillium rot: This infection usually begins on the wounds and causes mushy soft rots that are light tan in color. Later on, bluish green mold grows on the rots.

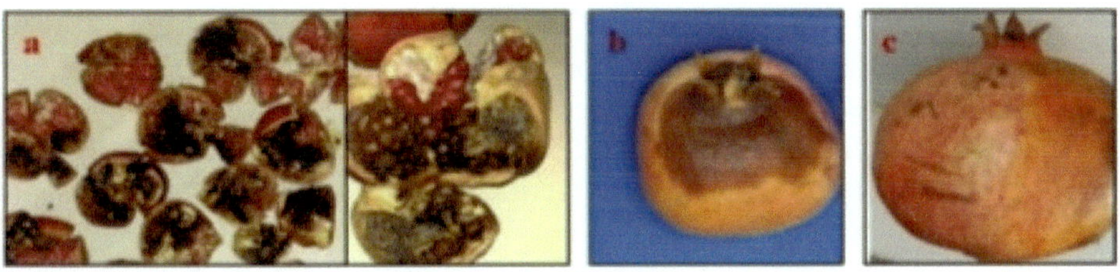

Symptoms of fruit rots on pomegranate fruits. **(a)** Alternaria Heart rot; **(b)** Aspergillusfruit rot and **(c)** Penicillium fruit rot.

8. Phomopsis rot: Rainy season is when the fungal disease is most common. Distinctive sharp-edged dry hard rot, dark brown in colour, usually originating at the calyx end and progressing towards the fruit's stem end. Young fruits and flowers are impacted and eventually wither. Larger fruits have patches that range in color from reddish brown to black. Certain other rots, such as Rhizopus species, cause inconsistent brown, fluid rot that quickly disintegrates tissue. Fungi-infected apples may leak water. When an injury occurs, sclerotium rots typically develop on any surface. They are soft, watery, and eventually turn white fungal growth on the surface.Under normal temperatures, the entire fruit quickly rots. Once fungal sporulation has occurred, the majority of rots may usually be detected.

9. Phytophthora blight/rot:*Phytophthora* species is a pathogen that primarily affects fruits, foliage, and seedlings in the warm, rainy season, when temperatures range from 25 to 30 degrees Celsius and humidity is high. The impacted seedlings exhibit signs of wilting and damping off, while the leaves and immature twigs have the typical blighted appearance. at addition, the fungus produces fruit rot and destroys fruits and flowers at all stages. Generally appearing first in fruits close to the ground, light brown tan mushy rot (after turning darker), which is neither firm nor extremely soft, spreads quickly with white sporulation in humid conditions. The fungus lives for a very long time in soil as thick-walled chlamydospores.

Phytophthora blight (Left); *Phytophthora* fruit rot (right)

Management

- Change flower regulation time from rainy to winter season in orchards where fungal spots, rots, or Phytophthora blight appear every year and negatively impact fruit quality or marketable output.
- Ensure the cleanliness of your orchard by clearing away any plant waste, trimming any dead, dry branches, soaking the top 0.5 to 1.0 inch of soil following rainstorms, applying 2.5% bleaching powder (33.3 % Cl) at 25 kg/1000 liters of water/ha, and sterilizing your pruning instruments with 2.5% sodium hydroxide.
- The majority of fungi responsible for fungal spot and rot are dormant on plant surfaces. When the right conditions arise, these fungi can become active and cause infection. To effectively prevent this, it is recommended to spray appropriate chemicals at recommended doses at intervals of 7 to 14 days, depending on the season and type of fungicide (systemic or non-systemic).
- During the rest period, a 1% Bordeaux combination alternating with copper or other cost-effective fungicides for 15 days helps lower the pathogen inoculum load for the following crop season.
- Sprays of Mandipropamid 23.4% SC @ 1 ml/L or Propiconazole 25% EC @ 1 ml/L + Azoxystrobin 23% SC @ 1 ml/L or Azoxystrobin 20% + Difenoconazole 12.5% SC @ 2 ml/L or Chlorothalonil 50% WP + Metalazxyl M 3.75% @ 2 ml/L or Bordeaux mixture @ 0.5% or Copper Oxychloride 45% + Kasugamycin 5% WP @ 2.5 g/L
- Both fruit quality and yield are enhanced by the combination treatments of Benomyl (0.1%) + Mancozeb (0.2%) and Carbendazim (0.1%) + Mancozeb (0.2%). Spraying a combination of 18% Tricyclazole and 62% Mancozeb WP at 2.5 g/L can effectively prevent anthracnose and calyx end rot in pomegranates caused by *C. gloeosporioides.*
- If there is a high incidence and favorable weather, provide two to three combination sprays at intervals of seven to ten days, depending on the weather.

References

Aditya Bhatia (2022). Major Diseases of Pomegranates. Diseases of Horticultural Crops and their Sustainable Management: Fruits, Plantation and Ornamental Plants. Nova Science Publisher. Hauppauge USA.

Arora, A. (2016). An overview of bacterial blight disease: A serious threat to pomegranate production. International Journal of Agriculture, Environment and Biotechnology, 9(4), 629-636.

Bagal, S. N., Patil, M. S., Thite, S. V., Chandanshive, A. A. andPise, N. M. (2022). Survey of Diseases and Pest on Pomegranate from Solapur District. Journal of Pharmaceutical Sciences and Research, 14(7), 818-824.

Gaikwad, D., Karande, K. and Deshpande, H. (2016, December). Pomegranate Fruit Diseases Identification and Grading. In International Conference on Communication and Signal Processing 2016 (ICCASP 2016) (pp. 672-679).

Gaikwad, D.S. and Karande, K.J. (2016). Image processing approach for grading and identification of diseases on pomegranate fruit: An overview. vol, 7, 519-522.

ICAR 2020. https://www.rvskvv.net/images/Disease_Hort_Crops_20.04.2020.pdf

Lamani, S. B. (2018). Pomegranate fruits disease classification with K means clustering. International Journal of Research Trends Innovation, 3, 74-80.

Managehyderabad2022.https://www.manage.gov.in/publications/eBooks/Hightech Pomegranate Production Practices.pdf

Mondal, K.K. and Mani, C., (2009). ERIC-PCR generated genomic fingerprints and their correlation with pathogenic variability of Xanthomonas campestrispv. Punicae, incitant of bacterial blight of pomegranate, Current Microbiology, 59: 616-620.

Mukesh, Satish Kumar Sharma and A. Sharma (2022). Etiology and Management of Cercospora Fruit and Leaf Spot Disease of Pomegranate through Fungicides and Systemic acquired Resistance Inducers. Biological Forum- An International Journal, 14(3): 1350-1356.

NRCP2011.https://krishi.icar.gov.in/jspui/bitstream/123456789/39845/1/Integrated Diseaseand Insect Pest. English.pdf.

Shilpa, Somasekhara Y M. Mahesh M. and Ravichandra (2022). Studies on Leaf Spot of Pomegranate (Punica granatum L.) caused by Cercospora punica Henn.Biological Forum-Ar International Journal, 14(3): 1571-1576

T.K. Bose (1988). Pomegranate. Fruits of India Tropical and Subtropical. NAYA Prokash Culcutta.22, 537-548

Xavier, Katia v., Achala N. KC. and Gary E. Vallad (2019). Diseases of Pomegranate (Punica granatum) in Florida, Electronics Data Information Source, 349.

9

Diseases of Apple and Their Management

***Ram Prakash Rai*[1]*, Roma Singh*[2]*, Sanjay Yadav*[3] *and Ankit Kr. Chaurasia*[4]**

[1]*Department of Plant Pathology, Banda University of Agriculture and Technology, Banda, Uttar Pradesh*
[2,3&4]*SardarVallabh Bhai Patel University of Agriculture and Technology Modipuram, Meerut, Uttar Pradesh*

Abstract

Apple is king of temperate zone and is a valuable and economically significant crop, is grown in India's northwest Himalayan region. After orange, banana and grapes, apple is the fourth most widely grown fruit crop in the world. The tree is native to Central Asia, where its wild ancestor Malus sieversii is still found today. The world largest apple grower is China with India at 5th position among the world. Himachal Pradesh is known as the apple bowl of India. They contain high amount of Vit C and act as a blood purifier and also have higher minerals and nutrients. There are many disease in apple which causes serious problem in apple yield reduction of which major disease are caused by bacteria, fungi and viruses and some through nematodes, viroids, phytoplasma and there are some physiological disorders. Most common disease in apple are caused by fungus and bacteria and viruses which are apple scab, powdery mildew, fire blight, soft rot, bitter rot, crown gall, black rot, cedar apple rust, collar rot, white rot, apple mosaic virus, apple dwarf and apple chlorotic leaf spot etc.

Out of which we have discussed some major disease of apple disease which are listed are apple scab, powdery mildew, alternaria leaf spot, black rot, fire blight, collar rot, their causal organism, symptoms, disease cycle and their management.

Keyword: *Apple, Symptom, Etiology, Management, Causal organism.*

DOI: 10.1201/9781042014460-9

Introduction

The apple (*Malus domestica*), a valuable and economically significant crop, is grown in India's northwestern Himalayan region. After Orange, Banana, and Grape, it is the fourth most widely grown fruit worldwide. The tree is native to Central Asia, where its wild ancestor *Malus sieversii* is still found today. Apples have been cultivated in Asia and Europe for thousands of years. The world's largest apple grower is China and in India it is grown in the hills of Uttarakhand, Himachal Pradesh, and Jammu & Kashmir in India. The average summer temperature during the active period is about 21-24°C growing season. Apple does this best in areas where trees are dormant. The color development will be better in winter when plenty of sunlight hits. It is possible even at an altitude of 1500 -2700 meters from sea level. 1000-1250 mm well distributed rainfall during the growing season which is most suitable for optimal growth and fertility of apple trees.

There are many diseases of apple which causes loss in the apple crop in which bacteria and fungi are most common pathogens to cause disease. The most common disease of the apple are listed below -

1. Apple Scab

Fungus *Venturia inaequalis* causes apple scab disease, also known as black spot in Australia (Mc Hardy,1996). *V. inaequalis* is spread throughout the world and is present in every region where apple is produced for businedd purpose. According to the expenses of controlling it economically, it is most significant apple disease in the world (Carisse and Bernier.2002). However in the early spring season the disease is more widespread in temperate regions with cool, humid weather (MacHardy,1996;Manktelow*et. al.*,1996).

Causal organism – *Venturia inaequalis*

Etiology

The most destructive disease in all apple-growing regions with strong spring and summer rainfalls is apple scab, which is caused by the haploid fungus *Venturia inaequalis* (Cooke) G. Wint (Ascomycotina: Pleosporales) (McHardy, 1996). *V. inaequalis* has one compulsory sexual generation and multiple asexual generations each year as part of its life cycle. Ascospores are created by sexual reproduction between strains of compatible mating types throughout the winter, when the fungus is saprotrophic on foliar litter. On immature apple leaves in the spring, ascospores generate primary lesions. Up to eight asexual generations of this phase, which lasts over the winter, produce conidia, which are predominantly dispersed by rain splashing on leaves and fruits (McHardy, 1996).

Symptoms

Symptoms are appeared on the upper ground parts of the tree and most parts of the parts. Tiny dull green lesions appear at the base of the petal on flowers also on the sepals and stem pedicel before and during flowering causing fruit drop.

Lesions appeared on the upper part of the leaves in the spring and later it also appeared on the top of the leaves due to older leaves being resistant to infection, lesions may seem fairly large (>1cm in diameter) on young leaves and modest in size on enlarged and older leaves. The early lesions are round and modest in size and grows bigger with time and become velvety and olive in colour due to production of the asexual spores. Damage to the leaves and the fruit appears as huge rankled and scabby lesions with direct edge. And velvety, green to olive-brown lesions are the symptoms on the fruits in the water soaked places.(Gupta,1990,1985).

Fruits and foliage that have been diseased typically fall from the tree.

The tree become more susceptible to freezing damage, insect and other disease infections if scab disease cause leaf defoliation for 2-3 years in a row.

Disease cycle

V. inaequalis is a hemibiotroph, and after a phase of saprobic development in fallen leaf tissues, it overwinters as pseudothecia (sexual fruiting bodies). Ascospores, which germinate and pierce the cuticle, make up the main inoculum. Although they originate on top of the epidermal cells, stromata do not invade them. Given that cell wall-degrading enzymes are only synthesized toward the end of the infection cycle, it is unclear how *V. inaequalis* obtains nutrients from the host. Throughout the growth season, conidia (secondary inoculum) are formed from the stromata's upper surface, starting a series of infections.

Management

Cultural Control

Apple scab disease can be managed through cultural practices, the planting of disease-safe varieties, the use of synthetic chemicals, and the use of natural control agents. In addition to cultural practices, cleaning up and throwing away diseased leaves that have fallen to the ground can be an efficient disease control strategy. Large orchards, however, cannot use it (Gupta, 1985). Use of urea solution splash sprayed to apple trees shortly before leaf drop or on infected fallen leaves on the ground lowers ascospore inoculum throughout the upcoming season. Regular tree pruning is a common method for preventing apple scab disease.

Chemical Control

Protective and curative chemical fungicides can be bused to control the disease. Before the disease preventive chemicals are applied to fruits and leaves, the ascospores and conidia which maybe present on the leaf surface.

Fungicidal application prevents the parasitic spores from germination or penetrating the host tissues.to avoid the infection on the newly developing tissues in the early season chemicals must be spayed at the interval of 7 days (Gupta and Lele,1976).

During the crop's growing period a spray program with 7 non systemic and systemic spray is adopted in protective mode.

Biological Control

In the future, a good alternative to chemical fungicides for controlling apple scab disease would be a biological control agent or natural microbial antagonist (Ellis *et. al*., 1997).

2. Powdery Mildew

The first case of powdery mildew disease on apple seedlings was documented in 1871 in Iowa, United States, and then spread to other apple-growing regions throughout the world (Bessey, 1877). Almost all of India's developing states have recorded cases of the disease. The primary victims of this illness are nurseries and seedling stocks. According to Glawe (2008), severe mildew infection in plants can impair production by up to 80%.

Causal organism – *Podosphaera lophotricha*

Powdery mildew is caused by *Podosphaera leucottricha* but the presence of Erysiphae heracleid and *Podosphaera oxycanthae* can also cause disease.

Etiology

Podosphaera leucotricha, an ascomycete fungus from the Erysiphaceae family. This obligate fungal parasite constantly produces asexual spores (conidia) on specialized short stalks known as conidiophores during the growth season. Conidia are 20-38 12 m in size, hyaline (transparent, colorless), and have prominent fibrosinbodies Conidia are wind-dispersed and can germinate without the need for free moisture. When they touch tissue that is vulnerable, they start an infection and form mycelium colonies. The first supply of inoculum the next spring comes from infected lateral and terminal apple buds, which also act as overwintering sites. However, *P. leucotricha*will be badly impacted by excessively cold winter temperatures because infected buds are more susceptible to winter mortality.

In sac-like asci that are encased in fruiting bodies, *P. leucotricha* also generates sexual spores, each ascocarp has a single ascus with eight elliptical ascospores. Ascocarps can be identified as distinct black specks on a mycelial mat's surface. The ascocarps of *P. leucotricha* have variously been referred to as cleistothecia, perithecia, and most recently, chasmothecia. These names refer to the ascocarps' closed nature and lack of a preformed opening, respectively. In order to identify powdery mildew ascocarps from other cleistothecia, the name chasmothecia has recently been created, because other ascomycetes cleistothecia lack a hymenium and instead have asci that are dispersed randomly throughout the enclosed structure.

Symptoms

Newly emerging shoots, young and mature leaves, flowers, and fruits all displayed symptoms. The first sign of an illness that manifested during the spring was a 3–4 day delay in the bud-opening process. The disease's symptoms included leaves, flowers, and fruits that were covered with a white to light-dark powdery mass that is made up of the pathogen's spores (Glawe, 2008).The spread is very quickly to other parts. The pathogen's dark earthy or earthy colored fruiting bodies completely cover the white to pale dim powdery mass on apple tree branches and the affected foliage is brittle and rigid.

Some leaves may twist, shatter, or develop disfigurement as a result of the disease. The size of the entire shoot may be reduced if disease affects succulent leaves and growing shoots, and leaves and shoots may turn brown in the middle of summer. Flowers fail to develop, turn greenish-white, and exhibit abnormal flower development (Hickey and Yoder, 1990).

Disease Cycle

The *P. leucotricha* mycelium infects apple flower and shoot buds during the poor year. Winter causes the mycelium to remain dormant. With the arrival of spring, fungi emerge from dormancy and begin to grow again. They colonize on developing young leaf shoots and tissue. These primary infections on conidiophores result in the production of asexual conidia, which are then disseminated by the wind.

Pathogen conidia flourished between 10 °C and 25 °C and at high humidity (>70%). Conidia develop into frames for hyphal outgrowths that crisscross the leaf's surface, develop, and finally straighten into distinct shapes. These structures emit chemicals that let parasite contamination pegs enter the plant's epidermal cells and subsequently spread out to form haustoria. Haustoria are specialized organs created within the living plant cells that continue to be the parasite while absorbing nutrients. As the mold state develops or as optional diseases cause the development of new settlements, the contamination process (hyphal outgrowth > appressorium > contamination peg haustorium) is repeated until helpless tissue is no longer accessible at this point. An unanticipated rise in mold action could result from development in the late season. Aside from helping to create quick inoculums, auxiliary disease cycles also contaminate horizontal and terminal buds that will harbor the organism throughout the winter. (Hickey and Yoder, 1990; Turechek, *et. al.*, 2004).

Management

Cultural Control

By removing the early sources of inoculum, the main source of contamination is controlled.

White terminal shoots that are infected need to be removed over the winter or early spring (Yoder, 1992).

There are cultivars that are resistant to accumulation and require control only in conditions of extreme disease stress. Jonafree, Prima, and Enterprise are apple resistant cultivars, but they are not widely cultivated.

Chemical Control

Foliar fungicide spraying helps prevent secondary infection and fruit infection.

From the tight-cluster stage to the conclusion of the terminal shoot growth in the middle of the summer, fungicides are often sprayed at intervals of 7 to 10 days.

The postbloom stage, when the leaf is growing, is when chemical sprays are most effective. For disease management, this may occasionally require 18 sprays on cultivars that are extremely vulnerable (Yoder *et. al.*, 1997).

The four compounds sulfur, fenbuconazole, myclobutanil, and strobilurins are the most useful for controlling mildew.

3. Alternaria Leaf Spot

Apples are the most widely produced and commercially important fruit in the world, producing around 81 million tonnes annually at a productivity of almost 18 tonnes/hectare (FAOSTAT, 2018).

These are vulnerable to more than 70 plant diseases, with pathogenic fungi being responsible for the most majority of these (Nabi *et. al.*, 2020).

Apple output and productivity are seriously threatened by the *Alternaria* leaf and fruit spot (ALFS) disease, which is brought on by *Alternaria* spp. In the majority of apple-growing locations worldwide over the past ten years, Alternaria illness in apples has resulted in significant losses for the industry (Filajdic*et. al.*, 1991). The Alternaria leaf blotch was originally noted in the USA in 1924 (Roberts, 1924) and has since been noted in the majority of nations that produce apples. In 2002, the Kashmir valley of J&K was reported to have the disease (*Alternaria malli*) (Shahzad *et. al.*, 2002).

Causal organism – *Alternaria alternata*

Etiology

Nees described the *Alternaria* genus in 1816, and according to Schoch *et. al.*, (2009), it is a member of the Pleosporaceae family and Pleosprinae subclass of the Dothideomycetes class of Ascomycota. Pathogens, endophytes, epiphytes, and saprophytes are all members of Dothidiomycetes (Schoch *et.al.*, 2006). Worldwide, *Alternaria* species are a frequent saprophyte and plant pathogen that are responsible for economically important plant diseases (Rotem, 1994).

The conidia of the genus Alternaria are distinctive and simple to distinguish from those of other fungal species. They are large and dark, catenate or solitary, ovoid or obclavate, frequently beaked, pale brown to brown, multi-celled, and have transverse or longitudinal septa (Ellis, 1971). In general, the genus Alternaria can be divided between species with large-spored and small-spored conidia based on conidial morphology. Due to their high degree of genetic and physical similarities, the smallspored Alternaria species are where the majority of the difficulty in species demarcation and nomenclature exists.

Symptoms

Small, round, purplish or black spots first form on leaves in the late spring or early summer. They eventually enlarge progressively to 14 inch in diameter and have a purplish border. Most lesions grow larger over time, becoming irregular, darker, and showing symptoms like frog eyes or crescent-shaped rings. Frog eye leaf spots are seen surrounding fruit mummies or dead wood and typically emerge early in the season. The leaves turn yellow and there may be early defoliation if the petioles get infected. Premature fruit drop can be caused by severe defoliation. When there is a severe mite infestation, defoliation is more severe (Filajdic*et. al*., 1995). Fruit symptoms include those that lenticels are accompanied by small, black, somewhat depressed patches.

Disease Cycle

The risk of disease and secondary infection increases throughout theb rowing season. Due to increasing temeperature and precipitation throughout spring, latent leaves produce more spores, which transport in the tree canopy is accelerated by wind and rain splashes. Ater one month of petal fall the primary infection occurs (Harteveld*et. al.,*2013). Humid and a temperature range between 25 and 31°C are ideal for the disease rapid progression, with symptoms beginning two days after infection (Filajdic*et. al*., 1992). About 20 days after bloom, the newly sprouting shoots become infected.

The risk of disease and secondary infections rises throughout the growing season, when fruit diseases are caused by a combination of high temperatures, heavy rainfall, and relative humidity 90 days after blossoming. The disease spreads more quickly on leaves and fruit until the end of summer or the beginning of autumn, when defoliation takes place.

Management

Using resistant cultivars will enable effective disease control. No resistance gene has been found as of yet.

Few research (Filajdic *et. al*., 1991; *Abe et. al*., 2010; Li *et. al*., 2011) have examined resistance to *Alternaria* leaf blotch.

A combination of fungicidal spray, mite treatments, and appropriate cultural practices could be used to suppress the ALFS outbreak.

Leaf debris on the orchard floor serves as the primary source of inoculums, and throughout the winter, spores can be found on twigs and buds so destroy the source of inoculum.

Cut down the branches infested with the disease and mites to prevent it from overwintering.

The source of inoculums in orchards may be decreased by applying urea to fallen leaves, mulching, pulling weeds, removing apple litter from orchards, using lime sulphur, covering plastic foil, and manually removing leaf residues.

Spay of copper based fungicides.

Selective canopy trimming also lessens the amount of inoculum in orchard twigs and buds. With the right time of application, fungicide sprays could help reduce infections.

4. Black Rot

Botryosphaeria obtusa (Schwein.) and *Diplodia seriata* are the pathogenic organisms. Shoemaker infections frequently cause apple tree branch canker, frog-eye leaf spot, black fruit rot, and dieback of shoots. Summer and fall types are severely affected by the disease. By removing the limb cankers and adhering to the spraying schedules, the disease is frequently entirely under control in the orchard.

Causal organism - *Botryosphaeria obtuse*

Etiology

The significant apple parasite *Botryosphaeria obtusa* (Schwein.) Shoemaker, which causes frog-eye spot, black rot, canker, and shoot dieback, has long been known by this name. *B. obtusa* has been isolated from at least 34 additional hosts in addition to apples (Punithalingam and Walker, 1973). It has recently been identified as a pathogen of *Vitis vinifera* in South Africa (van Niekerk *et. al*., 2004), Australia (Castillo-Pando *et. al*., 2001), and Portugal (Phillips, 1998, 2002).

The fungus was initially named *Sphaeriaobtusa* Schwein. by Schweinitz in 1832, and Cooke changed its name to *Physalospora obtusa* (Schwein.) Cooke in 1892. By combining this with Botryosphaeria Ces. & De Not., Shoemaker (1964) created the new name *Botryosphaeria obtusa*. Hesler (1916) first proposed the relationship between the teleomorph and the anamorph, which was later supported by Shear et al. (1925) and Stevens (1936).

The anamorph of *Botryosphaeria obtusa* definitely belongs in Diplodia Fr because of its brown, aseptate conidia developed on phialides, propagating by periclinal thickening or annellations, and lining the inner wall of pycnidial conidiomata. The precise name of the species has, however, generated a lot of debate and misunderstanding.

Symptoms

Fruit, leaves, and bark displayed indications of the *B. obtusa* pathogen-caused disease known as black fruit rot. A few weeks after the petals, start looking for signs of infection on your apple trees.

Fruit rot symptoms include:

Large portions of the fruit may be covered in diseased, decaying tissue with intense brown and black rings. On earlier infections, the fungal spore Pycnidia appears as little black dots. Some fruits remain tethered to the tree and turn to mush. Fruits that are infected early ripen 1 week earlier than they would under normal circumstances. Fruits with an infection have a rotting core.

Frogeye Leaf Spot signs were seen on infected apple leaves before the development of the black rot disease. Circular lesions with red or violet outside borders and light tan interiors can be seen on the leaves.

Symptoms as Branch Cankers: On infected limbs, symptoms presented as a sunken, reddish-brown area that occasionally had rough-looking or broken bark. Those cankers might not be immediately noticeable. Care should be taken to notice the canker on shoots if the plants have fruit rot or frogeye leaf spot on them.

Disease Cycle

The fungus cankerous, mummified fruits, and deadwood bark all through the winter. Conids and ascospores are released over the season by the black fungal fruiting bodies (pycnidia and perithecia). These spores spread the illness into sound wood, fruit, and leaves. Around blossom time, spore production is at its peak, but conidia assembly can go on all summer long under moist spells. At least a year will pass before the conidia become inactive. Typically, an infection with the leaf happens as the petals are falling. Conidia are affixed to the leaf and should germinate in 5–6 hours if the environment is moist. After conidia germination, they enter the leaves by natural apertures below the surface or through wounds caused by insects, hail, or other things.

RH (relative humidity) and rain are additional factors that facilitate the propagation of the virus and sickness. At a temperature of around 32 °C, spore germination and infection are most favorable.

Management

The best method for controlling apple disease is a combination of cultural activities and fungicidal treatment.

By removing preserved apples and trimming dead wood in the trees or on the plantation's ground surface, the inoculum can be decreased. Punched shoots and twigs serve as a disease repository for the following season. Reduced disease contamination is achieved through annual tree pruning and the use of customized composts on soil test rigs. When trees are stressed out and weak, cancer frequently occurs. To maintain healthy apple trees, specific social customs and behaviors must be observed.

5. Fire Blight

T.J. Burrill (1883) was the first to identify *Erwinia amylovora* bacterium as a disease-causing agent. In 1995-1996, reports on the causes of fireplace blight in apples came from Hungary, Romania, Northern Italy, and Northern Spain. More than 40 nations have identified *E. amylovora* as a disease source.

Causal organism – *Erwinia amylovora*

Etiology

In the early 1800 first time reported that *Eriwiniaamylovora*caused disease in apple plants.It is widely proved that North America is the place where first time this bacterial diease is seen. In addition to some regions of the United States of America, including Alabama, California, Colorado, Connecticut, Georgia, Illinois, Maine, Maryland, Massachusetts, Michigan, New York, North Carolina, Ohio, Oregon, Pennsylvania, Texas, Utah, Virginia, Washington, West Virginia, and Wisconsin, *E. amylovora* is currently present in all over the world in severe form.

The infection is reported in Northern Europe for the first time in the 1950s in which fruit containers are contaminated with bacterial ooze that were imported from the USA. Large portions of Germany and France are unaffected with this disease, but later 1990s,the disease was also found in Germany. The bacterium was first time discovered in remote areas of the Eastern Mediterranean in the 1980s, and from 1995 to 1996, occurrences of fire blight started to be reported in nations including Hungary, Romania, Northern Italy, and Northern Spain.

Symptom

The symptoms of fire blight are brown color with rotting leaves, twigs and flowers. The disease enter the apple tree from the tips of branches and then move downwards causing dieback. It affect younger leaf first and then infected leaf turn into brownish and the bark may became reddish during sunny days ooze may come out.In severe infestation the disease kills the shoots,blossom and entire trees. The scorched appearance of infected leaves give the name fire blight. Accordingly, this infection causes collar or rootstock blight, limb or trunk blight, fruit blight, blossom blight, and fruit blight (Van der Zwet and Keil, 1979; Van der Zwet and Beer, 1995).

(a) Blossom blight and spur blight symptoms include: Blight and blistering blight symptoms appear in the spring. Bacteria can infiltrate the trees through the flowers in addition to new shoots.

(b) Symptoms as Shoot Blight: The "shoot curse" starts from the tips of the shoots and swiftly spreads down to older twig segments. The initial symptoms were water-soaked, earthy or black-colored twigs. Infected shoots shrivel, and the branches curl at the point of growth, resembling a shepherd.

(c) Symptoms as Twig/Branch/Stem Cankers: The bark showed cracks along with infested areas when infection-causing bacteria spread from infected twigs to the largest branches, resulting in a prominent canker.

(d) Symptoms as Fruit Blight: Apple fruits may be contaminated with fire blight. Fruit rot was caused by a pathogen, and affected portions went from brown to black and were covered in whitish bacterial slime droplets.

(e) Symptoms as on Rootstock: Symptoms of fire blight that appear close to the graft union and resemble those of stem cankers. By girdling the rootstock, fire blight infections in rootstocks can quickly destroy the tree (Van der Zwet and Keil, 1979).

Disease cycle

Fire blight bacteria can be found on fruit that has been mummified as well as on living tissue on the trunk and major limbs of trees during the winter. Primary infection caused by raindrops, pollinating insects (such as bees, wasps that collect pollen, flies, and ants), or pruning (Paulin, 2000) during the growing season. The pathogen overwinter in bark.The pathogen is gram positive, rod shaped and peritrichous.The overwintering spores in bark became active and it have sufficient moisture. Secondary infections are caused by nectar pores and other natural openings. Shooting infections can also be caused by insect bites (such as those from aphids, leafhoppers, or domesticated plant bugs), wind, or rain. After the shoot infection the pathogen multiply fastly and oozing are seen in infested plant within three days. Until vegetative growth stops and the end bud is formed, the shooting is still very susceptible to infection (Thomson, 2000) the optimum temperature for pathogen growth is 27°C. Due to unfavorable condition the rate of infection and lesion extension are slowed down at high temperatures (5 °C or below, 31 °C and above) (Thomson, 2000).

Management

It is challenging dealing with the apple fire blight disease. Although managing a fire is difficult, applying integrated managementprinciples may help to reduce the damage and losses. The grower may be greatly assisted in controlling fire blight disease by implementing the suggested practices listed below.

Cultural Control

Proper distruction of plant pathogen and sanitation of infected trees.

Pruning blight-infected twigs and cankers prevents infection in the spring.

Carefully performed summer pruning of sick shoots inhibits the spread of pathogen inside the orchard.

Only the dry season is suitable for pruning infested shoots. Infected timber with healthy tissue cannot be cut below 12 to 15 inches. After each cut, pruning tools need to be sterilized.

Biological Control

Sprayed with a copper-based herbicide to remove that pathogen had accumulated on the bark and buds. At the silver tip (when the buds begin to swell), copper can be put to green tips, but not afterwards.

Antibiotic streptomycin eliminates pear blight-causing bacteria before they may infect the fruit trees.

Sucking insects that produce wounds that pathogens enter and causing the fire blight disease.. It is best to control these pests throughout the growing period. When flowers are in bloom, insecticides shouldn't be used to protect bees.

6. Collar rot of apple

Apples are frequently affected by Phytophthora collar rot, another disease. Spreading the illness also involves the production of unpolluted apples and overhead irrigation using tainted surface water. Many other tree fruits, especially nut trees, serve as the host plants for the causative organism.

Causal organism - *Phytophthora cactorum* is the disease's pathogenic organism, however *Pythium ultimum*, *Phytophthora megasperma, Phytophthora syringae,* and *Phytophthoracambivora* have all been reported to occasionally contribute to the onset of collar rot (Rana and Gupta, 1984).

Etiology

According to Judelson and Blanco (2005), the Oomycota genus Phytophthora is in charge of some of the most severe illnesses with significant economic consequences. By identifying *Phytophthora infestans* as the cause of potato late blight, Anton de Bary first used the term *Phytophthora* (plant destroyer) in 1876. A total of 51–54 new species have been discovered since the year 2000, adding to the 54 species that were discovered in the 20th century (Erwin and Ribeiro, 1996). They have a wide range of host plants, including fruit trees, citrus, woodland, and park species, and are widespread around the world.

Phytophthora can seriously harm apple, pear, cherry, and other fruit orchards and nurseries. Apple crown or collar rot is known to be caused by *P. cactorum*. Almost all apple-growing locations in the globe currently experience the disease, which is particularly serious in orchards with dwarfing root stocks (Harris, 1991). Early spring is when *Phytophthora* root and crown rot symptoms first show.

Symptoms

Around the crown, the *Phytophthora* disease spreads to the rootlets and causes a lesion along the soil line. If the scion cultivar is helpless, the disease may also spread to the bud association. The canker forms a characteristic shape as the bark dries up, and the callus tissues thicken at the borders. The *Phytophthora* organism can also infect fruits from cultivars that are susceptible. According to Rana and Gupta (1984), rotting and a solid, light tan spoil appear on or close to the ground in the case of contaminated apple trees.

At the lateral root intersection of the trees, contaminations frequently start. To define the new canker issue on a practical level. As they grow, they take on a dark, sad appearance and take on an oddly shaped, roughly oval shape. Earthy-colored, frequently delicate, fragile, or thin symptoms are caused by contaminated bark. Additionally, there are dark streaks that start close to the shift and continue past the canker's edge. If an infection persists for a considerable amount of time, just a small portion of the body exhibits the typical surface and coloring of freshly consumed tissue. Both in even and vertical bearings, the canker spreads swiftly.

One of the main effects of neckline damage is to support the affected appendage roots or trunk, resulting in the death of the affected organs or the entire tree. Declining vigor, poor terminal bud development, and the early emergence of small, sparse yellow leaves in the summer or rosy bronze leaves in early harvest are the prominent symptoms of the infection. Early fruit ripening and the production of fruits with extra colors are typical indications of the illness. Only the branches above the canker can experience the foliar symptoms, and badly affected trees finally perish.

Disease cycle

The pathogen mycelium endures the winter in trunk cankers as thick-walled oospores or as chlamydospores, and for several years in the soil. The highest concentrations of oospores and chlamydospores are found in the soil of old apple plantations. These long-lasting oospores with strong walls are resistant to harsh conditions like cold temperatures and dryness. Chemical therapy is largely ineffective for them as well. In late winter, oospores may develop in bark holes, or mycelium may enter the bark near the soil groundline. The many zoospores are produced by sporangia in the soil in the early summer or late spring. The lowest section of the trees can be accessed by zoospores, which float in irrigation water or rainwater. Rain splash also contaminates fruits that have grown close to the ground, infecting new branches and resulting in fruit deterioration. The Phytophthora fungus can generate vast quantities of spores with the aid of a small number of zoospores (Anonymous, 1989).

For fungus growth and infection, natural temperatures and moist soil are preferred. The pathogen *Phytophthora cactorum* develops at 13 °C and is viable up to 21 °C. Apple cultivars classified as "Midwest" vulnerable have a tendency to be harmed most frequently during the blossom cycle, again in late August or early September, and in October at the start of or during dormancy. During these periods, the fungus is most active in the soil, and the host is most vulnerable.

Management

Cultural control

The disease Phytophthora collar rot is exceedingly challenging to track and treat.

In soils with a high clay content or that are consistently moist and poorly drained, Phytophthora collar rot is typically prevalent.

Apple trees should be replanted on sloping land with sufficient drainage. Before planting apples on ridges, a reasonable slope should be made in the fields. Water will flow more smoothly as a result.

Improve internal soil drainage and disease management by tiling well-drained orchard wetlands. Avoid flooding and creating a soil ring around the trunks of apple trees. Apple trees should be planted as shallowly as possible inside the nursery.

Choose Resistant Cultivars and Rootstocks

Chemical Control

Disease can be controlled by saturating a 30 cm radius around the apple tree trunk with Blitox/fytolon/Blue Copper @0.1% or Dithane M45 @0.03% (Agrwal, 1970).

Conclusion

Apple being a staple fruit crop of the world among the fruit crops and it was originated in central Europe. It is rich in vitamins specially in vitamin C and it is fat and cholesterol free has good amount of carbohydrate and fiber. Apple is used in various industrial products, like apple jam, vinegar, pickles, wine and various sauce manufacturing companies but the disease caused in the apple fruit deteriorate the quality and quantity both which thus affect the apple producers and their product manufacturers.

Apple suffers from many fungal, viral, bacterial and phytoplasmas disease and their full knowledge and management should be known. Thus knowledge about the apple disease is important and their identification, and management should be known. Only chemical methods should not be adopted as it has residual effects on others organism and also has some harmful effect on the ecosystem, thus judicious use of the chemicals and biological and cultural methods should also be adopted to manage the diseases of the apple fruit.

References

Billing, E.; Berrie, A.M. (November 2002). "A Re-Examination of Fire Blight Epidemiology in England". Acta Horticulturae (590): 61–67.

Brisset, MN., Cesbron, S., Thomson, S.V. et. al.,(2000) Acibenzolar-S-methyl Induces the Accumulation of Defense-related Enzymes in Apple and Protects from Fire Blight. European Journal of Plant Pathology 106, 529–536.

Burrill, T. J. (1883). New species of Micrococcus (bacteria). Am. Nat., 17: 319.

Ellis, M. A., Chatfield, J., and Draper, E. (1997). Scab of Apple and Crab Apple. The Ohio State University Extension Factsheet. HYG-3003-94.

Ellis MB, (1971). Dematiaceous Hyphomycetes. Wallingford, United Kingdom: CABI Publishing.

Filajdić, N., & Sutton, T. B. (1991). Identification and distribution of Alternaria mali on apples in North Carolina and susceptibility of different varieties of apples to Alternaria blotch. Plant Disease, 75(10), 1045-1048.

Glawe, D. A. (2008). The powdery mildews: a review of the world's most familiar (Yet poorly known) plant pathogens. Annu. Rev. Phytopathol., 46: 27-51.

Gupta, G. K. (1985). Apple Scab (Venturia inequalis) (Cke) Went. E. Merck (India) Ltd., Bombay, 36 pp.

Gupta, G. K. (1990). Apple scab - a review. Indian J. Agric. Sci. 18; 157-165.

Gupta, G. K. and Lele, V. C. (1976). Morphology, physiology and epidemiology of the apple scab fungus (Venturia inequalis) (Cke) Went., in Kashmir valley. Indian J. Agric. Sci. 18;157-165.

Harteveld, D. O. C., Akinsanmi, O. A., & Drenth, A. (2013). Multiple Alternaria species groups are associated with leaf blotch and fruit spot diseases of apple in Australia. Plant Pathology, 62(2), 289-297.

Hickey, K. D. and K. S. Yoder. (1990). Apple powdery mildew. Pages 9-10 in A. L. Jones and H. S. Aldwinckle, Eds. Compendium of Apple and Pear Diseases. American Phytopathological Society, St. Paul, MN, USA.

Madhu, G. S., Nabi, S. U., Mir, J. I., Raja, W. H., Sheikh, M. A., Sharma, O. C., & Singh, D. B. (2020). Alternaria leaf and fruit spot in apple: Symptoms, cause and management. Eur. J. Biotechnol. Biosci, 8, 24-26.

Marine, S. C., Yoder, K. S., & Baudoin, A. (2010). Powdery mildew of apple. Powdery mildew of apple.

McHardy, W.E., (1996). Apple Scab: Biology, Epidemiology and Management. The American Phytopathological Society, St. Paul, MN.

Moriya, S., Terakami, S., Iwanami, H., Haji, T., Okada, K., Yamamoto, T. and Abe, K. (2013). Genetic Mapping and Marker-Assisted Selection of the Gene Conferring Susceptibility To Alternaria Blotch Caused by Alternaria Alternate Apple Pathotype in Apple. Acta Hortic. 976, 555-560.

Phillips, A.J.L. (1998). Botryosphaeria dothidea and other fungi associated with excoriose and dieback of grapevines in Portugal. Journal of Phytopathology 146: 327-332.

Phillips, A.J.L. (2002). Botryosphaeria species associated with diseases of grapevines in Portugal. Phytopathologia Mediterranea 41: 3-18.

Punithalingam E and Walker JM. (1973). Botryosphaeria obtusa. Descriptions of pathogenic fungi and bacteria, No. 394. Commonwealth Mycological institute, Kew, Surrey, England.

Rana, K. S., & Gupta, V. K. (1984). Occurrence of Pythiaceous fungi in collar rot affected apple soils of Himachal Pradesh [India]. Indian Phytopathology (India).

Rotem J, (1994). The genus Alternaria: Biology, epidemiology, and pathogenicity. St. Paul, MN, USA: American Phytopathological Society Press.

Schoch CL et.al., (2006). A multigene phylogeny of the Dothideomycetes using four nuclear loci. Mycologia98, 1041-52.

Schoch CL et.al.,(2009). A class-wide phylogenetic assessment of Dothideomycetes. Studies in Mycology 64, 1-15.

Shahzad, A., Bhat, G. N., & Mir, N. A. (2002). Alternaria mali-A new pathogen of apple in Kashmir. SKUAST J. Res, 4, 96-98.

Soldo, T., Svitlica, B., & Mesić, J. (2015). Phytophtoracactorumnajabukama-simptomi, biologijapatogena, epidemiologijaikontrolabolesti. GlasnikZaštiteBilja, 38(5), 24-30.

Turechek, W. W., Carroll, J. E. and Rosenberger, D. A. (2004). Powdery mildew disease of apple. Tree Fruit Factsheet, Cornell University.

Utkhede, Raj and Smith, Emmerson (1995). Control of Phytophthora crown and root rot of apple trees with fosetyl aluminium in new plantings. Pesticide Science, 45 (2): 117-122.

Van der Zwet, T. & Keil, H. L. (1979). Fire blight: a bacterial disease of rosaceous plants. United States Department of Agriculture (USDA) Handbook 510. Washington, DC.

Van dyZwet, T. & Beer, S. (1995). Fire blight: its nature, prevention and control. A practical guide to integrated disease management. USDA Agricultural Information Bulletin, No. 631.

Verma, K. D. and Gupta, G. K. (1988). Field reaction of apple germplasm to powdery mildew. Indian J. Agric. Sci., 58: 223-234.

Yoder, K. S. (1992). Powdery mildew of apple. In: J. Kumar, H. S. Chaube, U. S.. Singh, and A. N. Mukhopadhyay (eds.) Plant Diseases of International Importance, pp. 66-89, Vol. 3. Diseases of Fruit Crops, Prentice Hall, NJ.

Yoder, K. S., and Biggs, A. R. (1997). Apple cultivar susceptibility to the powdery mildew fungus. Biol. Cult. Tests Control Plant Dis., 12: 42-43.

10

Diseases of Brinjal and Their Management

Prince and Ankita

Department of Plant pathology, Sardar Vallabhbhai Patel University of Agriculture and Technology, Meerut, Uttar Pradesh

Abstract

Brinjal, scientifically known as Solanum melongena, holds great significance as a tropical vegetable cultivated throughout India. This vegetable is known by various name in different regions of India. India cultivates numerous varieties of brinjal, each differing in fruit size, shape, and color. The immature fruits are commonly used in various curries, and a wide range of dishes can be prepared using brinjal. Brinjal fruits are rich sources of vitamins, minerals like phosphorous, calcium, and iron, and their nutritional value varies across different varieties. However, the production and quality of brinjal crops have been negatively affected by climate change, leading to an increase in diseases. This crop is highly susceptible to attacks by various microbes, resulting in wilt, soft rot, and root rot. To address these issues and reduce the use of chemical fungicides, extensive experimentation has been conducted over the past decade using microbial inoculants to the plant diseases. Bacteria have been explored not only as biocontrol agents for plants, but also as promoters of plant growth and inducers of resistance to disease.In addition to improving the growth and plant health, these microbial agents contribute to the growing demand for low-input agriculture.

Keywords: *Brinjal, Phomopsis Blight, Bacteria*

Brinjal

Botanical Name	:	*Solanum melongena*
Family	:	*Solanaceae*
Chromosome number (2n)	:	*24*

Introduction

Brinjal (*Solanum melongena*), is a self-pollinated plant also known as eggplant, is an easily cultivated plant belonging to the Solanaceae family.Brinjal ishigh productive vegetable and easily grown everywhere.It is also known as poor's

DOI: 10.1201/9781042014460-10

man vegetable. Brinjal is cultivated widelyand grown as a vegetable crop in any environment. Brinjal grown large amount in India, Pakistan, Chinaand United States. The origin of the brinjal is India. In Brinjaleconomic parts are fruit. It is a high nutritional valuable food.The brinjal fruit and whole plant use as a medicine in some area. It is cultivated everypart of India except high attitude, all the year around. Depending upon consumer's preference and yield, different types of cultivars are cultivated.Somewild spp of brinjal are fond in northeast states of India and Malaysia. *Solanum insanum*, a wild cultivar of the plant, mainlyobserved in the dry and hilly region of west Bengal.Some yellow color brinjal varieties are found in Malaya.

The brinjal crop severely affected by various diseases, mainly attacks roots, leaves, fruits andsteams. The severity of the disease isdepending on the pathogen, location where its grown and time. The brinjal crop has a wild range of disease including fungal, Bacterial, and viral.

Fungal Disease		
S.N	**Name of the Disease**	**Causal organisms**
1	Damping off	*Pythium aphanidermatum*
2	Leaf spot (Cercospora)	*Cercospora solanimelongenae,*
3	Leaf spot (Alternaria)	*Alternaria solani*
4	Collar rot	*Sclerotium rolfsiis*
5	Fruit rot	*Phomopsis vexans*
6	Verticilium wilt	*Verticilium dahiae*
Bacterial disease		
1	Bacterial wilt	*Pseudomonas solanacearum*
Viral disease		
1	Tobacco mosaic disease	Tobacco mosaic virus

1. Damping off

Causal organisms:*Pythium aphanidermatum, P indicum, P parasitica, R solani and S rolfsi.*

- One of the most important soils borne disease of brinjal Damping-off. The disease mainly infectedin seedlings nursery.The seedlings that have been infected are decaying at ground level, causing the plants to topple over. The seedlings in the nursery are dying in certain areas. This pathogen, which is present in the soil, is accountable for significant losses in the early stages of plant growth.

Symptom

The nursery is affected by the disease, resulting in significant damage. This is particularly evident in areas with high soil moisture, moderatetemperature, and high moisture levels, especially during the rainy season, where the disease thrives. There are two distinct types of damping off symptoms that can be observed.

The first type is known as pre-emergence damping off, which occurs when the fungal pathogens attack the germinating seeds. These pathogens infect the main stem, hypocotyls, and primary roots of the seedlings, causing damage before they even emerge from the soil.

The second type, post emergence damping off is characterized by the discoloration of the upper part of the plant. This discoloration is a result of the infection juvenile tissue and the development of collar rot in seedlings. This type of damping off occurs after the seedlings have emerged from the soil.

Etiology

Pythium root rot is an important serious disease. The damping off, a common issue in field and greenhouses, refers to the killing of newly emerged or emerging seedlings by the fungal organisms pythium. Despite the widespread occurrence of this problem, no crop has been developing sufficient resistance to the pythium. Consequently, there is a significant interest in finding genetic host resistance. Pythium wilt, on the other hand, occurs when older plant infected by zoospores, resulting in biotrophic infections. The response of the plant to colonization/ reinfection pressures or environmental tress can lead to necrotrophic behavior, which in turn causes varying degrees of wilting due to impaired root functioning.

Disease Cycle

Pythium, a member of the Oomycetes or lower fungi, is classified as a lower fungus. It has the ability to withstand unfavorable conditions by existing as oospores, which are resting spores that possess resistance to dehydration. The growth of these oospores is stimulated by the exudates released from the roots of susceptible hosts. As the oospores develop, they undergo transformation into sporangia, which then generate zoospores. These zoospores are equipped with flagella, enabling them to move through water. Chemically attracted to the growing roots of young plants, they infect the roots through wounds or damaged areas where side roots are formed. Once the plant is infected, new sporangia containing zoospores are produced, facilitating the spread of the disease to neighboring plants. The zoospores disperse through water, making the disease particularly prevalent in wet conditions with high soil or substrate moisture content (70%). Additionally, oospores can also be dispersed by human or mechanical transmission.

Favorable Conditions

Damping off diseasecaused by *Pythium* spp thrives in conditions of high humidity and soil moisture, promoting its growth and development. The presence of high soil moisture encourages the sporulation of Pythium, while low temperatures between 10-20°C facilitate the production of zoospores, which contribute to severe secondary infections and aid in the progression of the disease (James 2012; Starkey and Enebak 2012). To survive in unfavorable temperatures, the pathogens resort to producing oospores.

Management

It is essential to carefully select healthy seeds for sowing. Prior to sowing, the seeds should undergo treatment with either Thiram or carbendazim at a rate of 2g per kilogram of seed. It is crucial to avoid continuously cultivating a nursery in the same area. To maintain the quality of the topsoil in the nursery, it is recommended to treat it with Thiram or Bavistin at a rate of 5g per square meter of soil. Additionally, it is advised to drench the nursery with the same chemical at a rate of 2g per liter of water every two weeks. Another effective approach to prevent damping-off is soil solarization. Moreover, the application of *Trichoderma viride* in the soil at a rate of 1.2kg per hectare has also proven to be significantly effective in controlling damping-off.

2. Leaf Spot (Cercospora)

Causal agent:*Cercospora solani-melongenae and C. solani*

Cercospora solani melongena is a significant obstacle for the extensive cultivation of brinjal due to leaf spot. Patel (2009) reported the presence of Cercospora spores in the brinjal field in Nashik. The airspora over the brinjal field has been documented by several researchers.

Leaf Spot of Brinjal

The stromata produced by the fungus are spherical in shape. The conidiophores exhibit a medium-dark color with a slight olive-brown hue, becoming lighter towards the tip. The conidia range from translucent to pale olive or yellowish in color.

Symptoms

The leaf spots exhibit characteristics of chlorotic lesions that are randomly distributed, with angular to irregular shapes. As the spots progress, they turn grayish brown and show abundant sporulation at the center. Severe infection leads to premature leaf drop, resulting in a decrease in fruit yield quality.

Etiology

The conidiophores of the fungus have an irregular width and appear grayish to brown. They also exhibit a highly branched structure. The conidia, on the other hand, are numerous and can range from light to dark in color. They are straight or slightly bent, cylindrical or obclavate in shape, and have distinct scars. The conidia can have a maximum of six septa. In terms of size, the spores measure between 16 to 59 cm in length and 4 to 6 cm in width. The teleomorph stage of the fungus has not been identified.

Disease cycle

Cercospora melongenae exclusively exists in its asexual phase, even when cultivated in a controlled laboratory environment. This particular disease exhibits an incomplete life cycle, classifying it as a deuteromycetes fungus. During the winter season, *Cercospora melongenae* survives within conidiophores, which are responsible for asexual conidia production. In late spring the conidia are released and disseminated through the air, wind, or rain-splash, facilitating their infection of suitable hosts, such as eggplants.

The conidia attach themselves to the underside of young leaves and enter the leaf through stomata and natural opening as well as wound of the plant. Once the infection takes hold, visible symptoms manifest on the leaves. The germ tube of Cercospora can penetrate both closed and open stomata, and multiple germ tubes from different conidia can enter the same stomata. It is important to note that this fungus primarily infects young leaves, while mature leaves exhibit a lower occurrence of infection.

Optimal Environmental Conditions

The optimal temperature range for the growth and spread of diseases is typically observed between 77°F and 95°F (25°C–35°C) during the daytime, while the night temperatures should remain above 61°F (16°C). Additionally, a relative humidity level of 90 to 95% is considered ideal for disease development. The primary source of disease-causing agents is the leftover residue from a previously infected crop. However, it is worth noting that the fungus responsible for the disease can also be found on seeds and various weed species. The spores of this fungus are dispersed through rain splashes and can also be carried by wind, affecting susceptible brinjal leaves.

Disease Management

To manage the disease effectively, it is advised to eliminate and dispose of the affected plant parts and apply Bavistin @ 0.1% or Chlorothalonil @ 0.5% to the affected plants. Cultivating resistant varieties is also a viable approach to control the disease. Furthermore, it is crucial to eradicate infected plants from the field to prevent further spread.

3. Leaf spot (Alternaria)

Causal organisms: *Alternaria melongenae and A.solani*

The only known reproductive stage of the disease. The conidia and mycelium of the fungussurvive in the soil and on the crop residues during the winter, causing primary infection of the leaves. The pathogen can survive severalyears, up to ten years.

Symptoms

The presence of concentric rings on the leaves is a distinctive indication of the disease, which is accompanied by the eventual shedding of affected leaves. Furthermore, the pathogens responsible for this ailment can also invade fruits, resulting in premature yellowing and dropping. In the case of brinjal plants, early blight symptoms manifest as small lesions measuring 1-2 mm in black or brown coloration on the foliage. These lesions tend to enlarge under favorable environmental conditions and are often encircled by a yellow halo, forming a target-like pattern referred to as a Bullseye lesion. As the lesions continue to expand and new ones emerge, the affected leaves may undergo a yellowing process and shed, leading to a significant loss of foliage.

Etiology

The fungal mycelium the pathogens is characterized by its septate instructure, ranging in color from light brown to dark brown. It can be found both within and between cells. The emergence of dark-colored conidiophores through stomata is a notable feature. These conidiophores produce single-celled, muriform conidia that are arranged in chains.The conidia typically have 4-10 transverse septa and somelongitudinal or oblique septa.

Disease Cycle

Under optimal temperatures (20-30 °C), the pathogen undergoes sporulation, with an average time of 13 hours. The dispersal of conidia occurs through water and wind after sporulation is complete. When the conidia come into contact with a leaf surface in favorable environmental conditions, germination takes place. Germination is most effective at higher temperatures. The ideal temperature range for germination is between 21 and 28 °C. Moisture (>90%) is essential for germination. Once germination occurs, the pathogen enters the host by penetrating the leaf surface using an appressorium and infection peg. This leads to the infection of leaf cells and the subsequent appearance of symptoms. Mature lesions produce conidiophores that produce conidia. The dispersal of conidia through water or wind is accompanied by the emergence of these new conidiophores. Dispersal can occur within a single leaf or over long distances between plants. The spores can be carried by wind for approximately 1.1 miles. Both *'A. melongenae' and 'A. solani'* can survive in the form of microsclerotia and chlamydospores, which develop after infected leaves partially decay. These survival structures are most effective at low temperatures (30°C) and are resistant to desiccation and freezing. Eventually, the microsclerotia and chlamydospores produce conidiophores, allowing for the production of conidia and the continuation of the cycle.

Favorable Environmental condition

The course and severity of epidemics are primarily determined by weather conditions. Therefore, the study of different weather parameters such as temperature, relative humidity, and rainfall is crucial in understanding the infection and progression of leaf spot. The disease severity increased as the temperature rose (15.1–34.4 and 13.2–32.0 °C) and rainfall & RH decreased during the period from 24th September to 28th October.

Management

To control early blight, it is important to practice crop rotation and remove and destroy crop debris from previous crops. Other methods such as staking and mulching can also be effective.

Fungicides are frequently utilized to manage early blight. Extensive field experiments have proven the efficacy of fungicides such as maneb, chlorothaloniland mancozeb. These fungicides are conveniently accessible at different trade names. When applied in accordance with the instructions on the label and during the early stages of the season, they offer effective control against early blight.

4. Collar Rot

Causal organisms: *Sclerotium rolfsii*

Symptoms

This illness manifests sporadically in a severe manifestation. The lower segment of the stem is impacted by the soil borne inoculum (sclerotia), leading to decortications as primary indication. The exposure and necrosis of underlying tissue may of result in the demise of the plant. The mycelia and sclerotia can be observed near the ground surface on the stem. Several factors, including diminished plant vitality, water accumulation surrounding the stem and injuries play a role in the progression of the ailment.

Disease Cycle

The fungus undergoes its life cycle as a sclerotium, which is a compact mass of hyphae surrounded by a tough outer shell. Under warm and humid conditions, the sclerotia “germinate” and infect various parts of the plant. Unlike systemic pathogens, this pathogen does not spread throughout the entire plant. Therefore, removing the infected plant components can effectively control the disease.

5. Fruit Rot

Causal organisms: *Phomopsis vexans*

Symptoms

Phomopsis vexans affects the ground parts of the plants. Typically, spots appear on the stems or leaves of seedlings. These spots girdle the seedlings, leading to their death. The spots on the leaf are well-defined, circular and range in diameter up to approximately 1 inch. They exhibit a light brown to gray color with a margin. On the other hand, on the fruit's spots are bigger. Initially, the affected fruit becomes soft, watery, and smoggy.

Etiology

In the impacted region, the central area undergoes a color change, turning gray and black, while pycnidia begin to form. The causative agent, *Phomopsis vexans,* produces pycnidia that may or may not have a rupture. These pycnidia can exhibit a spherical or irregular shape. Inside the pycnidium, the conidiophores are transparent and can be either simple or branched. The conidia themselves consist of transparent, single-celled structures that are cylindrical in shape. Furthermore, the ascospores of Phomopsis vexans are transparent and possess an ellipsoid to bluntly fusoid shape, featuring a single septum.

6. Verticillium Wilt

Verticillium species are plant pathogens that do not have a specific host range. Many of these species are saprophytes, but the most significant plant pathogens are *Verticillium albo-atrum* and *Verticillium dahliae*. *Verticillium dahliae* specifically causes verticillium wilt and early dying of Brinjal plants.

Life Cycle

Verticillium dahliae has a long-life cycle, surviving in the soil for many years as microsclerotia, which are small, resilient structures that can withstand extreme temperatures and dehydration. Additionally, the fungus (mycelium)survives in crop residues. It infects plant roots, either directly or through wounds caused by nematodes, and then colonizes the plant's vessels, producing spores that are transported through the transpiration stream. As a defense mechanism, the plant generates gum in the vessels. The presence of both the fungus and the gum leads to blockage of the xylem vessels, resulting in wilting of the plant. Spore can be dispersed by air to infect new hosts, while the remaining mycelium or microsclerotia are deposited in senescent tissue. The spread of the fungus is limited by the mycelium, and dispersal from plant to plant requires water, soil particles, tools, and other components. Verticillium can also be transmitted through infected seeds and tubers.

Etiology

The optimum temperature for the*Verticillium albo-atrum* is 21°C, and its growth ceases above 30 °C. On the other hand, *Verticillium dahliae* thrives at temperatures around 25-27 °C.

7. Bacterial Wilt

Causal organisms:*Pseudomonas solanacearum*

This wilt disease can manifest at any growth stage of crops, from the seedling stage to maturity. However, plants are most susceptible during the pre-flowering and early fruiting stages. The plant pathogen *Ralstonia solanacearum*, previously identified as Pseudomonas solanacearum, is responsible for inducing rapid wilting and eventual demise of the host plant (Hayward AC, 1991).

Symptoms

Bacterial wilt is characterized by common signs such as wilting, stunted growth, yellowing of theleaves, and eventual complete collapse of the plant. In the early stages, the lower leaves may sag before wilting takes place. The plant's vascular system becomes darkened, and bacterial ooze is discharged from the affected areas. Although the plants may show wilting symptoms during the day, they may temporarily recover at night, only to ultimately succumb to death. The bacterium responsible for this disease is non-acid fast and non-spore forming.

Etiology

The disease-causing bacterium is known for its non-acid fast,non-spore forming,non-capsulated, and motile nature, withpolarflagellum. It possesses the capability to generate acid, yet it does not produce gas when exposed to dextrose, sucrose, lactose, and glycerol. Moreover, it can break down starch and slightly liquefy gelatin through hydrolysis.

8. Little Leaf of Brinjal

The occurrence of Brinjal little leaf first documented at the central farm in Coimbatore by Thomas and Krishnaswamy (1939), and subsequent studies have explored various biological aspects of the disease (Varma et al., 1969; Mitra, 1993). The insect vector responsible for transmitting this disease is *Hishimonusphycitis*, as reported by Bindra and Singh (1969). This disease is transmitted by the *Hishimonusphycitis.* The infected plants experience a complete loss in yield.

The causal agent of this disease is phytoplasmas, which are cellwallless mollicutes that inhabit the phloem sieve tube elements of plants and the gut wall of insects. They are notorious for causing severe damage to crop and natural ecosystems worldwide. Leaf hoppers, plant hoppers, and psyllids are responsible for transmitting phytoplasmas (Kumar et al., 2012b).

Symptom

Little leaf was initially believed to be caused by a viral infection, but in 1969, it was discovered that aMLO actually causes the disease closely related to aster-yellows and curly top. One of the notable symptoms of this disease is the reduced size of the leaves, which are attached to the stem due to elongated petioles. These leaves are slender, soft, smooth, and exhibit a yellow coloration. The newly formed leaves display even pronounced symptoms, with shortened spaces between the stems. The lateral buds become larger, but their petioles and leaves are shorter, resulting in a compact and bushy appearance of the affected plant. If the infection occurs during immature growth stages.

Disease Cycle

This disease is transmitted through sap. It affects plants such as Datura, tomato, and tobacco. The presence of *Datura fastuosa* and *Vinca rosea* contributes to the spread of the disease. *Cestiusphycytis* (*Eutettixphycytis*) and *Empoascadevastans* are both vectors that naturally transmit the disease. The organism thrives in weed hosts (Sohi et al, 1974).

Favorable Environmental Condition

For the conidia to germinate, a relative humidity of 80-100% and a temperature between 20 and 26 °C are ideal and necessary. Conidial infection occurs best at temperatures between 25 and 30 °C. If the temperature drops below 12 °C, no conidia will be produced (Premila et al., 2015).

Management

To prevent the disease, controlling the insect vector through insecticides and cultivating resistant varieties are the main methods of management.

9. Tobacco Mosaic Virus (TMV)

Symptoms

The distinctive signs of potato virus Y include mosaic mottling on the leaves and stunted growth of brinjal plants. Initially, the mosaic symptoms are mild, but they progressively worsen over time. The infected leaves become deformed, small, and leathery. Additionally, contaminated plants produce numerous fruits. On the other hand, tobacco mosaic virus primarily manifests as swollen leaves. In advanced cases, leaves are swollen and discolored.

Plants that are infected at an early stage tend to experience stunted growth. PVY can be easily transmitted through sap. In the field, aphids transmit it, specifically *Aphis gossypii* and *Myzus persicae*, and it can also be propagated on weeds such as *Solanum nigrum* and *S. xanthocarpum*. On the other hand, TMV transmitted through sap, contaminated materials and clothing, soil debris, and the hands of laborers. It can cause harm to cultivated plants such as tomato, tobacco as well as weed host. The virus can persist in plant debris present in the soil.

Management

To minimize disease incidence, it is important to collect seeds from plants that are free from viruses. Additionally, reducing the population of aphid,removing and destroying infected plantsand eradicating suitable host can also help in minimizing the spread of diseases. In the nursery, aphids can be controlled by applying carbofuran @1 kgat the time of sowing seeds. Furthermore, 3-4 foliar sprays of Phosphamidan @ 0.05% can effectively control aphids. It is recommended to start spraying Phosphamidan (0.05%) at 10-day intervals, beginning from 20-25 days after transplanting.

Conclusion

Eggplants are a crucial component of the human diet, providing essential nutrients. They are widely cultivated in Asia and Africa and face various diseases that negatively impact their production and quality. This chapter focuses on the management of microbial diseases i.e., damping off, soil-borne fungi like Pythium and *Rhizoctonia*, which result in seedling rot. Additionally, diseases like leaf spot (*Cercospora* and *Alternaria*) and fruit rot, caused by fungi, reduce fruit production and quality. Little leaf disease and tobacco mosaic virus, transmitted by insect vectors, are also significant threats to eggplants. Bacterial wilt, which blocks vascular tissues and causes leaf drying and drooping, ultimately leads to plant death. This chapter provides detailed information on disease management.

References

Balai, L., Ahir, R. and Yadav, S. 2010.Varietal screening of brinjal genotypesagainst leaf spot of disease caused by Alternaria alternata. Environment and Ecology 31: 1276-1278.

Bhattacharya P, Samajpatti N, Bhattacharya P (2003). Characteristics of Ralstonia (Pseudomonas) solanacearum from Gangetic West Bengal. J.Mycopathol. Res. 41: 15-19.

Daskalov, H. and Murtazov, T. 1937. News Inst. Pl. Indtr. Sofia. 4: 65-72. Cited in: Som, M. G. and Malty, T. K. 1986. Brinjal In: Vegetable Crops in India. Eds. Bose, T.K. and Somm, M.G. pp. 293-342. Noya Prakash Calcutta India.

Dingar, S. M. And Singh, M. (1985). Role of weather in development of leaf spot disease of brinjal. Indian Phytopathology,38(4):721–726.

Fageria MS, Choudhary BR, Dhaka RS. Vegetable Crops, ProdoctionTecnology. Kalyani Publishers, Ludhiana, 2003, 25-74.

Gupta VK, Paul YS. Fungal diseases of Brinjal, Chilli and Brinjal. In: Diseases of Vegetable crops. (Ed.) Gupta and Paul, 2001, 87-101.

Hayward AC (1991). Biology and epidemiology ofbacterial wilt caused by *Pseudomonas solanacearum*. Annual Review of Phytopathology 29: 65-87.

https://agritech.tnau.ac.in/crop_protection/dried%20plant.JPG.

https://www.google.com/imgres?imgurl=https%3A%2F%2Fcontent.peatcloud.com%2Fw400%2Flittle-leaf-of-brinjal .

James R.L., (2012). Fusarium root and stem diseases. In: CramMM, Frank MS, Mallams KM (eds) For. Nurs. Pests. USDA Forest Service. Agriculture Handbook, Washington DC, pp 117–120.

Pandey, A. (2010) Studies on fungal diseases of eggplant in relation to statistical analysis and making of a disease calendar. Research in Science and Technology, 2(9): 1 – 3.

Pandey, A. and Pandey, B. N. (2001). Fungal diseases of brinjal in Bareilly region. Advances in Plant Sciences, 14(1):99 –104.

Patel, S.I. (2009). Observationon the airborne Cercospora spores causing leaf spot disease of brinjal at Nashik, India. Floraand Fauna, 15(1):31 -33.

Premila, A. and Sophiarani, Y. (2015). Downey mildew of onion: epidemiological aspects. International Journal of Innovative Research and Development,4(13):257 –259.

Sohi, A.S., Bindra, O.S. and Deal, G.S. (1974). Studies on the control of the brinjal little leaf disease and insect pests of brinjal. Indian Jour. Ent. 36:362-364.

Starkey T., Enebak S.A., (2012). Rhizoctonia blight of southern pines. In: Cram MM, Frank MS, Mallams KM (eds) For. Nurs. Pests. USDA Forest Service. Agriculture Handbook, Washington DC, pp 63–65.

Sutton, John Clifford; Sopher, Coralie Rachelle; Owen-Going, Tony Nathaniel; Liu, Weizhong; Grodzinski, Bernard; Hall, John Christopher; Benchimol, Ruth Linda (1990-01-06). "Etiology and epidemiology of Pythium root rot in hydroponic crops: current knowledge and perspectives". Summa Phytopathologica. 32 (4): 307–321.

Vijaya, T., M. Nagalaxmidevamma, and A. Sreeramulu. "Epidemiology of Cercospora Leaf Spot Disease of Brinjal - Process of Penetration." Environment & Ecology 9.2 (1991): 396-98.

11

Diseases of Chilli and Their Management

Anshika Sharma[1], Sibte Sayyeda[2] and Shivanshu Mishra[1]

[1]*Department of Plant Pathology, Sardar Vallabbhai Patel University of Agriculture and Technology, Meerut, Uttar Pradesh*
[2]*Department of Plant Protection, Aligarh Muslim University Aligarh, Uttar Pradesh*

Abstract

Chilies have been reported to be in use since the infancy of agriculture. It is one of the oldest domesticated crops in the Western Hemisphere. Long before the arrival of Columbus in the Americas, the local people already used peppers as food, as war artefacts and in religious rituals (Chiou et al., 2014). Common hot chilli (Capsicum annuum L., Solanaceae, 2n=2x=24) is one of the vegetables or spices known and used all over the world for its green fruits and pungency.

Introduction

Chili peppers, also spelled chile or chilli, are varieties of the berry-fruit of plants from the genus Capsicum, which are members of the nightshade family Solanaceae, cultivated for their pungency. Chili peppers are widely used in many cuisines as a spice to add "heat" to dishes.

Botanical Name : *Capsicum annum*

Family : Solanaceae

Chromosome Number (2n) : 24

Diseases of Chilli

1. Powdery Mildew

The disease was first reported from Morocco in 1937.

The powdery mildew caused by *Leveillula taurica*. Is a major constraint in chilli production in India causing heavy yield loss ranging from 14 to 30%, due to severe defoliation and reduction in photosynthesis, size and number of fruits per plant (Mathur *et al.,* 1972; Sivaprakasam *et al.,* 1976; Gohokar and Peshney, 1981).

DOI: 10.1201/9781042014460-11

Causal organism: *Leveillula taurica*

Symptoms

First disease symptoms are noticed on the older leaves which progress to younger leaves. The infected leaf is covered with white to grey powdery masses of the fungus where lower leaf surfaces also turn necrotic. The heavy infection leads to leaf shedding resulting in heavy losses in yield. Shedding of foliage. (A. Sudha and P. Lakshmanan, 2007.)

Powdery mildew of chilli

Epidemiology

Cultivated and wild hosts such as *Cynara cardunculus*, *Cicer Arietinum*, Lucerne, Oxalis cernua are the source of the pathogen during off-season. The optimum and maximum temperature for disease development is 20 °C.

Temperature above 30°C appears to be lethal for the fungus. (A. Sudha and P. Lakshmanan, 2007).

Disease Cycle

The powdery mildew disease cycle (life cycle) starts when spores (known as conidia) land on a chilli leaf. Spores germinate much like a seed and begin to grow into the leaf. Chilli powdery mildew parasitizes the plant using it as a food source. The fungus initially grows unseen within the leaf for a latency period of 18-21 days. Then the fungus grows out of the breathing pores (stomates) on the under surface of the leaf, producing spores which are borne singly on numerous, fine strands or stalks (conidiophores). These fungal strands become visible as white patches or mildew colonies on the underside of the leaf. Repeated cycles of powdery mildew can lead to severe outbreaks of powdery mildew that economically damage the crop.

Management

Spray with Wettablesulphur@0.25% or Dinocap @ 0.05%2-3 times at 10 -15 days interval. Varieties such as B 15, R 7, Padasali, Tinwari, Golden Superior have been reported as moderately resistant.

2. Fusarium Wilt

Causal organism: *Fusarium oxysporum f.sp.capsici*

Symptoms

Fusarium wilt is characterized by wilting of the plant and upward and inward rolling of the leaves. The leaves turn yellow and die. Generally appear as localized areas of the field where a high percentage of the plants wilt and die, although scattered wilted plants may also occur. By the time above-ground symptoms are evident; the vascular system of the plant is discoloured, particularly in the lower stem and roots. (Chavdarov, Liliya Krasteva, Nikolaya Velcheva, Stefan Neykov, 2013)

Fusarium Wilt

Epidemiology

High temperature and wet soil conditions favour disease development and are more severe in poorly drained soils. The organism grows and sporulates best at 20 to 30 °C and 5 to 8 pH.

Disease Cycle

When healthy plants grow in contaminated soil, the germ tube of spores or the mycelium penetrates root tips directly or enters the roots through wounds or at the point of formation of lateral roots. The mycelium advances through the root cortex intercellularly, and when it reaches the xylem vessels it enters them through the pits. The mycelium then remains exclusively in the vessels and travels through them, mostly upward, toward the stem and crown of the plant. While in the vessels, the mycelium branches and produces microconidia, which are detached and carried upward in the sap stream. Microconidia germinate at the point where their upward movement is stopped, the mycelium penetrates the upper wall of the vessel, and more microconidia are produced in the next vessel. The mycelium also advances laterally into the adjacent vessels, penetrating them through the pits. (G.N. Agrios 2005 edition)

Management

Use of wilt resistant varieties. Drenching with 1% Bordeaux mixture or Blue copper 0.25% may give protection. Seed treatment with 4g *Trichoderma viride* formulation or 2g Carbendazim per kg seed is effective. Mix 2kg *T. viride* formulation with 50kg FYM, sprinkle water and cover with a thin polythene sheet. When mycelia growth is visible on the heap after 15 days, apply the mixture in rows of chilli in an area of one acre.

3. Cercospora Leaf Spot

Causal organism: *Cercospora capsici*

Cercospora capsici was shown to be consistently associated with leaf spot of chilli, reducing appreciably the photosynthetic activity of infected plants leading to losses in yield (Rangaswami, 1979)

Cercospora leaf spot of chilli is one of the major problems of chilli cultivation in

Bangladesh (Meah and Khan, 1987). It is caused by *Cercospora capsici* luxuriantly less than 28 °C temperature, 92% relative humidity and pH 5-6. Below 90% relative humidity, the disease does not develop. The fungus survives in plant debris, primary infection coming from air-borne spores derived from it. The disease is more severe in wet weather than in dry weather and becomes destructive in high relative humidity (Cerkauskas, 2004).

Symptoms

Leaf lesions typically are brown and circular with small to large light grey centres and dark brown margins. The lesions may enlarge to 1cm or more in diameter and sometimes coalesce. Stem, petiole and pod lesions also have light grey centres with dark borders, but they are typically elliptical. Severely infected leaves drop off prematurely resulting in reduced yield.

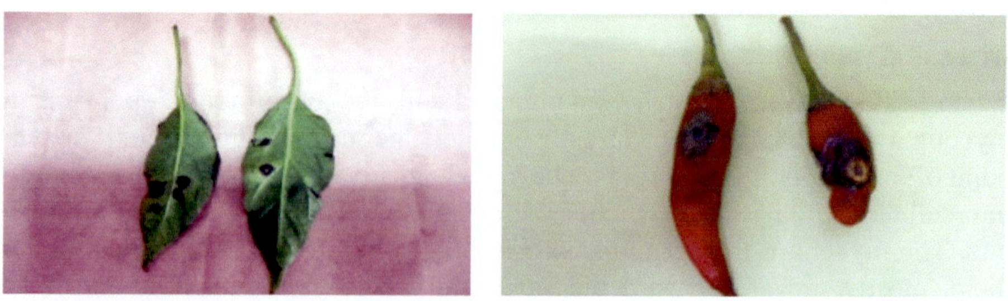

Cercospora leaf spot

Epidemiology

The fungus survives on seed and in crop residue. The conidia are disseminated by wind, rain or intercultural operation tools. The fungus sporulates abundantly at 20-30 °C, but reduces at 10 °C and stops at 40 °C.

Management

Some fungicides such as Bavistin-50 DF, Rovral-50 WP etc. have been found to give reasonable control of the disease in Bangladesh. Seed treatment with fungicides (53 °C, 10 minutes) reduced seed borne infection and higher seedlings vigour (Meah and Wick, 2006). Application of Bavistin-50 DF @ 0.02% three times sprays with 10 days intervals are most effective to control the Cercospora leaf spot of chilli (Meah, 2006).

3. Damping off

Causal organism: *Pythium aphanidermatum*

Chilli crop is attacked by several diseases of biotic and abiotic nature leading to great loss to cultivators. Damping off disease is one of them that causes heavy losses to the growers. It is responsible for 90 per cent of plant death either as pre or post-emergence damping off in the nurseries and fields (Sowmini, 1961).

Pythium aphanidermatum (Edson) Fitz. is a causal organism causing damping off of disease in chillies. *P. aphanidermatum* is cosmopolitan in distribution and one of the most common plant parasitic pathogens of a number of other crop plants in warmer parts of the world. (Waterhouse and Waterston, 1964).

Symptoms

Damping off occurs in two stages, i.e. the pre-emergence and the post-emergence phase. In the pre-emergence phase the seedlings are killed just before they reach the soil surface. The young radical and the plumule are killed and there is complete rotting of the seedlings. The post-emergence phase is characterized by the infection of the young, juvenile tissues of the collar at the ground level. The infected tissues become soft and water soaked. The seedlings topple over or collapse (Agrios, 2005).

Management

Cultural management practices are the primary strategies used to address disease problems (Sullivan, 2004). Site selection, such as choosing land without history of soil borne diseases, and planting in fields with well-drained soil, is important for preventing damping-off (Koike et al., 2000). The use of resistant cultivars can be the most important weapon for damping off disease control. However desirable cultivars with resistance to important soil borne diseases are not always available (Koike *et al.*, 2000).

5. Fruit Rot, Die Back and Anthracnose

Causal agent : *Colletotrichum capsici*

Anthracnose, derived from a Greek word meaning 'coal', is the common name for plant diseases characterised by very dark, sunken lesions, containing spores (Isaac, 1992).

The *Colletotrichum capsici* is reported to be most virulent in causing higher fruit rot of chilli (Amusa et al., 2004; Than et al., 2008; Ratana Cherdchai *et al*., 2010; Akhtar *et al*., 2017).

The optimum temperature for growth of Colletotrichum capsici is 28 to 32 °C. However, Sawle (2016) reported that temperature more than 30 °C would inhibit the growth of *Colletotrichum capsici. Colletotrichum capsici* can be identified on the basis of colony colour, growth pattern and pattern of acervuli formation on PDA medium (Smith and Black, 1990). The *C. capsici* will be fairly white to light grey colour with circular fluffy mycelia and; black coloured acervuli scattered all over the colony (Gupta et al., 2017). *Colletotrichum capsici* is reported to be seed and debris borne (Richardson, 1990) and also airborne (Asalmol *et al.,* 2001).

Symptoms

On attainment of favourable conditions, characteristic symptoms appear on the ripe chilli fruit, which appear as sunken circular or angular lesions. Often multiple lesions coalesce to form severe fruit rot. Generally, the lesions are characterised by the presence of black coloured spots in concentric rings at maturity. Initially, orange to pink conidial masses may be visible on the fruit surface. The dark spots when observed under a microscope are the acervuli structures containing setae hairs entrapping the conidia of the pathogen. Further, the pathogen forms micro sclerotia in plant debris or seed soil, which is the mode of survival under unfavourable conditions. The pathogen infects all parts of the host plant, including stems and leaves . Lesions on stems and leaves appear as small sunken greyish brown spots with dark margins, further on which development of acervuli in concentric rings could be easily seen.(Amrita *et.al*., 2016)

Disease cycle and epidemiology of anthracnose

Environmental factors play a major role in the development of disease epidemics. The relationships among rainfall intensity, duration and crop geometry and the dispersal of inoculum possibly lead to different levels of disease severity (Dodd *et al*., 1992). The effects of temperature often interact with other factors, such as leaf surface wetness, humidity, light or competitive microbiota (Royle and Butler, 1986). The duration of the surface wetness, however, ap-pears to have the most direct influence on the ger-mination, infection and growth of the pathogen on the host. Generally infection occurs during warm, wet weather. Temperatures around 27 °C and high humidity (a mean of 80%) are optimum for anthracnose disease development (Roberts *et al.,* 2001).

Management

Effective control of Colletotrichum diseases usually involves the use of a combination of cultural control, biological control, chemical control and intrinsic resistance (Wharton and Diéguez- Uribeondo, 2004).

The fungicide traditionally recommended for an-thracnose management in chilli is Maganese ethyl-enebisdithiocarbamate (Maneb) (Smith, 2000), although it does not consistently control the severe form of anthracnose on chilli fruit. The strobilurin fungicides azoxystrobin (Quadris), trifloxystrobin (Flint), and pyraclostrobin (Cabrio) have recently been la-beled for the control of anthracnose of chilli, but only preliminary reports are available on the efficacy of these fungicides against the severe form of the disease (Alexander and Waldenmaier, 2002; Lewis and Miller, 2003).

Trichoderma species have been applied to control *Colletotrichum species* in chilli (Boonratkwang et al., 2007).

5. Bacterial Leaf Spot

Causal agent:*Xanthomonas campestris pv. vesicatoria*

TheGram-negative, motile, aerobic, short rod-shaped bacteria can infect leaves, fruits, andstems, causing necrotic lesions and defoliation (Goode and Sasser,1980).

Symptoms

BLS-associated leaf lesions are often surrounded by a yellow halo of variable size. Approximately 24 h post-inoculation, hypersensitive cultivars exhibit a tissue collapse reaction from dark green to grey, while susceptible reactions lead to the appearance of water-soaked lesions, 3 days after inoculation. (Cook and Stall,1969)

Epidemiology

BLS epidemiology is induced by environmental factors which promote the broader dispersal of pathogens. This leads to the question of why susceptible plants are infected in some areas, while plants grown from the same seed in another area may not develop symptoms. (Potnis et.al., 2020). It was reported that *X. euvesicatoria pv. euvesicatoria* can survive in plant debris, inoculated sandy loam soil, and the rhizosphere of chilli and non-hostplants. In addition, the pathogen was found to survive at least 18 months either in soil or in seed in a field previously planted with chilli diseased with BLS. The pathogen can also be spread from infected plants to healthy plants by rain and wind. (Bashan *et.al.,*1982)

Management

Copper products have historically been used as the main chemical prevention measure for BLS. The use of copper was started by Professor Millardet in 1882, who developed Bordeaux mixture (copper sulphate and lime), which was effective at controlling grape downy mildew. (Johnson,1935).

Copper-tolerant strains were killed by the combination of copper and mancozeb, while the sprays of xedcopper controlled only sensitive strains (Marco *et.al.*,1983)

Biological control offers an environmentally sustainable alternative to traditional control methods that does not impact beneficial organisms and is a low-cost strategy that can enjoy sustained utility if applied judiciously. (He *et.al.*, 2021)

6. Leaf Curl Disease

Chilli is susceptible to various pathogens including viruses, which can cause heavy production losses. So far 65 viruses have been reported, including begomoviruses causing chilli leaf curl virus disease (ChiLCVD) infecting chilli throughout the world (Nigam *et al.* 2015). ChiLCVDis the most destructive virus in terms of incidence and yield loss. In severe cases, 100 percent losses of marketable fruit have been reported (Senanayake *et al.* 2007; Kumar et al.2011a,b; Senanayake *et al.* 2012).

Symptoms

The symptoms of ChiLCVD consist of abaxial and adaxial curling of the leaves accompanied by puckering and blistering of intravenous areas and thickening and swelling of the veins. In advanced stages of the disease, auxiliary buds are stimulated to produce clusters of leaves which were reduced in size. The whole plant assumed a bushy appearance and stunted growth. Fewer flowers and fruits were developed on the diseased plants, and those that were formed were much reduced in size and curled at the stylerend (Mishra *et al.* 1963). Two distinct types of symptoms were observed during a survey of *C. annuum* at Indian Agricultural Research Institute (Dhanraj *et al.* 1968). Symptoms in the firrst type were similar to those earlier reported (Mishra *et al.* 1963). In another type of symptoms, the affected plants became erect and bushy and the leaves were dark green. Upward curling, puckering and reduced size of leaves were also observed (Senanayake *et al.* 2007). Typical upward leaf curling, crinkling, puckering and reduction in leaf area along with stunting of whole plants were observed (Kumar *et al.* 2012).

Etiology

It is a polyphagous pest and is noted on more than 600 plant species spreading more than 60 plant viruses (Rishi 2004). ChiLCVD caused by begomoviruses is transmitted by whitefly *B. tabaci* in a persistent manner. Whiteflies are small piercing and sucking insects of the family Aleyrodidae, order Hemi-ptera, which have been consociated with agriculture and with the transmission of plant viruses for many years (Czosnek *et al.* 2002).The non-viral aetiology of leaf curl disease complex was experimentally proved, and it was concluded that it was the result of damage caused by thrips (upward curl) and mites (downward curl) (Amin 1979).

Cultural control

Use seeds from healthy plants of the previous season. Growing nursery in protected structures. Removal of infected seedlings and weed hosts from nursery as soon as seen. Treatment of seedlings with proper systemic fungicides to control vector.Use of yellow sticky traps just above the plants to control insect vectors. Destroying previous year susceptible crops, particularly solanaceous weeds and volunteer plants. Good weed control in the crop that may be alternative host to virus and vectors. Transplanting dates should be adjusted to avoid peak season of the vector population. Use of reflective (silver colour) plastic mulch. Use of live mulches, border crops or hedges which are more attractive to the vectors than pepper crop (Kenyon *et al*. (2014))

Biological Control of Vector

Predators- *Coccinella septempunctata, Clitostethus arcuatus, Orius spp, Chrysoperla carnea, Chrysopa spp., Sinea confusa* (Gerling et al.(2001))

Parasitoids- *Eretmocerus emeritus, Eretmocerus eremicus, Encarsia accenta, Encarsia adusta*, etc.

Pathogens (Fungi)- *Verticillium lecanii* and *Paecilomyces fumosoroseus, Paecilomyces farinosus*

Mycoinsecticides- BotaniGard, Bea-Sin, Boveril PM, Mycotal, Ago Biocontrol Verticillium, Pae-Sin (Faria and Wraight (2001))

Chemical control of vector (Synthetic)

Difenthiuron 50 WP @ 0.75 g per litre (Heathcote (1973)).Spraying diazinon, malathion, metasystox at 10 days interval (Devi and Reddy (1995)).0.07% monocrotophos with 0.25% wettable sulphur (Bhattiprolu and Rahman(2006)). Diafenthiuron @ 200 ml/litre (Hussain et al. (2017)). Imidacloprid 17.8 SL (0.003%) (Pandey et al. (2010)). Imidacloprid (0.05%), acephate (0.1%) and malathion (0.05%) (Ahmed and Ram (2016)).

References

A. Sudha and P. Lakshmanan. Solanum nigrum, a new host for powdery mildew disease of Capsicum annuum in the Madurai district of Tamil Nadu, India. Australasian Plant Disease Notes, 2007, 2, 97–98

Ahmed AA, Ram MR (2016) Medicinal plant extracts for the management of leaf curl disease of chilli (C. annuum Linn).World J Pharm Pharm Sci 5:1916-1924

Alexander, S.A., Waldenmaier, C.M., 2002. Management of Anthracnose in Bell Pepper. Fungicide and Nematicide Tests [Online]. New Fungicide and Nematicide Data Committee of the American Phytopathological Society. Vol. 58, p.49. Available from http://apsjournals.apsnet. org/doi/abs/10.1094/PDIS.2004.88.11.1198 (Accessed 25/12/2007). [doi:10.1094/FN58]

Amin PW (1979) Leaf curl disease of chilli peppers in Maharashtra,India. PANS 25:131134

Amrita Saxena, Richa Raghuwanshi,Vijai Kumar Gupta and Harikesh B.Singh:Chilli Anthracnose: The Epidemiology and Management, Front. Microbiol., Sec. Plant Pathogen Interactions Volume 7 - 2016

An, S.; Potnis, N.; Dow, M.; Vorholter, F.; He, Y.; Becker, A.; Teper, D.; Li, Y.; Wang, N.; Bleris, L.; et al. Mechanistic insights into host adaptation, virulence and epidemiology of the phytopathogen Xanthomonas. FEMS Microbiol. Rev.2020,44, 1–32.

Bashan, Y.; Diab, S.; Okon, Y. Survival of Xanthomonas campestris pv. vesicatoria in pepper seeds and roots in symptomless and dry leaves in non-host plants and in the soil. Plant Soil. 1982,68, 161–170.

Bhattiprolu SL, Rahman MA (2006) Management of insect borne viral diseases of chilli using nylon net, chemicals, neem products and barrier crop. Karnataka J Agric Sci 19:154–157

Boonratkwang, C., Chamswarng, C., Intanoo, W., Juntharasri, V., 2007. Effect of Secondary Metabolites from Tricho-derma Harzianum Strain Pm9 on Growth Inhibition of Colletotrichum Gloeosporioides and Chilli Anthracnose Control. Proceedings of the 8th National Plant Protection Conference. Naresuan University, Phisanulok, Thailand, p.323-336

Cerkauskas, R. 2004. Pepper Diseases: Cercospora Leaf Spot Published by AVRDC–The World Vegetable Center, Shanhua, Taiwan. pp. 741.

Chavdarov, Liliya Krasteva, Nikolaya Velcheva, Stefan Neykov Phytopathogens Causing Wilt in Pepper – Distribution, Symptoms and Identification Petar Institute of Plant Genetic Resources - Sadovo, Bulgaria 10.7251/AGREN1304549C

Chiou, K. L., Hastorf, C. A., Bonavia, D., & Dillehay, T. D. (2014). Documenting cultural selection pressure changes on chile pepper (Capsicum baccatum L.) seed size through time in Coastal Peru (7,600 B.P. - Present). Economic Botany, 68(2): 190–202. Doi:10.100712231-014-9270-y

Cook, A.A.; Stall, R.E. Differentiation of pathotypes among isolates of Xanthomonas vesicatoria. Plant Dis. Rep.1969,53, 617–622

Czosnek H, Ghanim M, Ghanim M (2002) Circulatory pathway of begomoviruses in the whitefly vector B. tabaci—insights from studies with Tomato yellow leaf curl virus. Ann Appl Biol140:215–231

Devi S, Reddy HR (1995) Effect of insecticides on aphid transmission of pepper vein banding virus and cucumber mosaic virus on chilli (C. annuum L). Mysore J Agric Sci 29:141–148

Dhanraj KS, Seth ML, Basal RC (1968) Reactions of certain chilli mutants and varieties to leaf curl virus. Indian Phytopathol 21:342–343

Dodd, J.C., Estrada, A., Jeger, M.J., 1992. Epidemiology of Colletotrichum Gloeosporioides in the Tropics. In: Bailey, J.A., Jeger, M.J. (Eds.), Colletotrichum: Biology, Pa-thology and Control. CAB International, Wallingford, p.308-325.

Faria M, Wraight SP (2001) Biological control of B. tabaci withfungi. Crop Prot 20:767–778

Gerling D, Alomar O, Arno J (2001) Biological control of Bemisia tabaci using predators and parasitoids. Crop Prot 20:779–799

Goode, M.; Sasser, M. Prevention-The key to controlling bacterial spot and bacterial speck of tomato. Plant Dis.1980,64, 831–834.

He, D.C.; He, M.H.; Amalin, D.M.; Liu, W.; Alvindia, D.G.; Zhan, J. Biological control of plant diseases: An evolutionary andeco-economic consideration. Pathogens 2021,10, 1311.

Heathcote GD (1973) Control of viruses spread by invertebrates toplants. In: Gibbs AJ (ed) Viruses and invertebrates. Elsevier,Amsterdam, pp 587–609

Isaac, S., 1992. Fungal Plant Interaction. Chapman and Hall Press, London, p.115.

Johnson, G.F. The early history of copper fungicides. Agric. Hist. 1935,9, 67–79

Kenyon L, Kumar S, Tsai WS, Hughes JA (2014) Virus diseases of peppers (Capsicum spp.) and their control. Adv Virus Res 90:297–354

Koike, S.T., Gaskell, M., Fouche, C., Smith, R. and Mitchell, J. 2000. Plant disease management for organic crops. ANR Publication 7252, University of California, Davis, CA.

Kumar RV, Singh AK, Chakraborty S (2012) A new monopartite begomovirus species, Chilli leaf curl Vellanad virus and associated beta satellites infecting chilli in the Vellanad region of Kerala, India. New Dis Rep 25:20

Kumar S, Kumar R, Kumar S, Singh AK, Singh M, Rai AB, Rai M(2011a) Incidence of leaf curl disease on capsicum germplasmunder field conditions. Indian J Agric Sci 8:187–189.

Kumar Y, Hallan V, Zaidi AA (2011b) Chilli leaf curl Palampur virusis a distinct begomovirus species associated with a betasatellite.Plant Pathol 60:1040–1047.

M. S. Islam, K. Fatema , K. M. B. Alam and M. B. Meah .Diagnosis and prescription for Cercospora leaf spot of chilli.J. Bangladesh Agril. Univ. 13(2): 191–196, 2015

Marco, G.M.; Stall, R.E. Control of bacterial spot of pepper initiated by strains of Xanthomonas campestris pv. vesicatoria that differin sensibility to copper. Plant Dis. 1983,67, 779–781.

Mathur, R.L., G. Singh, and Gupta, R.B.L. 1972. Chemical control of powdery mildew of chilli (Capsicum annum) caused by Leveillula taurica. Paper presented in the first disease on plant disease problems of Rajasthan in Dec 16-17. Udaipur.

Meah, M. B. 2006. Plant disease diagnostic manual. Plant Disease Diagnostic Clinic. IPM-laboratory. Department of Plant Pathology, Bangladesh Agricultural University, Mymensingh. Bangladesh. pp. 6-11.

Meah, M. B. and Khan, A.A. 1987. Check list of fruit and vegetable diseases in Bangladesh. 1st Edition, Department of Plant Pathology, Bangladesh Agricultural University, Mymensingh, Bangladesh. pp. 13.

Meah, M. B. and Wick, .LR. 2006. Diagnostic plant pathology. Dept. of plant soil and insect sciences, University of Massachusetts. Amherst MA, USA. pp. 5-71.

Mishra MD, Raychaudhuri SP, Jha A (1963) Virus causing leaf curl of chilli (Capsicum annuum L.) Indian. J Microbiol 3:73–76

Moghe PG (1977) Investigations into the causes of churda murda(malformation) disease of chilli in Vidarbha. Curr Sci 46:831–832

Nigam K, Suhail S, Verma Y, Singh V, Gupta S (2015) Molecular characterization of begomovirus associated with leaf curl diseasein chilli. World J Pharm Res 4:1579–1592.

Pandey KS, Mathur AC, Srivastava M (2010) Management of leaf curl disease of chilli (Capsicum annuum L.). Int J Virol 6:246–250

Ramesh Chand, Chhattar Pal, Vineeta Singh, Manoj Kumar, Vinay Kumar Singh and Pallem Chowdappa,Draft genome sequence of Cercospora canescens: a leaf spot causing pathogen.CURRENT SCIENCE, 109, NO. 11, 10 2015.

Rangaswami, G. 1979. Diseases of vegetables. In: Diseases of crop plants in India. 2nd Ed. McGraw Hill Book Company, New Delhi. pp. 345-347

Rishi N (2004) Current status of begomoviruses in the Indian Subcontinent. Indian Phytopath 57:396–407

Roberts, P.D., Pernezny, K., Kucharek, T.A., 2001. Anthrac-nose caused by Colletotrichum sp. on pepper [Online]. Journal of University of Florida/Institute of Food andAgricultural Sciences. Available from http://edis.ifas. ufl.edu/PP104 (Accessed 25/12/2007).

Royle, D.J., Butler, D.R., 1986. Epidemiological Significance of Liquid Water in Crop Canopies and its Role in Disease Forecasting. In: Ayres, P.G., Boddy, L. (Eds.), Water, Fungi and Plants. Cambridge University Press, Cam-bridge, p.139-156.

Senanayake DMJB, Mandal B, Lodha S, Varma A (2007) First reportof Chilli leaf curl virus affecting chilli in India. Plant Pathol 56:343.

Senanayake DMJB, Varma A, Mandal BJ (2012) Virus–vector relationships, host range, detection and sequence comparison of chilli leaf curl virus associated with an epidemic of leaf curl disease of chilli in Jodhpur. Indian Phytopathol 160:146–155

Smith, K.L., 2000. Peppers. In: Precheur, R.J. (Ed.), Ohio Vegetable Production Guide. Ohio State University Ex-tension, Columbus, Ohio, p.166-173.

Sowmini, R. 1961. Studies on phycomycetes in agricultural soils with special reference to Pythiaceae. M.Sc. (Agri.) Thesis, University of Madras, pp. 160.

Sullivan, P. 2004. Sustainable management of soil- borne plant diseases. ATTRA, National Sustainable Agriculture Information Service, California, U.S.A.

Waterhouse, G.M. and G.M. Waterston. 1964. Pythium aphanidermatum. [Descriptions of Fungi and Bacteria].CAB International. 4: Sheet 36

Wharton, P.S., Diéguez-Uribeondo, J., 2004. The biology of Colletotrichum acutatum. Anales del Jardin Botanico de Madrid, 61:3-22.

12

Diseases of Okra and Their Management

Abhay Pratap Singh[1], Pankaj Singh[2], Shivanshu Mishra[2] and Ankit Kumar Chaurasiya[2]

[1]*Department of Plant Pathology, Banda University of Agriculture and Technology Banda- 210001, Uttar Pradesh, India*
[2]*Department of Plant Pathology, Sardar Vallabhbhai University of Agriculture & Technology, Meerut, Uttar Pradesh*

Abstract

The well-known vegetable crop okra (Abelmoschus esculentus (L) Moench) is a member of the Malvaceae family which is extensively found across the world's tropical, subtropical, and mild temperate climates. It is native to Ethiopia. It has a significant impact on human nutrition and a rich source of nutrients such as protein, carbohydrates, vitamins, calcium, potassium, enzymes, and other minerals that are frequently deficient in impoverished nations' diets. Haemorrhoids and ulcer treatment both have been credited to its medical benefits. The yield of okra is being decreased due to a variety of biotic and abiotic reasons. In this chapter we have discussed several major diseases viz. damping off, wilt, Cercospora leaf spot, powdery mildew and yellow vein mosaic disease along with their causal agent, symptomatology, etiology, favourable environmental conditions, survival and spread and some management strategy with the aim of reducing disease inoculum and spread which ultimately enhance the fruit production.

Keywords: *Okra, Symptoms, Pathogen, Diseases, Etiology etc.*

Okra

Botanical Name	:	*Abelmoschus esculentus* (L) Moench
Family	:	Malvaceae
Chromosome number (2n)	:	130

DOI: 10.1201/9781042014460-12

Introduction

Okra (*Abelmoschus esculentus* (L).is a cross pollinated vegetable crop. Okra originated near about Ethiopia& its cultivation became widespread in North Africa and the Middle East (Lamont, 1999; Tindal, 1983). Okra is grown commercially in many nations, such as Ethiopia, Cyprus, India, Bangladesh, Afghanistan, Pakistan, Turkey, Iran, Western Africa, Yugoslavia, Myanmar, Japan, Malaysia, Thailand, Brazil, Southern United States, and Ethiopia (Purseglove, 1987; benjawan et al. 2007; Qhureshi, 2007). It is a widely used, high-valued vegetable crop that is commercially grown for its tender pods during the spring, summer, and kharif seasons. Okra has high moisture content, making it a good source of vitamins, minerals, and mucilage. Okra is cultivated and consumed commercially throughout the year in many different places of India. Gujarat is the state that produces the most okra in India out of all the others, followed by West Bengal and Bihar (Anonymous, 2022) (Ndunguru & Rajababu, 2004). Okra has also been used medicinally to treat ulcers and haemorrhoids (Adams, 1975). Additionally, it is an excellent source of iodine, which cures the simple goitre, and other compounds that are valuable in medicine (Moaward et. al. 1984).

Numerous diseases, including *Fusarium* wilt, powdery mildew, *Cercospora* leaf spot, yellow vein mosaic, enation leaf curl, and others, have a devastating impact on okra plants, resulting in substantial economic losses due to production reductions.

1. Damping off

One of the most significant and deadly diseases affecting okra nowadays is damping off, which is brought on by soil-borne pathogens. It causes significant losses in the first plant stand in the field, especially in summer-grown crops. The economic loss due to damping off disease can be represented in two ways, first is to destruction of planting material (seeds) and seedlings and another is extra cost due to replanting and lower yield due to late planting. Damping off of okra is known to occur worldwide and caused by *Pythium* spp. (Jukte*et al.,* 2016), *Phytophthora nicotianae* (Matny, 2013), *Rhizoctonia solani* and *Macrophomina phaseolina* (Jiskani *et al.,* 2007) under different climatic conditions.

Symptoms

From the time of planting until the fourth or sixth week after sowing, damping-off symptoms might be seen (Horst 2013). Depending on when they first appeared, the disease's symptoms might be separated into two phases.Seeds become fragile, rotten, and do not germinate as a result of pre-emergence damping off, which also causes distinctive water-soaked lesions to appear at soil surface. (Cram 2003, Landis 2013). Post emergence damping off in which seedlings becomes thin and tough, water-soaked sunken lesions and rotting at or slightly below the soil level that leads to decay, wilt and final collapse of seedling after emergence (Boyce 1961, Horst 2013). Plant shows stunted growth with mould like growth at or near soil level and leaves becomes greyish to brown in colour.

Pre emergence damping off

Post emergence damping off

Pathogen

Damping off disease of okra is caused by many soil borne pathogens including *Pythium* spp. (Jukte*et al.,* 2016), *Phytophthora nicotianae* (Matny, 2013), *Rhizoctonia solani* and *Macrophomina phaseolina* (Jiskani *et al.,* 2007). *Pythium* spp are most frequently causes post and pre-emergence damping off in okra.

Etiology

Pythium spp are oomycetes soil borne pathogen that grow in soil as saprophyte and parasitically in okra tissues. It belongs to order peronosporales having hyaline and aseptate mycelium and hyphae grows inter or intracellularly. Globose or oval sporangia are produced terminally on the hyphae which produces reniform biflagellate zoospore and oospore is also produced as sexual resting spore through gametangial contact.

Disease Cycle

Damping off disease in okra caused by *Pythium* spp and other soil borne fungal plant pathogens. Since the pathogens are soil borne so they do not require any natural opening to enter host plant. As the favourable environmental condition comes, the fungal spores or mycelia present in the soil germinate to form appressoria or infection peg to enter the host tissues. After the penetrating the host tissues it starts to colonize and exhaustion of host nutrition and the pathogen starts to reproduce and produces enormous number of sporangia that contains enormous number motile zoospores that leads to secondary infection to nearby seeds or seedling. At the last of cycle, they reproduce sexually through gametangial contact and produce thick-walled oospore that serve as resting spore that remains dormant in the soil for a long period until the favourable conditions come.

Favourable Conditions

Damping off disease caused by *Pythium* spp are favoured by poorly drained soil, high humidity cool and wet soil moisture. High soil moisture favoured the zoospore movement and low temperature (10-20°C) also favoured the production of zoospore that cause serious secondary infection and helps in the disease development. (Wright 1957; Papavizas and Davey 1961;Duniway 1983; James 2012; Starkey and Enebak 2012).

Management

Damping off disease is devastating threats to okra plant that is caused by a variety of soil borne pathogen including *Pythium* spp., *Rhizoctonia* spp. etc. it invades seeds and seedlings of the plant and leads to death of plant. Following combination of preventive and curative measures are required to manage damping off disease in okra.

Cultural Methods

Seed bed Preparation

In order to prevent soil crusting, use pest-free growing media or soil by adding compost, plant residues, or microbial amendments that inhibit soil-borne pathogens. You can also perform soil solarization and bio-fumigation, choose a mixture of particle sizes and good porosity, install drainage tiles, subsoiling, and by adding composted organic matter to improve the soil texture, water-holding capacity, and net production of the soil.(Dürr and Aubertot 2000; Diab *et al.* 2003; Deadman *et al.* 2006; Njoroge *et al.* 2008; Pane *et al.* 2011; He *et al.* 2011; Landis 2013; Bahrami sharif *et al.* 2013; Vitale *et al.* 2013)

Soil pH

Use low-pH (4.5–6.0) somewhat acidic soils; raise the pH of the soil with organic amendments; use sulphur, aluminium sulphate, or acid peat. (Russell 1990; Davey 1996; Cram 2003).

Sowing Time

Avoid warm or damp conditions while planting, seed when it's not too early or too late, and irrigate soils well enough to reach the developing root's depth without overflowing them. (Hwang *et al.* 2000; Cram2003).

Crop Rotation

Mono culturing of okra plant should be averted and a long duration crop rotation with *Brassica* crops should be acompanied because their root exudates contain antagonistic substances to minimise the pathogen inoculum. (Hwang *et al.* 2008; Abdel-Monaim and Abo-Elyousr 2012).

Physical Methods

Soil Solarisation

In severely affected areas soil solarisation method should be applied to kill the pathogen by sterilising the soil. Raising the soil temperature by covering the soil after a slight irrigation (up to 10 cm) with a transparent polythene sheet during summer month that kill the resting inoculum of the pathogen.

Chemical Control

Seed Treatment

- Especially when applied to weaker seeds or ones with damaged seed coats, seed treatments can effectively boost the emergence of seedlings. (Mancini and Romanazzi 2014).
- Seed treatments with chemicalsare mostly followedto manage damping-off diseases (Leisso *et al.* 2009; Dorrance *et al.*, 2009; Rothrock *et al.*, 2012; Kandel *et al.* 2016). Seed treated with Thiram (2-3 g /kg of seed) or *Trichoderma viride* (3-4 g/kg of seed) and soil drenching with Dithane M 45 (0.2%) or Bavistin (0.1%) provide best management against the disease. In spite of this, a variety of chemicals can be used to eliminate the pathogen from the seed coat, including fungicides, bleach, hydrogen peroxide, and ethanol. (Dumroese and James 2005; Mancini and Romanazzi 2014).

Foliar Application

For the control of damping off disease Azoxystrobin, Metalaxyl, and Pyraclostrobin alone or in combination with *P. Fluorescence* is found effective. Metalaxyl, Azoxystrobin and Pyraclostrobin @ 5 & 10 um ml^{-1} provide protection to post&pre-emergence mortality and yield healthy seedlings. When these fungicides are applied along with *P. Fluorescence* gives an excellent control over damping off. (Salman &Abuamsha, 2012)

2. Fusarium Wilt

Wilt disease of okra is commonly caused by soil borne pathogen *Fusarium oxysporum* f.sp. *vasinfectum*. *Fusarium oxysporum* have a wide host range including agricultural and horticultural crops such as tomato, potato, egg plant, pepper, pea, banana, pegion pea, cotton, lentil etc. *Fusarium oxysporum* f.sp. *vasinfectum* is a vascular wilt pathogen that invades vascular system of okra plant leading to death of entire plant and cause huge economic loss. The crops sown in May-June were more susceptible to wilt in the region of Punjab and Haryana and leads to significant loss of 25-35% by *F. oxysporum f. sp. vasinfectum*. (Grover RK and Singh G. 1970).

Symptoms

Fusarium wilt is known to occur widely and is a most devastating disease of okra. Young plants that are impacted by the disease eventually perish from symptoms such as yellowing, loose turgidity, and drooping leaves. Older plants have abrupt leaf wilting and yellowing or browning of the vascular bundles in the collar area.

a- Drying and drooping of leaves, b- Presence of dark brown discolouration underside of bark, c- Mechanical blockage of vascular tissues by physical presence.

By invading the roots and colonising the vascular system, the fungus limits the movement of water. A dark woody section and a dark brown stripe on the underside of the bark are visible at the longitudinal section base of the stem. (Nene *et al*., 1984, Anon. 2011).

Pathogen

Wilt disease of okra is caused by soil borne *Fusarium oxysporum* f. sp. *vasinfectum* (Atk) Snyder & Hansen. *F. oxysporum f. sp. vasinfectum* also Seed-borne in nature (Gangopadhyay S and Kapoor KS. 1977).

Etiology

Fusarium oxysporum is a soil borne anamorphic fungi. It is a facultative parasite having hyaline, intracellular mycelium that asexually produces three kinds of spores viz. macroconidia, micro conidia and chlamydospores. Macro conidia are colourlesstwo to many celled sickles shaped and micro conidia are one to two celled colourless ovoid shaped and smaller than macro conidia (Champawat and Pathak 1989).

Disease Cycle

The pathogen is soil borne and survive in the soil during off season in the form of resting spore or in infected plant debris. As favourable condition comes, the fungus comes in close contact with the roots and chlamydospores or conidia starts to germinate and penetrate the host cortical tissues. Pathogen starts to proliferate extensively in the xylem vessels and leads to cause discoloration and mechanical blockage of the xylem vessels and leads to death of plants.

Favourable Environmental Conditions

Wilt disease caused by *Fusarium oxysporum* is adored by relatively high temperature (28-30°C) and humidity. Warmer, slightly acidic and moist soils favour the disease development (Mahesh *et al.* 2006).

Management

Wilt of okra is a soil borne as well as seed borne infection so the management tactics should tackle the pathogen's life cycle. Some of the common strategy to manage this problem are as follows.

Sanitation/ Crop hygiene

The disease is soil borne so it is important to remove the old infected plant debris to prevent the further disease progress. The tools used in inter cultivations should be clean and free from contaminations.

Soil Improvement

The pathogen is most favoured by moist condition so the crop should be cultivated in well drained and amended with some green and organic matters such as compost or well rotten FYM to enhance the soil fertility to promote the growth of the plant and also reduces the inoculum in soil.

Soil Solarisation

This practice involves the use of transparent polythene sheet to cover the moist soil during hot summer month to raise the soil temperature that kills the resting structures of the pathogen and ultimately reduces the disease inoculum.

Crop Rotation

Avoid the mono cultivation of okra plant in the same land. Adopt a long (2-3 years) crop rotation such as with sorghum to reduce the initial inoculum of the disease.

Soil Fumigation

Soil fumigation with 2, 3-bromopropionitrile and trichloronitroethylene is recommended for effective management of the disease. (Ribeiro R De L D et. Al. 1971).

Seed Treatment

Seed treatment with *P. fluorescence and T. viridae* (10g/kg) give better control over wilt of okra. Razeena and Ahmad (2007). Fusarium wilt can be managed by treating the seed with 0.3 % Ziram (Gangopadhyay and Kapoor (1977).

Chemical Control

The disease is controlled by administering three sprays of either Bavistin (1g/liter) or Karathane (6g in 10 litres of water) as soon as the first symptoms occur, five to six days apart.

3. Powdery Mildew of Okra

Powdery mildew is one of the severe diseases of okra which is caused by an obligate parasite *Erysiphe cichoracearum*. A wide host range of crops such as beans, southern peas, squash, cucumbers, muskmelons, pumpkins etc are affected by powdery mildews and leads huge economic loss. There have been reports of the disease from Mexico (Diaz-Franco, 1999). The disease has been documented to occur in India in the following states: Himachal Pradesh (Raj et al., 1992), Delhi (Prabhu et al., 1971), Karnataka (Sohi and Sokhi, 1974), and Maharashtra (Jambhale and Nerkar, 1983). Okra is most important vegetable crop and almost all cultivars of okra are vulnerable to Powdery mildew and yield loss ranges between 17% to 86.6% which is totally depend the age of the plant at the time of infection. (Sridhar & Poonam 1989).

Symptoms

Plants infected with powdery mildew are characterized by the presence of white powdery mass of microscopic spores on the upper surface of leaves, stems, petioles etc. There is a characteristic yellow halo can be seen on the lower surface of leaves. In the severe infestation infected plant parts gets stunted and distorted. Leaves becomes yellow, wither and flowers become distorted and reduced in number which leads to reduction of fruit yield and quality (Khalikar*et al*., 2011).

White powder like spore mass on the upper surface of leaves

Pathogen

Powdery mildew disease of okra is caused by wind borne obligate pathogen *Erysiphe cichoracearum.*

Etiology

The pathogen *Erysiphe cichoracearum* produces ectophytic branched mycelium and conidiophores are unbranched and septate with hyaline single celled oval to barrel shaped conidial mass which disappeared by producing dark brown to black coloured cliestothecia with hyphae like appendages (Dahivelkar *et al.,*2017).

Favourable Environmental Condition

The disease development and spread are favoured by relatively low temperature (11-28°C), cool and dry weather, high humidity (55-85%) and 6.5 mm rainfall with 4.5 km/hr wind speed. (kavyashre e*et al.*, 2022).

Disease Cycle

Powdery mildew fungi of okra are ectophytic in nature. Ascospores or cleistothecia or conidia are the source of primary inoculum and comes in contact with the host tissue and produce germ tube to form appressoria which penetrate the cell wall of the host.

After that a specialized feeding structure haustoria is formed to absorb the nutrient for growth and development. After that pathogen goes to asexual cycle to produce enormous number of conidia that serve as source of secondary infection and the cycle continue to spread of the disease.

Management

Biological Control Measures

Spray of bio dewcon @ 0.75 kg/acre that contains *Ampellomyces quisqualis* 1.15% WP. gives effective control of powdery mildew. Spray of *Trichoderma harzianum* and *Bacillus subtilis*and 0.03% of neem oil gives better control over powdery mildew of okra. (Khalikar *et al.*,2011).

Chemical Control

Powdery mildew of okra is successfully managed by foliar application of benomyl (0.1%), wettable sulfur (0.2, 0.3 and 0.5%) (Sridhar SS 1985, Ramakrishnan et. al. 1973)). Applications of potassium silicate K_2SiO_3 (200 mg/L) and the alternation of systemic fungicide Bayleton WP 5 Triadimenol (25 mg/L) and thiovit 80% WP (sulphur (2.4g/L) with 12 days interval is also highly effective against powdery mildew (Yanar *et al.*, 2011).

4. Cercospora Leaf Spot of Okra

A multitude of biotic and abiotic stress affect the okra crop. Fungal infections are one type of biotic factor that can seriously hinder the production of okra (Jha and Dubey, 2000). Okra leaf spots have been found to caused by two species of *Cercospora* in India: *C. malayensis* Stev. and Solh. and *C. abelmoschi* Ell. and Ev. (Sridharan and Rangaswamy, 1968). It is also known as black mould disease which is usually causes maximum damage to the crop particularly in humid warm areas.

Symptoms

The disease primarily affects the foliar part of the plant that results in higher reduction in plant health and production.

Both the spp. of *Cercospora* cause different symptoms on host plant.

C. abelmoschi first appears on older leaves and produces light brown coloured spots initially. Later turned to purplish in colour of different size.

Light brown coloured spot that coalesces to form larger spots giving blighted appearance

Later the spots coalesce to form large sized spots and gives blighted appearance in infected leaves. The fungus produces sooty mould growth on the lower surface of leaves and severely affected leaves rolled up, wilt and fall down to the ground (Deighton 1976).*Cercospora malayensis* produces long fairly large spots with brown streaks between the veins which extending from edge to midrib and produce definite spots with red to purple border. (Stev. and Solh. 1931).

Pathogen

Cercospora leaf spot of okra is caused by two spp. of *Cercospora* viz. *C. malayensis* Stev. And Solh. and *C. abelmoschi* Ell. and Ev. (Sridharan and Rangaswamy, 1968).

Etiology

The pathogen *Cercospora* spp produces septate branched intercellular, hyaline to yellowish-brown mycelium. Conidiophores are olive brown to brown geniculate, straight to flexuous and arising from loose to compact stroma produces 2-8 septate hyaline acicular obclavate conidia. (Solheim and Stevens, 1931).

Favourable Environmental Condition

Temperatures between 25 and 30 °C are beneficial for conidial infection, whereas relative humidity in the range of 90 to 100% and temperatures between 20 and 26 °C are excellent and necessary conditions for conidia germination. Conidia do not form at temperatures lower than 10 °C. (Windels *et al.,* 1998).

Disease cycle

The pathogen sustain oneself for a long period of time in the infected plant tissues through conidia, dormant mycelium and perithecia in soil. The subjected okra plant also provide shelter to the pathogen. Primary infection caused by ascospores or conidia from infected plant tissues or infected seed thus infects roots and lower leaves of the plant. Secondary infection is caused by air borne conidia. Rainfall also plays a crucial role in secondary spread of the disease.

Management

Crop Rotation

Avoid the mono cultivation of okra in same land because the pathogen survives in soil during off season so crop rotation with non-familiar crop plays crucial role in reducing the build-up of *Cercospora* spores in soil.

Fertilizer Management

This is a foliar disease infecting leaves of okra so avoid the excessive use nitrogenous fertilizer because nitrogenous fertilizer enhances the lush, succulent growth of the plant which more susceptible to the pathogen. So, it is advised to use a balanced dose of fertilizers to promote healthy plant development.

Seed Treatment

During the first stage of pod development, seeds are treated with thiophanate methyl or carbendazim at 0.1% at 2 g/kg and two sprays spaced 15 days apartgives effective control over the same disease. (Satya Prasanth and Patil, 2007). Application of Trichoderma viride in a furrow at a rate of 4.0 kg, supplemented with 50 kilogramme of farm yard manure per hectare +The most effective and cost-effective methods for managing the illness were determined to be Tebuconazole 2 DS @ 1.5 g/kg and two foliar applications of Tebuconazole 25.9% EC @ 1.0 ml/l, the first applied at the onset of the disease and the second applied at intervals of 15 days. (Kumhar *et al.*, 2018).

Chemical Management

Foliar spray of Tebuconazole (0,15%)(Nath *et al.*, 2013; Kumar *et al.*, 2015), carbendazim @ 0.05% or propiconazole @ 0.1% (Jahagirdar andHundekar, 2010) give effective control of Cercospora leaf spot of okra.

5. Yellow Vein Mosaic Disease

Viruses are major threat in okra growing region of the world and pose a major constrains to production. Yellow vein mosaic viral disease is one of them causing huge economic loss. Yellow vein mosaic disease of okra was first described by Kulkarni in 1924 in India and further studied by many researchers (Capoor and Verma, 1950; Verma, 1952). Marketable yield loss caused by yellow mosaic disease assumed in the range of 50-94% depending on the stage of plant being infected (Sastry and Singh 1974; Pun and Doraiswamy 1999). Early infection within 20 days of germination causes up 94% which lowers with the late infection after 50-60 days (Nath and Sakia, 1992).

Symptoms

The most characteristic symptoms of okra yellow vein mosaic disease are the formation of light to dark yellow-coloured veins in the infected leaves and remaining leaf lamina remains green. Typically yellowing occurs only along with the vein and shows a yellow-coloured interwoven appearance of veins. At the later stages there is alternate pattern of green and yellow coloured patches on whole leaf giving mosaic appearance and curling, vein clearing which leads to reduction in photosynthetic area that results in reduced growth of the plant and ultimately reduction in fruit production (Venkatara vanappa *et al.*, 2012).

a- initial infection of OYVMV, b- severe infestation of OYVMV showing leaf curling & vein clearing

Pathogen

The disease is caused by okra yellow vein mosaic virus which is belongs to genus begomovirus under the family *geminivirirdae*. The virus is monopartite with ssDNA genome. The genome of virus consists of two particles *i.e.*DNA A & DNA B in which former is associated with viral replication and later with coat protein. The virus is transmitted by sap sucking white fly (*Bemisia tabaci)* in persistent circulative manner. (Brown 1994; Duffus 1994).

Disease Development

White fly is the only mode to spread the disease. When virus free white fly inserts their stylet to an infected plant to suck the sap then fly also suck some virus particle present in the parenchymatous tissue along with the sap and fly becomes viruliferous. Now fly again insert their stylet and secrete some saliva to next healthy plant during feeding then it transmits the viral particle to healthy plant which becomes infected. The cycle continues to spread the disease very rapidly.

Favourable Conditions

Population of white fly is directly proportional to disease spread. Higher temperature without rain favours the white fly population so there is high disease incidence (Mahanta *et al.,* 2020)

Management

Management of okra yellow vein disease is totally relied on the population of its vector (Denholm *et al.* 1998; sani*et al*, 2020). Population of white flies can be managed by changing in sowing date, follow intercropping with pulses.

Foliar application of foratox, methyl demeton 0.025% or three applications of phosphamidon 0.02% effectively manage the white fly population in the field (Munir *et al.,* 2020). Spray of different plant extracts also gives better management of white fly such as seed extract of neem and leaf extract of tobacco (2%) (Bhagat *et al.* 2001).The disease can be effectively managed by using some plant defence activator such as KH_2PO, benzoic acid, salicylic acid, citric acid etc. (Ryals *et al.,* 1996).

Biocontrol Agent for White Fly

In nature, there are a lot of parasitoids and fungal pathogens that prey on whiteflies and cause their population to decline.

- *Eretmocerus* spp. and *Encarsia formosa* (Capinera, 2001)
- *Beauveria bassiana, Paecilomyces* spp., and *Verticillium lecanii* (Ioannou, 1997)
- Another significant nymphal parasitoid of *B. tabaci* is *Prospaltella flava* (Arneja, 2000).
- *B. tabaci* eggs and nymphs are consumed by predators such as *Chrysoperla carnea, C. scelestes, Geocoris bicolor, and Mallada boninesis* (Dhawan et al., 2007).

6. Conclusion

Vegetables are the most important component of human diet to fulfil the nutritional requirement. Okra is renowned for its nutritional and medicinal properties. Okra is widely cultivated crop across the world mainly Asia and Africa. There are several diseases causing negative effect on the production of okra. Damping off of okra caused by several soil borne fungi such as *Pythium, Rhizoctonia* etc. causes pre and post emergence seedling rot. Wilt of Bhindi caused by *Fusarium oxysporum* f.sp. *vasinfectum* causes mechanical blockage of vascular tissues resulting in drying and drooping of leaves which leads to death of the plant. Powdery mildew is caused by obligate parasite *Erysiphe cichoracearum* produces white powdery spores on upper surface of leaves, stem, flowers, petioles etc. that reduces the photosynthetic area of the leaves which results in stunting and distortion of entire plant. Another important disease okra is *Cercospora* leaf spot caused by *Cercospora abelmoschi* and *Cercospora malayensis* causing light brown to dark coloured large sized spots with reddish to purplish margin giving blighted appearance from distance. Other and most important disease of okra is yellow vein mosaic disease caused by a begomovirus (*okra yellow vein mosaic virus*) producing bright yellow-coloured interwoven veins which results in vein clearing leaf curling and stunting of the plant at later stage. The chapter emphasize on the common understanding of management of these major diseases of okra.

References

Abdel-Monaim, M. F., & Abo-Elyousr, K. A. (2012). Effect of preceding and intercropping crops on suppression of lentil damping-off and root rot disease in New Valley–Egypt. Crop Protection, 32, 41-46.

Adams C. F., (1975). Nutritive value of American foods in common units, U.S. Department of Agriculture, Agric Handbook. 425, pp 29.

Anonymous, (2011). Biology of Abelmoschus esculentus(okra), Ministry of Biotechnology and Ministry of Environment and Forest, Government of India. Pp. 1-26.

Arneja, A. K. (2000). Biology of whitefly, Bemisiatabaci (Gennadius) on American cotton (Doctoral dissertation, M. Sc. Thesis, Punjab Agricultural University, Ludhiana, India).

Bahramisharif, A., Lamprecht, S. C., Calitz, F., & McLeod, A. (2013). Suppression of Pythium and Phytophthora damping-off of rooibos by compost and a combination of compost and nonpathogenic Pythium Taxa. Plant disease, 97(12), 1605-1610.

Benjawan, C., Chutichudet, P., & Kaewsit, S. (2007). Effects of green manures on growth, yield and quality of green okra (Abelmoschus esculentus L.) Har Lium Cultivar. Pakistan journal of biological sciences: PJBS, 10(7), 1028-1035.

Bhagat, A. P., Yadav, B. P., Prasad, Y. (2001). Rate of dissemination of okra yellow vein mosaic virus in three cultivars of okra. Indian Phytopathology, 54(4), 488–489

Boyce J.S., (1961). Forest pathology, third. McGrawHill, USA, New York.

Brown, J. K. (1994). Current status of Bemisiatabaci as a plant pest and virus vector in agroecosystems worldwide. FAO Plant Protection Bulletin, 42(1/2), 3-32.

Capinera, J.L., (2001). Handbook of vegetable pests. Academic Publishers, San Diego.

Champawat, R. S., & Pathak, V. N. (1989). Cultural, morphological and pathogenic variations in Fusarium oxysporum f. sp. cumini. Indian Journal of Mycology and Plant Pathology, 19(2), 178-183.

Cram M.M., (2003). Damping-Off. Tree Plant Notes 50:1–5.

Dahivelkar, P., Atre, G. E., Gawande, P. V., & Mate, G. D. (2017). Management of powdery mildew of okra caused by Erysiphe cichoracearum. Int. J. Curr. Microbiol. App. Sci, 6(8), 3183-3189.

Davey CB (1996). Nursery soil management-organic amendments. In: Landis TD, South DB (eds) Natl. Proceedings, For. Conserv. Nurs. Assoc. Portland (OR), p 6–18.

Deadman, M., Hasani, H. A., & Sa'di, A. A. (2006). Solarization and biofumigation reduce Pythium aphanidermatum induced damping-off and enhance vegetative growth of greenhouse cucumber in Oman. Journal of Plant Pathology, 335-337.

Deighton F. C., (1976). Mycological papers. 140:138.

Dhawan A.K., Butter N.S., Narula A.M., (2007). The cotton whitefly Bemisiatabaci (Gennadius). Technical Bulletin, Department of Entomology, Punjab Agricultural University, Ludhiana.

Diab HG, Hu S, Benson D.M., (2003). Suppression of Rhizoctonia solani on impatiens by enhanced microbial activity in composted swine waste-amended potting mixes. Phytopathology 93:1115–1123.

Diaz Franco, A. (1999). Okra (Abelmoschus esculentus) powdery mildew in Mexico. Revista Mexicana de Fitopathol. 17: 44-45

Dorrance AE, Robertson AE, Cianzo S et al (2009). Integrated management strategies for Phytophthora sojae combining host resistance and seed treatments. Plant Dis 93:875–882.

Duffus, J. E. (1994). Diseases vectored by whiteflies: etiology, geographical distribution and possible control measures. Arab Journal of Plant Protection.

Dumroese, R. K., & James, R. L. (2005). Root diseases in bareroot and container nurseries of the Pacific Northwest: epidemiology, management, and effects on outplanting performance. New Forests, 30(2-3), 185-202.

Duniway, J. M. (1983). Role of physical factors in the development of Phytophthora diseases.

Dürr C, Aubertot J.N., (2000). Emergence of seedlings of sugar beet (Beta vulgaris L.) as affected by the size, roughness and position of aggregates in the seedbed. Plant Soil 219:211–220.

Gangopadhyay, S., & Kapoor, K. S. (1977). Control of Fusarium wilt of okra with seed treatment. Indian Journal of Mycology and Plant Pathology, 7(2), 147-149.

Grover, R. K., & Singh, G. (1970). Pathology of wilt of okra (Abelmoschus esculentus (L.) Moench) caused by Fusarium oxysporum f. vasinfectum (Atk.) Snyder & Hasen, its host range and histopathology. Indian Journal of Agricultural Science, 40, 989-96.

Hamm, P. B., Campbell, S. J., & Hansen, E. M. (1990). Growing healthy seedlings: Identification and management of pests in Northwest forest nurseries (Vol. 19). Forest Pest Management, US Department of Agriculture, Forest Service, Pacific Northwest Region.

He, M., Tian, G., Semenov, A. M., & van Bruggen, A. H. (2012). Short-term fluctuations of sugar beet damping-off by Pythium ultimum in relation to changes in bacterial communities after organic amendments to two soils. Phytopathology, 102(4), 413-420.

Horst R.K., (2013). Damping-off. Westcott's plant disease handbook. Springer Netherlands, Dordrecht, p 177.

http://www.apnikheti.com/upload/crops/crops_thumb/2202idea99okra_cercospora_leaf_spot_(okra).jpg

https://infonet-biovision.org/res/res/files/1528.400x300.jpeg

https://www.agrifocuszambia.com/wp-content/uploads/2020/06/Yellow-Vein-Mosaic-Disease-Bhendi-Yellow-Vein-Virus-BYVMV-okra.jpg

https://www.pioneer.com/content/dam/dpagco/pioneer/na/us/en/images/agronomy/crops/soybeans/weather-effects/IMG-photo-soybean-seedlings-damping-off-symptoms.jpg

https://www.thedailygardener.com/wp-content/uploads/2019/07/Yellow-vein-mosaic-disease.jpg

Hwang, S. F., Ahmed, H., Turnbull, G. D., & Kutcher, H. (2008). Effect of crop rotation on canola seedling blight and soil pathogen population dynamics. Canadian Journal of Plant Pathology, 30, 369.

Hwang, S. F., Gossen, B. D., Turnbull, G. D., Chang, K. F., Howard, R. J., & Thomas, A. G. (2000). Seeding date, temperature, and seed treatment affect pythium seedling blight of field pea. Canadian Journal of Plant Pathology, 22(4), 392-399.

Ioannou, N. (1997). Current Status of the Whitefly-Virus Complex in the Near East.

ISLAM, H. R., Bhardwaj, M. L., Sharma, I. M., & Sharma, N. K. (1993). Performance of commercial okra (Hibiscus esculentus) varieties in relation to diseases and insect-pests. Indian Journal of Agricultural Science, 63(11), 747-748.

Jahagirdar, S., &Hundekar, A. R. (2010). Management strategies against frog eye leaf spot pathogen (Cercosporanicotianae) in India. Indian Phytopathology, 63(1), 63.

Jambhale, N.D. and Nerkar, Y.S. (1983). Inheritance of resistance to powdery mildew in okra. Nat. Sem. Breeding crop plants for resistance to pests and diseases, May 25-27, 1983, Coimbatore, Tamil Nadu.

James R.L., (2012). Fusarium root and stem diseases. In: CramMM, Frank MS, Mallams KM (eds) For. Nurs. Pests. USDA Forest Service. Agriculture Handbook, Washington DC, pp 117–120.

Jha, A.K. and Dubey, S.G.,(2000). Occurrence of collar rot of (Abelmoschus esculentus) in the plateau region of Bihar. Journal of Research- Birsa Agricultural University. 12(1): 67-72.

Jiskani, M. M., Pathan, M. A., Wagan, K. H., Imran, M., & Abro, H. (2007). Studies on the control of tomato damping-off disease caused by Rhizoctonia solani Kuhn. Pakistan Journal of Botany, 39(7), 2749-2754.

Jukte, S. R., Badgujar, S. L., Suryawanshi, A. P., Utpal, D., &Kuldhar, D. P. (2016). Symptomatology, isolation, identification and pathogenicity test of damping off disease in okra. International Journal of Plant Protection, 9(1), 358-361.

Kandel, Y. R., Wise, K. A., Bradley, C. A., Chilvers, M. I., Tenuta, A. U., & Mueller, D. S. (2016). Fungicide and cultivar effects on sudden death syndrome and yield of soybean. Plant disease, 100(7), 1339-1350.

Kavyashree, S., Naik, G., Ravindra, H., Adivappar, N., &Salimath, S. B. (2022). Correlation of weather parameters with powdery mildew disease severity on okra (Abelmoschus esculentus L.). J. Pharma Innov. J., 11(1), 314-316.

Khalikar, P. V., Jagtap, G. P., & Sontakke, P. L. (2011). Management studies of okra powdery mildew (Erysiphe cichoracearum) using bio-agents, plant extracts and chemical fungicides. Indian Phytopathology, 64(3), 286.

Kumhar, D. R., Meena, A. K., & Meena, P. N. (2018). Efficacy of bio agents and fungicides against collar rot and early leaf spot of groundnut. Indian Phytopathology, 71, 549-554.

Lamont, W. J. (1999). Okra-A versatile vegetable crop. Horttechnology-Alexandria VA-, 9, 179-184.

Landis, T . D. (2013). Forest nursery pests: damping-off. Forest Nursery Notes, 2, 25-32.

Leisso, R. S., Miller, P. R., & Burrows, M. E. (2009). The influence of biological and fungicidal seed treatments on chickpea (Cicer arietinum) damping off. Canadian journal of plant pathology, 31(1), 38-46.

Mahesh, M., Saifulla, M., &Byregowda, M. (2006). Nutritional and physiological studies on different isolates of pigeonpea wilt pathogen (Fusarium udum Butler). PLANT Disease Research-Ludhiana-, 21(2), 219.

Mancini, V., &Romanazzi, G. (2014). Seed treatments to control seedborne fungal pathogens of vegetable crops. Pest management science, 70(6), 860-868.

Matny, O. N. (2013). First report of damping-off of okra caused by Phytophthora nicotianae in Iraq. Plant disease, 97(4), 558-558.

Moaward F.G., Abdelwhab B.M., Abdelnahun F.M., Shehaya F.W., (1984). Annual of Agricultural science, 21: 603 – 613.

Mohanta, R., Nath, R., Jena, N. K., Mishra, A. (2020). Effect of different dates of sowing and weather parameters on vector whitefl (Bemisiatabacigennadius) and incidence of yellow vein mosaic disease in okra (Abelmoschus esculentus L. Moench). Journal of Entomology and Zoology Studies, 8(2), 1774–178.

Munir, A., Ali, S., Zeshan, M. A., Ghani, M. U., Khan, A. A. (2020). Evaluation of organic amendments and insecticides against okra yellow vein mosaic virus and its vector. Pakistan Journal of Agriculture, Agricultural Engineering and Veterinary Sciences, 36(1), 13–20.

Ndunguru J., Rajabu A.C., (2004). Effect of okra mosaic virus disease on the above-ground morphological yield components of okra in Tanzania. Scientia Horticulturae 99: 225-235.

Nene, Y. L., Sheilla, V. R. and Sharma, S. B., 1984, A world list chickpea and pigeonpea pathogens. In: ICRISAT Pulse Pathology Progress Report, 8: 9.

Njoroge, S. M., Riley, M. B., & Keinath, A. P. (2008). Effect of incorporation of Brassica spp. residues on population densities of soilborne microorganisms and on damping-off and Fusarium wilt of watermelon. Plant Disease, 92(2), 287-294.

Pane, C., Spaccini, R., Piccolo, A., Scala, F., &Bonanomi, G. (2011). Compost amendments enhance peat suppressiveness to Pythium ultimum, Rhizoctonia solani and Sclerotinia minor. Biological control, 56(2), 115-124.

Papavizas C.G., Davey C.B., (1961). Saprophytic behavior ofRhizoctonia in soil. Phytopathology 51:693–699.

Prabhu, A. S., Phatak, K. D., & Singh, R. P. (1971). Powdery Mildew of 'Bhindf'(Abelmoschus Esculentus (L.) Moench) in Delhi State. Indian Journal of Horticulture, 28(4), 310-312.

Purseglove J.W., (1987). Tropical crop dicotyledons. Language Book Society Longman, London, UK.

Qhureshi, Z. (2007). Breeding investigation in bhendi (Abelmoschus esculentus (L.) Moench). University of Agricultural Sciences.

Ramakrishnan, L., Thamburaj, S., Kamalanathan, S., & Krishnamurthy, C. S. (1973). Control of powdery mildew of bhendi (Abelmoschus esculentus L.) with systemic fungicides. Madras agricultural journal.

Razeeena B.P.M. and Ahmad R,. (2007). Control of seed-borne fungi of okra with Pseudomonas fluorescens and aqueous leaf extract of Henna (Lawsoniainermis L.) Indian Journal of Mycology and Plant Pathology, 37(3), 485-487.

Ribeiro R. De L. D., Robbs C.F., Akiba F., Kimura O., and Sudo S., (1971). Studies on pre- and post-emergence rot of okra (Hibiscus esculentus L.) Arquivos da Universidade Federal Rural do Rio de janeird, 1(1), 9-13.

Rothrock, C. S., Winters, S. A., Miller, P. K., Gbur, E., Verhalen, L. M., Greenhagen, B. E., ... & Garber, R. H. (2012). Importance of fungicide seed treatment and environment on seedling diseases of cotton. Plant disease, 96(12), 1805-1817.

Ryals, J. A., Neucnsch Wander, U. H., Willits, M. G., Molina, A., Steincr, H. Y., (1996). Systemic acquired resistance. Plant Cell, 8, 1809–1819. DOI 10.2307/3870231

Salman, M., &Abuamsha, R. (2012). Potential for integrated biological and chemical control of damping-off disease caused by Pythium ultimum in tomato. BioControl, 57, 711-718.)

Satya Prasanth P, Patil P.V., (2007). Purple seed stain of soybean - its incidence, efect on seed quality and integrated management. Indian Phytopathology 60(4):482–488.

Sohi, H.S. and Sokhi, S.S. (1974). Behaviour of okra varieties to damping off, powdery mildew and cercospora blight. Indian Phytopathology. 27: 91-92.

Solheim, W. G., & Stevens, F. L. (1931). Cercospora Studies—II. Some Tropical Cercosporae. Mycologia, 23(5), 365-405.

Sridhar, T. S., & Poonam, S. (1989). Assessment of loss caused by powdery mildew (Erysiphe cichoracearum) of okra (Hibiscus esculentus) and its control. Indian Journal of Agricultural Sciences, 59(9), 606-607.

Sridharan R. and Rangaswamy G. (1968). Studies on two Cercosporaleaf spot of Abelmoschus esculentus. Indian Phytopathology 21:37-42.

Sridhar S.S., (1985). Disease thresholds for powdery mildew of okra. Seeds and Farms 11: 36-38.

Starkey T., Enebak S.A., (2012). Rhizoctonia blight of southern pines. In: Cram MM, Frank MS, Mallams KM (eds) For. Nurs. Pests. USDA Forest Service. Agriculture Handbook, Washington DC, pp 63–65.

Tindall H. D., (1983). Vegetable in the tropics Macmillan press ltd London first edition.

Vitale A, Castello I, D'Emilio A, Mazzarella R, Perrone G, Epifani F, Polizzi G (2013). Short-term effects of soil solarization in suppressing Calonectria microsclerotia. Plant Soil 368:603–617.

Windels C E, Lamey H A, Hilde D, Winder J and Knudesen T (1998). A cercospora leaf spot model for sugar beet. Plant Disease 82: 716 – 726.

Wright E., (1957). Influence of temperature and moisture on damping-off of American and Siberian elm, black locust, and desertwillow. Phytopathology 47:658–662.

Yanar, Y., Yanar, D., & Gebologlu, N. (2011). Control of powdery mildew (Leveillulataurica) on tomato by foliar sprays of liquid potassium silicate (K2SiO3). African Journal of Biotechnology, 10(16), 3121-3123.

13

Diseases of Potato and Their Management

Ankit Kumar Chaurasia[1], Shivanshu Mishra[1], Sanjay Kumar Maurya[2] and Abhay Pratap Singh[3]

[1]Department Plant Pathology Sardar Vallabhbhai Patel University of Agriculture & Technology, Meerut – 250110, Uttar Pradesh
[2]Department Horticulture, Chaudhary Charan Singh University, Meerut Uttar Pradesh
[3]Department Plant Pathology, Banda University of Agriculture & Technology, Banda, Uttar Pradesh

Abstract

Potato (Solanum tuberosum) is a widely cultivated and consumed crop with significant global importance due to its versatility, adaptability, and nutritional value.From a nutritional perspective, potatoes offer a range of essential nutrients, including carbohydrates, vitamins (particularly vitamin C and several B vitamins), minerals (such as potassium and magnesium), and dietary fibre. However, their nutritional value can vary depending on the variety and cooking method employed.Despite their significance as a food source, potatoes face challenges related to disease susceptibility, environmental concerns, and evolving consumer preferences. Many pathogens, such as bacteria, fungus, viruses, and nematodes, can cause a variety of diseases that affect potato crops. These diseases pose substantial threats to potato production, resulting in yield losses, decreased quality, and economic hardships for farmers. Understanding potato diseases, their causes, and effective control measures is vital for potato growers, researchers, and policymakers. Addressing these challenges is crucial for sustaining potato production and meeting the growing global demand for this essential crop.

Keywords-*nutritional value, diseases, pathogens.*

Potato

Botanical Name	:	*Solanum tuberosum*
Family	:	Solanaceae
Chromosome Number (2n)	:	48

DOI: 10.1201/9781042014460-13

Introduction

Potato *(Solanum tuberosum)* is a widely cultivated and essential staple crop globally, known for its versatile culinary applications and nutritional value. With the exception of Kerala, potatoes are grown in practically every state in India under a variety of climatic circumstances. During the brief winter months, which span from October to March, 82% of potatoes are grown in plains. In West Bengal, potatoes are the most widely grown crop after grains (Chakraborty, 2012). Originating from the Andes region of South America, potatoes have become a crucial part of diets worldwide, providing carbohydrates, vitamins, and minerals. With a rich history dating back to ancient civilizations, including the Irish potato famine, this crop has played a significant role in shaping agricultural practices and food security. Potatoes are adaptable to various climates and soil types, making them a vital source of food for both progressing and developed nations. Their economic and nutritional importance is underscored by their ranking as the 4th most consumed crop after rice, wheat & corn, with aworldwide production of over 370 million metric tons in 2020 [FAOSTAT]. Furthermore, ongoing research and breeding efforts continue to enhance potato yields, disease resistance, and nutritional content, ensuring its continued significance in the global food supply chain. Potatoes are prone to a number of diseases brought on by a variety of pathogens, including fungi, bacteria, and viruses.

1. Late Blight of Potato

Late blight is one of the main diseases that damages potato crop. It is known for its role in the Irish Potato Famine of the 1840s, late blight continues to pose a significant threat to the world's potato crop.

Causal Organism-*Phytophthora infestans*

Symptoms

The disease first manifests as sporadic, pale green lesions that are drenched in water and mostly located on the tips and margins of leaves. These lesions rapidly become big necrotic patches that range in color from brown to purplish-black. The stems are surrounded by light to dark brown spots. At such locations, the damaged stems and petioles weaken and may even collapse.Tubers infected with late blight, irregular reddish brown to purplish patches penetrating interior tissues are observed.In the field and in warehouses, bacteria that cause soft rot may attack the sick tubers, which are often hard, dry, and firm (Arora *et al.*, 2014).

Symptom of Late Blight

Etiology

Phytophthora infestans (Mont.) de Bary is the organism that causes late blight of potatoes (Jones, 1998). According to Shaw and Khaki (1971), the Oomycetes have no relation to actual fungi, include the genus *Phytophthora*. Phytophthora (A1 and A2 mating types) contains heterothallic species, such as *P. infestans* (Fry *et al*., 1993). The mycelium produces branching sporangiophores with lemon-shaped sporangia at their tips. Sporangiophores, unique to this fungus, produce swellings at the sites where sporangia are generated (Agrios 2005). Sexual spore called oospore is produced by *P. infestans* when two kinds of mating, A1 and A2, come in contact with each other (Kirk, 2009).

Disease Cycle

Infection Process

On host plants, sporangia and zoospores can both initiate a new infection. Infection through zoospores occurs faster than infection through the direct germination of sporangia. Zoospores are still motile for up to 22 hours at 5–6 °C, but only for 19 minutes at temperatures above 24-25 °C (Melhus, 1915).Small necrotic flecks typically start to form 24 to 48 hours after the zoospores are injected, usually causing infection 2.5 hours later.

Sporangia Dispersal

The sporangia of pathogenstart developingform night when there is a lot of humidity and disperse in the morning when the temperature rises. and the RH decreases. Instead of temperature fluctuations, sporangia detached from sporangiophores on leaves is mostly caused by changes in humidity (Hirst, 1958).Although they are hydrophilic, sporangia easily disperse into water that falls on them. Splash droplets carrying spores could cause the infection to be transported across short distances.

Development of An Epidemic

Temperatures between 7 and 27 °C are ideal for the growth of blight. Humidity is one of the primary elements in the emergence of an epidemic. High RH is necessary for the long-term survival of sporangia (Martin, 1923).Abiotic factors like the distribution and length of time in the crop canopy of nearly-saturated or saturated air at a favourable temperature and biotic factors like cultivar, age, dietary habits, and the quantity of inoculum on the host interact to determine the rate at which a disease spreads.However, lower humidity inhibits sporulation, their viability, and the infection process, which delays the spread of an epidemic (Easton, 1982).

Management

The initial step for controlling late blight is reducing the main sources of inoculums. The control of contaminated sources, such as volunteer plants, disease in neighboring fields, sick tubers, rubbish piles, and regrowth following haulm removal, can help manage this disease better (Turkensteen and Mulder, 1999). Oospores are problematic primary inoculum source, mainly when crop rotation timevis short. According to Turkensteen *et al.* (2000), oospore-contaminated sandy and clay soils persisted as infection causing soil for 48 to 34 months, respectively. If infected seed is the source of the initial infection, then application of systemic fungicides early in the season can be an effective management strategy for disease (Hermansen and Naerstad, 2009).A late blight year has been observed to generate a larger yield when phosphate and potassium dosages are increased. (Roy *et al.*, 2001). Traditionally, the damaged region has been sprayed with an effective fungicide to manage late blight. Bordeaux combination has been the recommended fungicide for a long time. Following that, the Bordeaux combination was swapped out for fungicides made organically, primarily carbamates, which managed the illnesses known as early and late blight(Arora *et al.,* 2014). But the oomycete-specific phenylamide group of fungicides Metalaxyl transformed the control of late blight (Bruck *et al.*, 1980).

2. Early Blight of Potato

In regions with favourable weather, early blight, also known as Alternaria blight and it is one of the most significant leaf diseases (CIP, 1996). Worldwide, potato crop is susceptible to early blight, which is brought on by two different *Alternaria species* (*A. solani* and *A. alternata*), and is most common in areas with high temperatures, alternating dry and humid spells, irrigation, andlow-organic matter, sandy and light-textured, potato soils (Gudmestad and Pasche, 2007).

Causal Organism- *Alternaria solani*

Symptoms

The symptoms on fading potato leaves were initially described by Ellis & Martin in 1882. The 'target spot' effect is characterized by blemishes on leaves that are dark brown or black and have concentric circles (Rands 1917a). This zonation, which is frequently lacking in greenhouse-inoculated plants, is brought on by variations in temperature, moisture, and radiation from day to night. Every host has similar lesions. (Pscheidt 1985). Compared to their usual oval shape, lesions can occasionally take on an angular shape that fits into the interveinal spaces under unfavourable conditions (Rands 1917a). The leaf finally dies from the lesions once they grow larger and strengthen (Pscheidt 1985). Dry rot, which appears as discrete, erratic, sunken lesions on the surface of infected tubers, develops (Pscheidt 1985). On stems and, to a lesser degree, on leaves, brown, angular necrotic patches that are internally identified by a series of concentric rings appear. Because of the broader leaf veins that enclose them, leaf lesions are rarely round. Lesions usually appear around the time of flowering and multiply as plants get older. On lower leaves, lesions initially develop. They could unite and bring to widespread yellowing, leaf loss, or early demise. (Binyam Tsedaley, 2014).

Symptom of Early Blight

Etiology

The dark-colored *Alternaria species* mycelium forms short, simple, upright conidiophores in older infected tissue that carry single or branched chains of conidia. Conidia are large, black, oblong, or multicellular organisms with transverse and longitudinal cross walls. Conidia can be easily separated and moved by wind currents. Many *Alternaria species* are mostly saprophytic, which means they can only develop on extinguished or senescent plant parts like petals, fruits, and leaves, or on dead or decaying plant material. They are therefore unable to infect the tissues of living plants. Determining if an Alternaria fungus present on sick tissue is the primary or secondary cause of the illness can be difficult (Binyam Tsedaley, 2014). After tubers have been exposed to spores during harvest, lesions may still develop while they are being stored (Schultz and French, 2009).

Disease Cycle

Overwintering and Survival

Fungal pathogen *A. solani* is polycyclic since it can cause numerous cycles of infection in a single season. (Shuman 1995). In order for the infection to endure, regarding soil temperatures, chlamydospores have also been reported to be the source of the early blight overwintering inoculum (Basu 1971; Patterson 1991). For 5-8 months, in uncultivated soil, the inoculum might endure in debris. The dark color of the hyphae makes them more resistant to lysis (Lockwood, 1960). The most common places for spore survival are infected seed and detritus. Biological, edaphic, and climatic factors dominate in determining survivability in seed and debris. In dry, unused fields, the fungus thrives (Rotem 1994).

Dispersal

Conidia produced by the primary inoculum are blown or splashed onto the lower leaves of the plant in the spring, when they germinate and spread the infection (Rotem 1994). The main methods for dispersing *A. solani* are wind, rain, and insects. Before infection reaches the point at which full leaves start to dry up and plants start to die, a rapid rate of spread does not begin to occur. The wind velocity curve resembles the spore dispersion curve. However, spore dispersal and other climatological variables don't seem to be significantly correlated (Rotem 1964).

Infection

According to Bashi and Rotem (1974), spore growth could start at 20 °C after just two hours of wetting time. A longer wetting period is needed as the germ-tube lengthens. The epidermis can be directly penetrated, or through wounds or stomata, by germ-tubes that generate appressoria. Rotem (1994) andWaggoner & Horsfall (1969) showed that a temperature of 10 °C might be the lowest temperature for infection, 35 °C can be the highest, and 20 °C to 30 °C is the ideal range. Age and plant susceptibility affect incubation periods, which measure the interval between the start of an infection and the appearance of symptoms (Rowell 1953 in Rotem 1994).

Sporulation

Temperature between 5 to 30 °C, with ideal range being 20 °C, are necessary for potato plant sporulation (Pscheidt 1985). The greatest sporulation happens after a significant precipitation or dew. The daytime is when spore production begins, although spores build up over a period of 7-14 days before being distributed (Bashi & Rotem 1975b).

Epidemics

Early blight develops mainly in the presence of moisture. Depending on the situation, *A. solani* responds to meteorological conditions differently (Waggoner & Horsfall, 1969 and Rotem, 1994). Weatherrelated elements may occasionally have an indirect effect by affecting the host's vulnerability (Rotem, 1994). Usually, epidemics begin at the conclusion of the growth season, when plants are at their most susceptible. However, geographical location and local climatic conditions have an impact on how quickly a disease spreads.

Management

Cultural Practices

Use resistant cultivars and verified pathogen-free seed. Before digging potatoes, wait for at least 3 to 4 days (and probably 2 weeks) following vine dying. This method raises resistance in tuber to the fungus that causes early blight. After harvesting, turn over all leftover plant material and stray potatoes. Lesion-free tubers should be stored at 40°F in an atmosphere that is dry, sterile, dark, and well-ventilated. To avoid bruises, handle tubers carefully. Avoid thick plant stands and extended overhead irrigation, instead choose well-drained and well-aerated fields and encourage air circulation. Irrigation should be avoided or arranged when the weather is cold and cloudy to give plants enough time to dry before dusk. Tubers ought to be stored in a way that encourages quick suberization since unbroken periderm is incapable of spreading *A. solani*.

Chemical Control

The most common and effective method of treating early blight is to apply foliar fungicides early in the growth season until vine mortality (Wharton and Kirk, 2012). In the early stages of the disease, it is recommended that contact fungicides be used often to prevent infection. Applying a systemic or contact fungicide in three to four applications should begin at blossoming. It is necessary to apply a systemic fungicide treatment as soon as symptoms start to show. There is no need to significantly limit the quantity of sprays if the initial and subsequent fungicide applications are scheduled appropriately (van der Waals, 2002). Spray a 7–10 day schedule of protectant-type fungicides containing the active ingredient (chlorothalonil, maneb, or macozeb) beginning at bloom or in accordance with a weather-timed spray schedule (such as Blitecast). This treatment should be continued until the foliage either naturally dies or is artificially killed by applying a “vine-killing” chemical.

3. Wart disease of Potato

Black wart, black scab, potato tumor, potato cancer or canker, cauliflower disease, warty disease, and many more descriptive terms in various languages and cultural backgrounds where potato is grown and the disease is present are a few examples of significant diseases that affect cultivated potatoes (Frank 2007). Due to warty exudations that were occasionally formed on stems, leaves, and flowers as well as on tubers, *S. endobioticum's* disease was known as potato wart (Hooker 1981).

Causal Organism-*Synchytrium endobioticum*

Symptoms

Galls, which can range in size, typical symptoms range from pea-sized to outgrowths that are as big as or bigger than the tubers that gave rise to them. Although it might be uneven, the shape is typically spherical. Subterranean warts are white to brown in color, whereas above-ground galls range in color from green to brown. When wart tissue reaches maturity, it turns black and causes complete tuber decomposition (Melodie and Sindermann 1994). Young, originating tubers that have been infected early acquire deformities and sponginess, rendering them unrecognizable. Older tubers' eyeballs get infected and develop into warts that resemble cauliflower. Stolons grow warts, but roots are unaffected. The aerial buds near the base of the stem can develop little green warts. Attacks on leaves occasionally occur. These warts are softer than tuber warts and have a pulpy appearance when touched.

Symptom of Wart Disease

They look like a mass of hyperplastic tissue made up of twisted, proliferating branches and leaves. Its host plant is not killed by a potato wart. Before harvest, indications of a below-ground infection might not be noticeable. Plant vitality may be noticeably diminished (Anon 2004).

Etiology

S. endobioticum is a member of the basal fungal lineage known as Chytridiomycota, which evolved between 1000 and 1600 million years ago (Heckman *et al.*, 2001The absence of hyphae or mycelium and the motile, flagellated spores, or zoospores, are characteristics that set the chytridiomycota apart. Chytrids are free-living saprophytes, or diseases, that infect and kill moist, aquatic land plants and animals (Barr, 2001; Sparrow, 1960).Initially, it was given the name *Chrysophylyctis endobiotica* and assigned to a recently constituted genus under the order Chytridiales (Schilberszky, 1896). It was later called *Synchytrium endobioticum* based on similarities in morphology and cytology to other species in the genus *Synchytrium* (Curtis, 1921; Percival, 1910).

Disease Cycle

The complexity of the life cycles of *Synchytrium* species varies greatly, and *S. endobioticum* is one of the long-cycled species (Karling, 1964). A new inner wall layer called a prosorus forms inside the resting spore during the resting phase, when the warted host tissue breaks down and the resting spores become dispersed in the soil. During germination, the outer spore wall cracks, releasing the newly produced sorus (also known as a zoosporangium), which then produces a large number (200–300) of uninucleate haploid zoospores. A single or a number of zoospores can infect a host cell (Lange & Olson, 1981b). The host nucleus grows and takes on an atypical shape; it is often found next to the fungal thallus that is budding. Five-minute chromosomes were reportedly seen during this repeated mitosis, according to Curtis (Curtis, 1921). Light microscopy may now show that the host tissue has developed a brown discoloration. Within the cells surrounds the infected host, hypertrophic expansion (cell enlargement) is generated. Meristematic tissues keep rising concurrently with hyperplasia (cell multiplication), increasing the likelihood that the host would become reinfected (Hampson, 1993).When the larger host cells split apart, the individual zoosporangia are liberated. These zoospores are morphologically identical to the zoospores that develop from resting spores. These zoospores have the ability to regenerate an asexual reproductive cycle by reentering the host tissue. As an alternative, the zoospores might behave as isogametes and combine to create a diploid biflagellate zygote (sensu Curtis), which infects the host in a manner akin to that of haploid zoospores (Curtis, 1921).

Management

The first acts were published in Ireland in 1908, followed by Scotland, England, and Germany a few years later. The first piece of law mainly dealt with banning the importation of potatoes contaminated with *S. endobioticum* from foreign nations. The prospect of the sickness spreading across the nation was not considered at the time the crimes were committed. It was the cause of the rising number of outbreaks in so many nations in such a short period of time. The emergence of research into the biology governing the fungus was a result of the widespread spread of the potato wart disease. Because of *S. endobioticum's* extended persistence in soil and the significant losses it causes to potato crops, EPPO has added it to the A2 quarantine list. The fungus is listed on the official list of selected agents and toxins maintained by the United States Department of Agriculture (USDA). Council Directive 69/29/EC of December 8, 1969, on the control of potato wart disease, as well as Directive 2000/29/EC of May 8, 2000, on protective measures against the introduction into the Community of organisms harmful to plants or plant products and against their spread within the Community (Obidiegwu *et al.* 2014).

4. Black Scurf

Black scurf and stem canker are two commercially important diseases that caused by *Rhizoctonia solani* Kuhn (teleomorph: *Thanathephorus cucumeris* Frank Donk) causes in the potato crop (*Solanum tuberosum* L.).

Causal organism- *Rhizoctonia solani*

Symptoms

According to Banville (1989), Plants that are infected frequently produce a few large tubers or a large number of little (3 cm diameter) progeny tubers. In plants with severe infection, the axils of the leaves may grow tubers (Hartill, 1989). Although they have very little influence on haulm yield and stem number, severe stem and stolon attacks have a considerable impact on tuber fresh yield, dry matter yield, and dry matter content. They also increase the amount of large and malformed tubers.

Symptom of Black Scurf

Etiology

A damaging soil-borne pathogen that affects numerous plant species globally, *Rhizoctonia solani* causes disease in those plants. *R. solani* is a pathogen that can be either soil- or seed-borne. In the form of mycelium, which is found in decaying plant materials, the fungus lives in soil. Additionally, it can live forever in the soil or on the surface of tubers as seed-borne sclerotia. In the absence of a susceptible host, *R. solani* population declines; however, the rate of decline is influenced by soil type, crop rotation, and potentially soil organic matter content (Martn, M. P., & El Bakali, A. M., 2006).

Disease Cycle

The pathogen *Rhizoctonia solani* is carried by soil. In reaction to exudates released by the root and shoot, *R. solani* colonizes the surface of the underground potato plant (Jeger *et al*., 1996). It multiplies to form a vast network of anastomosing hyphae on the root/stolon system. During the colonization phase, the host plant shows no symptoms as long as the infection structures do not form. According to Jongebloed and Hofman (1988; Keijer, 1996), the runner hyphae repeatedly branching, which results in the development of short, swollen cells that give birth to infection cushions, initiates the initial stages of infection.According to Keijer *et al.* (1997), infection cushions act as a supplemental food source for subsequent colonization of the plant and are thought to be necessary for causing stem and stolon lesions. In a network of interconnected hyphae, AG-3 condenses into relatively small infection cushions (Keijer, 1996).

Management

Cultural Control

The importance of agricultural techniques in managing soil-borne infections was acknowledged far in advance of the widespread availability of fumigants and fungicides. The incidence and severity of potato rhizoctoniasis are greatly influenced by a wide range of agronomic factors, including plant type, cultivar, crop rotation, soil management, tillage, irrigation, pesticide use, harvesting, crop residues, volunteer plants, and storage (Jeger et al., 1996).

Crop rotation

Since monoculture is known to be a source of additional inoculums for diseases that are spread through the soil, it should be avoided. The frequency and severity of stem canker increased with the number of potato cropping cycles due to an increase in the density of soilborne inoculums (Scholte 1992; Honeycutt et al., 1996). In order to minimize the harm produced by *R. solani*, three-year rotations or longer are frequently necessary (Banville et al., 1996).

Biological Control

A potential strategy for controlling harmful organisms is biological management. The most commonly employed bioagent for controlling the pathogen *R. solani* is the Trichoderma species (Beagle et al., 1985). According to Hussain et al. (2014), *Trichoderma viride* and *Trichoderma harzianum* both strongly restriscts the mycelial growth of R. solani isolates. However, *Trichoderma harzianum* is more efficient than *T. viride.*

Chemical Control

Numerous fungicides, including azoxystrobin, benomyl, carbendazim, fluazinam, fludioxonil, imazalil, iprodione, mepronil, pencycuron, and propiconazole, have been studied as seed or soil treatments against stem canker and black scurf.

5. Common Scab

The recurrent disease known as common scab compromises the quality of potatoes cultivated worldwide. Common scab has been reported everywhere potatoes are grown, including the Americas, India, Asia, and Africa. Scabs also affect other tuber and root crops, including sweet potatoes, turnip, radish, beet, and carrot.

Causal organism-*Streptomyces scabiei*

Symptoms

All of the symptoms of common scab can be present in tubers from the same plant and range from light to deep-pitted or elevated lesions. Lesions can be isolated or might combine to cover extensive areas of the tuber surface. They frequently have a dark, brown, or corky color. Brown corky lesions on the tuber periderm's surface have been identified as signs of netted scab (Bang 1979; Bang 1995; Scholte and Labruyere 1985). All of a potato's underground portions, including the roots, stolons, and stems, are susceptible to necrosis from streptomyces, and it can also slow the growth of roots from seed tubers (Han *et al.* 2008). By causing necrosis, this infection can also kill potato seedlings grown from real potato seed. Root necrosis caused by Streptomyces may reduce plant vigor, but it does not always result in foliar symptoms right away(Charkowski*et al.* 2020)

Symptom of Common Scab

Etiology

Aerobic, filamentous, spore-forming, gram-positive Streptomyces belongs to the Streptomycetaceae family of Actinobacteria. Cross walls are either absent or scarce in filamentous mycelia. At the ends of hyphae, spores develop in spiral chains. The largest genus of Actinobacteria, Streptomyces, contains more than 600 species. Certain species of Streptomyces coexist peacefully with eukaryotes, such as plants, although the majority of them are saprophytes that dwell in the soil. Potatoes can develop pitted, netted, or common scabs due to at least 12 different species of Streptomyces. Scabies is just one of several grammatically erroneous original names for pathogenic *Streptomyces species*. The scientific community has just recently begun to use corrected nomenclature for these species, which includes *S. scabiei*(Charkowski *et al.*2020).

Survival and Dispersal

In comparison to other bacterial diseases, the life cycle of streptomyces is merely complex. It develops vegetatively as cells that resemble filamentous mycelia. When resources ran away nutrients are transported to aerial reproductive hyphae, which produce spores, and the vegetative cells go through programmed cell death. Inside scab sores, these hyphae are occasionally apparent without magnification. Pathogenic streptomyces flourish in soils that are between pH 5.2 and 8.0 and temperature 20 to 22 °C. These conditions are also favorable for the growth of potatoes. The cylindrical spores that are formed at hyphal terminals are the main means by which Streptomyces spreads and survives. Spores have the ability to spread through seed tubers, soil-dwelling invertebrates, and water bodies. Streptomyces spores are heat-resistant and can survive in soil for up to 20 years. As the pathogen spores germinate, they enter the plant through lenticels or other naturally occurring holes like wounds. Tubers are particularly susceptible to colonization by Streptomyces in the first month of development. On developed tubers, streptomyces is unable to produce lesions, and lesions stay the same size and intensity. However, severely pitted tubers may become dehydrated and will not germinate the next season. (Charkowski *et al*,2020).

Management

The most suitable choice is disease resistance or tolerance, but there are currently few potato cultivars with high common scab tolerance available. Chemical treatments are the least sustainable way to manage disease; while they may work for a season, they are often costly and detrimental to the soil. Among the often used, efficient compounds are pentachloronitrobenzene for in-furrow use, chloropicrin as a soil fumigant, and fludioxonil for treating seed pieces (Al-Mughrabi et al. 2016; Powelson and Rowe 2008). Common scab may also be somewhat controlled by fluazinam (Santos-Cervantes *et al.* 2017).Crop rotation strategies may decrease the severity of common scab. These crop rotations often include allelopathic green manures that help control a number of soil-borne potato diseases, as well as brassica crops that are utilized as a biofumigant. Soil additives like rice bran, chelated iron, or peat can decrease the incidence of common scab by boosting the number of nonpathogenic streptomycetes (Tomihama *et al.* 2016; Sarikhani *et al.* 2017). Common scab will worsen with the addition of some soil amendments, such as manure, which elevates the pH of the soil.

6. Bacterial Wilt of Potato

R. solanacearum species complex (RSSC) is the culprit behind bacterial wilt, one of the most common and destructive bacterial illnesses that affect vegetables in tropical climates (Buyela *et al.,* 2017; Traore *et al.,* 2020). Around the world, regions with tropical, subtropical, and moderate climates are home to *R. solanacearum*, a soil-borne pathogen. More than 450 species from 54 different botanical families are infected by it, including dicots (Solanaceae, Potato, Pepper, Tomato, and Eggplant) and monocots (Musaceae).

Causal Organism- *Ralstonia solanacearum*

Symptoms

The immature plant is more prone to the sudden wilting of foliage that characterizes *Ralstonia solanacearum*. One of the signs is a discoloration of the vascular system that goes from pale yellow to dark (Gota, 1992). Even in typical climatic conditions that are perfect for the infection, the *R. solanacearum* infected plant may show all or none of these symptoms; this is often a well acknowledged stage known as latency. Through wounds from transplanting, culture, nematodes, insects, and naturally occurring wounds, the pathogen gets into the roots. Finally, the xylem components are teeming with bacteria and slime as it begins to expand quickly inside the vascular system (Kelman and Sequeira, 1965).

Symptom of Bacteria Wilt

Etiology

The soil-borne bacterial pathogen *R. solanacearum,* earlier known as *Pseudomonas solanacearum,* is a significant constraint on crop output. A strictly aerobic, gram-negativebacterium andit is rod-shaped with a size of 0.5–0.7 x 1.5–2.0 micrometers is known as *Ralstonia solanacearum*. Low quantities of sodium chloride (NaCl) (2%), which inhibit this pathogen in culture, are effective against it because it is extremely sensitive to desiccation (Fegan M. and Prior P. 2005). For optimum growth, most strains need a temperature of between 82 and 90 ºF, but some just require 80.5 degrees. Growing media, both liquid and solid (agar), are commonly used in the production of bacteria (CD Lambert, 2002).

Survival and Dispersal

In contaminated or diseased soil, *Ralstonia solanacearum* is a soil- and water-borne pathogen that can endure and proliferate for a long time. An environment with highs between 85 and 95 °F is more conducive to the growth and progression of illness. Furthermore, a variety of other factors, including salt concentration, pH, moisture content, soil structure, and type, influence the development and survival of diseases(Jones JB *et al.*, 1991). The bacterium can occasionally persist outside the plant, a condition known as the epiphyte or external phase. The pathogen's epidemiology does not place much emphasis on this stage because the bacteria does not survive outside for very long in hot conditions or when the relative humidity is lower than 95%.*R. solanacearum* can survive in soils, irrigation water that has been contaminated by the illness, and infected weeds for days to years. These sources operate as inoculum and are transferred from infested to healthy areas by machinery and surface runoff from irrigated water (Coutinho TA, 2005). The pathogen can also spread from infected water sources, such as ponds or rivers, to uninfected areas after a rainstorm or when infected water sources are used for irrigation. When bacteria were discharged from infected semi-aquatic weed roots into irrigation waters, this was a major contributor in the spread of the infection (Saddler GS, 2005).

Management

In the soil, bacterial wilt disease is challenging to manage (Jones, 1997). Biological control and host-plant resistance are two of the several management techniques that have been created to control and eradicate this illness (Dalal *et al*., 1999). The most significant and practical strategies are:

Cultural practices: Sowing a sensitive crop such as mungbean before cultivating it by crop rotation, intercropping, or applying green manure (Hartman *et al*., 1993). Crop rotation with a non-susceptible crop provides some control, but because of the pathogen's wide host range, this method is difficult to implement (Kelman, 1953).By utilizing non-susceptible crops and rotating crops every five to seven years, infection of potato plants can be considerably decreased (Smith *et al*, 1995).

Resistance cultivars: Developing resistance cultivars through breeding is the most efficient way for preventing *R. solanacearum*-induced bacterial wilt (Persley, 1991). A virulent mutation of *R. solanacearum* that is pathogenic has been employed as a biocontrol agent for the disease (Trigalet & Trigalet, 1990).

Biological Control: Bacteriophages, which can attack the bacterial wilt pathogen, have been utilized as a biocontrol agent, and Wall and Sanchez (1992) discovered that this has the effect of suppressing *R. solanacearum's* population dynamics. To control *R. solanacearum*, strains of the plant growth-promoting bacterium (PGPR) have been found as a promising bio-control agent.,

Chemical control: Since many years ago, many chemical types have been employed to treat bacterial wilt. Unfortunately, no approach has been demonstrated to be effective when used alone due to the complexity of *Ralstonia solanacearum* (Nion YA, Toyota K., 2015). Chemical control techniques for soil-borne plant diseases, insect pests, and weeds employ agricultural chemicals. To stop bacterial wilt from starting, many treatments have been employed, including fumigants like meta sodium, 1,3-dichloropropene, and chloropicrin, algicides like 3[3-indolyl] butanoic acid, and plant activators as Val doxylamine and validamycin A.Plant activators such as validamycin A and Val doxylamine, as well as compounds like 3[3-indolyl] butanoic acid, have been employed.

7. Black Leg or Soft Rot

Blackleg and soft rot in potato tubers are caused by bacteria belonging to the Dickeya and Pectobacterium genera(Pe'rombelon & Kelman, 1980; Pe'rombelon, 2002). These conditions are economically second only to the bacterial wilt brought on by *Ralstonia solanacearum* in the yield of seed potatoes.

Causal organism-*Erwinia carotovora* **subsp.** *Carotovora*

Symptoms

Both Dickeya and Pectobacterium species can cause potato blackleg, which is most frequently recognized by a slimy, moist lesion of black rot that rises up the stems from the rotting mother tuber, particularly during rainy seasons. In dry conditions, stems and leaves are more likely to stunt, yellow, wither, and desiccate(Pe'rombelon & Kelman, 1980). Extensive stem rot that develops late in the summer under continuously rainy weather and moves from the top of the plant to the base can be mistaken for blackleg symptoms, but it is typically exclusively caused by Pcc (Pe'rombelon & Kelman, 1980).

Symptom of Soft Rot

Etiology

The principal bacteria responsible for blackleg, which impairs the developing plant, and tuber soft rot in potatoes include *Dickeya spp., P. carotovorum* subsp. carotovorum (Pcc), and *Pectobacterium atrosepticum* (Pa) (Wolf & De Boer, 2007). Previously, these bacteria (*E. carotovora* subsp. carotovora, *E. carotovora* subsp. atroseptica, and *E. chrysanthemi*) were grouped together under the name Erwinia (Dye, 1969). They are facultative anaerobic, pectinolytic, straight rods with peritrichous flagellae that are motile and nonsporing (Charkowsky, 2006). They are grouped in the Enterobacteriaceae family and are members of the γ-Proteobacteria subgroup (Charkowsky, 2006). They often produce a range of enzymes that break down cell walls, facilitating the penetration and maceration of the plant tissues they ingest (Barbas *et al.,* 1994).

Survival and Dispersal

In the soil, microorganisms that cause soft rot do not hibernate. Bacteria can live in soil for one week to six months, depending on climatic conditions including pH, moisture content, and soil temperature. In combination with plant material, including volunteers, survival may be prolonged. But in a crop rotation system that lasts between three and eight years, the bacteria cannot thrive in the soil (Anilkumar & Chakravarti, 1970; Rangarajan & Chakravarti, 1970; Lim, 1975; Pe'rombelon & Hyman, 1988). The latently infected seed (mother) tubers are now widely acknowledged to be the primary source of blackleg infection (Pe'rombelon, 1974). As a mother tuber decays, the bacteria are released into the soil, where they are then transferred by soil water to neighboring progeny tubers.According to Czajkowski *et al.* (2010), potato roots can also become infected by soil-borne bacteria, which can subsequently spread through the vascular system to develop tubers. After entering the stems, the bacteria stay dormant and may not always cause stem rot (blackleg). Long-distance spread of flying insects carrying soft rot bacteria from damaged plants can contaminate adjacent potato crops. They might also be found in the aerosols that are created when rain impaction on blackleg plants and harvest preparation pulverizes the crop. A layer of water on the surface of the tuber creates anaerobic conditions inside the mother tubers, which encourages the growth of bacteria and the start of rotting (Pe'rombelon *et al.*, 1989b).

Management

Blackleg and tuber soft rot have been studied using a variety of methods, with differing degrees of success. Techniques that prioritize preventing contamination and relying on seed certification protocols are commonly employed and have shown some degree of effectiveness. Two outcomes of better store management are decreased tuber rotting and the bacterial load on tubers. Though their efficacy has varied, chemical and physical methods have been studied (especially in the context of hot water treatment). It was originally believed that streptomycin might effectively treat soft rot and potato blackleg. Before planting, seed tubers can be submerged in a solution containing streptomycin and oxytetracycline hypochloride or streptomycin and mercury compounds to lessen the risk of blackleg in the field and tuber infections in storage (Bonde & de Souza, 1954).

8. Potato Leaf Roll

When it comes to the financial impact on potato crops around the world, the potato leafroll virus (PLRV) is among the most significant viruses. The virus, which is a member of the luteovirus family, causes infected plants' leaves to become yellow or red in addition to dwarfing and stunting them (Matthews 1982; Waterhouse *et al.* 1988).

Causal Organism-Potato leaf roll virus (PLRV)

Symptoms

Chlorosis, reddening, rolling, and leathering of leaves, reduction in potato yield, and net necrosis of potato tubers are symptoms specific to PLRV (Alani *et al.*, 2002). In potato fields, PLRV infection occurs in circular pattern, usually surrounding the contaminated seed piece that is the source of the viral inoculums. Technically referred to as a "aphid's hole in the field," aphids cause direct damage and even kill potato plants (Robert *et al.*, 2000). If the virus infects the plants late in the growing season, even foliar symptoms might not show up; plants infected early in the growing season would be stunted.

Symptom of Potato Leaf Roll

Mode of transmission

Woodford *et al.* (1994) stated that, The intricate process of virus transmission by insects entails interactions between the virus, vector, and host plant.Aphids are the only mechanical means of transmission for luteoviruses, which are RNA-containing, phloem-restricted viruses (Casper, 1988). PLRV is only spread by aphids in nature. At least eleven distinct aphid species have the ability to disseminate it, with Myzus persicae Sulz., the peach-potato aphid, being the most efficient vector (Kennedy *et al*. 1962; Kostiw 1980). Stem grafting can artificially transfer PLRV.

Management

Right now, growing resistant potato types is the best and most efficient way to combat PLRV. Nevertheless, only few potato cultivars in use now are somewhat resistant to PLRV (Beekman 1987). According to Olubyo *et al.* (2010), combination of insecticides and mineral oils could be the very much efficient way to reduce aphids, although synthetic pesticides were thought to be more effective at controlling PLRV than mineral oils. The study added that imidacloprid and thiamethoxam are the most efficient insecticides for managing different aphid types.

Conclusion

Effective management of potato diseases is crucial for maintaining crop health and maximizing yields. This chapter has highlighted various strategies for disease management, including cultural practices, disease-resistant varieties, chemical control, and integrated pest management. It's important to implement a combination of these approaches to mitigate the risks associated with potato diseases and ensure sustainable potato production. Regular monitoring, timely interventions, and a proactive approach are key to success in managing potato diseases and safeguarding this important global food crop.

References

Agrios, G.N. (2005). Plant Pathology. 5th Edition. Academic Press, London, New York, 922pp.

Al-Mughrabi KI, Vikram A, Poirier R, Jayasuriya K, Moreau G (2016) Management of common scab of potato in the field using biopesticides, fungicides, soil additives, or soil fumigants. Biocontrol Sci Tech 26(1):125–135

Alani RA, Alessawi UN and Almashaikhy SA, 2002.Isolation of proteins from Datura stramonium has ability to inhibition the multiplication of potato virus Y (PVYn). Jerash J. Res. Studies. 7: 9-21.

Anilkumar TB, Chakravarti BP, 1970. Factors affecting survival of Erwinia carotovora, causal organism of stalk rot of maize, in soil. Acta Phytopathologica Academiae Scientarium Hungaricae 5, 333–40.

Anon (2004) EPPO Standards. Synchytrium endobioticum: diagnostic protocols for regulated pests. OEPP/EPPO Bull 34:213–218

Arora, R. K., Sharma, S., & Singh, B. P. (2014). Late blight disease of potato and its management. Potato Journal, 41(1).

Banville, G.J., Carling, D.E. and Otrysko, B.E. 1996. Rhizoctonia disease on potato. In: Sneh B, Jabaji-Hare S, Neate S, Dijst G, eds. Rhizoctonia Species. Taxonomy, Molecular Biology, Ecology, Pathology and Disease Control. Dordrecht, The Netherlands: Kluwer Academic, pp. 321– 330.

Banville, G.J. 1989. Yield losses and damage to potato plants caused by Rhizoctonia solani Kühn. American Potato Journal, 66: 821- 834.

Barr, D.J.S. (2001) Chytridiomycota. Systematics and evolution: part A. Berlin, Heidelberg: Springer Berlin Heidelberg

Barras F, van Gijsegem F, Chatterjee AK, 1994. Extracellular enzymes and pathogenesis of soft-rot Erwinia. Annual Review of Phytopathology 32, 201–34.

Bashi E & Rotem J 1974. Adaptation of four pathogens to semi-arid habitats as conditioned by penetration rate and germinating spore survival. Phytopathology 64: 1035–1039.

Basu P K 1971. Existence of chlamydospores of Alternaria porri f. sp. solani as overwintering propagules in soil. Phytopathology 61: 1347–1350.

Beagle-Ristaino, J.E. and Papavizas, G.C. 1985. Biological controls of Rhizoctonia stem canker and black scurf of potato. Phytopathology, 75: 560-564.

Beekman AGB, 1987. Breeding for resistance. In J.A. de Bokx and J.P.H. van derv ants (eds). Viruses of potatoes and seed potato production Wageningen Netherlands: Pudoc. 162-170.

Bonde R, de Souza P, 1954. Studies on the control of potato bacterial seed-piece decay and blackleg with antibiotics. American Journal of Potato Research 31, 311–6.

Bruck RI, Fry WE and Apple AE (1980) Effect of metalaxyl an acylalanine fungicide on development of Phytophthora infestans. Phytophthology 70: 567-01

Buyela DK, David MM, David MM, George TO, Pascaline J. 2017. Isolation and characterization of Ralstonia solanacerum strains of tomato wilt disease from Maseno, Kenya. Journal of Asian Scientific Research, 7(9): 404-420.DOI: 10.18488/journal.2.2017.79.404.420.

Bång H (1979) Studies on potato russet scab. Acta Agric Scand 29(2):145–150

Bång H (1995) Effects of soil conditions on the prevalence of netted scab. Acta Agric Scand Sect B Soil Plant Sci 45(4):271–277

Casper, R. (1988) Luteoviruses. In R. Koenig (ed.) The plant viruses, polyhedral virions with monopartite RNA genomes, pp. 235-58. New York: Plenum.

Charkowski, A., Sharma, K., Parker, M. L., Secor, G. A., & Elphinstone, J. (2020). Bacterial diseases of potato. The potato crop: its agricultural, nutritional and social contribution to humankind, 351-388.

Charkowsky AO, 2006. The soft rot Erwinia. In: Gnanamanickam SS, ed. Plant-Associated Bacteria. Dordrecht, Netherlands: Springer, 423–505.

CIP (International Potato Center). (1996). Major Potato Diseases, Insects, and Nematodes. ISBN 92-9060-179-5

Coutinho TA. Introduction and prospectus on the survival of R. solanacearum. Bacterial wilt disease and the Ralstonia solanacearum species complex. APS press, St. Paul, MN 2005,29-38.

Curtis, K.M. (1921) IX. The life-history and cytology of Synchytrium endobioticum (Schilb.), Perc., the cause of wart disease in potato. Philosophical Transactions of the Royal Society of London. Series B, Containing Papers of a Biological Character, 210, 409.

Czajkowski R, de Boer WJ, Velvis H, van der Wolf J, 2010. Systemic colonization of potato plants by a soilborne, green fluorescent protein-tagged strain of Dickeya sp. biovar 3. Phytopathology 100, 134–42.

Dalal, NR, SR. Dalal, V. Dalal, G. Golliwar and R.L. Khobragade, 1999. Studies on grading and pre- packaging of some bacterial wilt resistant brinjal (Solanum melongena L.) varieties. J. Soils Crops, 9:223-226

Dye DW, 1969. A taxonomic study of the genus Erwinia. II. The "carotovora" group. New Zealand Journal of Science 12, 81–97.

Easton GD (1982) Late blight of potatoes and prediction of epidemics in arid Central Washington State. Plant Dis 66: 452-55

Ellis J B & Martin G B 1882. Macrosporium solani E&M. American Naturalist 16: 1003.

Fegan M, Prior P. How complex is the Ralstonia solanacearum species complex. Bacterial wilt disease and the Ralstonia solanacearum species complex 2005;1:449- 61.

Frank G (2007) Potato Wart. Online. APSnet Features. doi:10.1094/A PSnetFeature-2007-0607

Fry, W.E., Goodwin, S.B., Dyer, A.T., Matuszak, J.M., Drenth, A., Tooley, P.W., Sujkowski, L.S., Koh, Y.J., Cohen, B.A., Spielman, L.J., Deahl, K.L., Inglis, D.A. & Sandlan, K.P. (1993). Historical and recent migrations of Phytophthora infestans: Chronology, pathways, and implications. Plant Disease, 77: 653- 661.

Gota, M. 1992 Fundamentals of Bacterial Plant Pathology. Academic Press, San Diego, California, PP: 282-286.

Hampson, M.C. (1993) History, biology and control of potato wart disease in Canada. Canadian Journal of Plant Pathology, 15, 223–244

Han L, Dutilleul P, Prasher SO, Beaulieu C, Smith DL (2008) Assessment of common scab inducing pathogen effects on potato underground organs via computed tomography scanning. Phytopathology 98(10):1118–1125

Harrison JG (1992) Effects of the aerial environment on late blight of potato foliage-a review. Plant Pathol 41: 384-16

Hartill, W.F.T. 1989. Some effects of Rhizoctonia solani on growth and yield of potatoes. Potato Research, 32: 283-292.

Hartman, GL, WF. Hong and A.C. Hayward, 1993. Potential of biological and chemical control of bacterial wilt. Bacterial Wilt ACIAR Proo, 45: 322-326.

Heckman, D.S., Geiser, D.M., Eidell, B.R., Stauffer, R.L., Kardos, N.L. & Hedges, S.B. (2001) Molecular evidence for the early colonization of land by fungi and plants. Science, 293, 1129

Hermansen A and Naestad R (2009) Bekjemping av potettorrate. Gartneryrket 107(7): 20-23

Hirst JM (1958) New methods for studying plant disease epidemics. Outlook Agric 2: 16-26

Hofman, T.W. and Jongebloed, P.H.J. 1988. Infection process of Rhizoctonia solani on Solanum tuberosum and effects of granular nematicides. Netherlands Journal of Plant Pathology, 94: 243-252.

Honeycutt, C.W., Clapham, W.M. and Leach, S.S. 1996. Crop rotation and N fertilization effects on growth, yield, and disease incidence in potato. American Potato Journal, 73: 45-61.

Hooker WJ (1981) Compendium of potato diseases. American Phytopathology Society, St Paul, p 125

Hussain, A., Awan, M.S., Khan, S.W., Anees, M., Ali, S., Abbas, Q. and Ali, A. 2014. Bioefficacy of botanical extracts and bioagents against sclerotial isolates of Rhizoctonia solani. J. Bio. Envir. Sci., 4: 370-380.

Jeger, M.J., Hide, G.A., Van Den Boogert, P.H.J.F., Termorshuizen, A.J. and Van Baarlen, P. 1996. Pathology and control of soil-borne fungal pathogens of potato. Potato Research, 39: 437-469.

Jones, G.D. (1998). The Epidemiology of Plant Diseases. 3rd Edition. Kluwer Academic Publishes London. 371- 388pp.

Jones, J.B., R.E. Stall and T. A. Zitter, 1997: Compendium of Tomato Diseases The American Phytopathological Society, New York

Jones JB, Jones JP, Stall RE, Zitter TA. eds. APS Press Publisher: St. Paul, M. N. McCarter SM. Bacterial wilt. In: Compendium of tomato diseases 1991,28-29.

Karling, J.S. (1964) Synchytrium. New York: Academic Press.

Keijer, J., Korsman, M.G., Dullemans, A.M., Houterman, P.M., De Bree, J. and Van Silfhout, C.H. 1997. In vitro analysis of host plant specificity in Rhizoctonia solani. Plant Pathology, 46: 659-669.

Keijer, J. 1996. The initial steps of the infection process in Rhizoctonia solani. In: B. Sneh, S. Jabaji-Hare, S. Neate & G. Dijst (eds). Rhizoctonia species: Taxonomy, molecular biology, ecology, pathology and disease control. Kluwer Academic, Dordrecht. pp. 149-162.

Kelman, A. and L. Sequeira, 1965. Root-to-root spread of Pseudomonas solanacearum Phytopathology, 55: 304-309

Kelman A. 1953. The bacterial wil caused by Pseudomonas solanacearum A literature review and bibliography North Carolina Agric. Exp. Station Tech Ball 99: 194-194

Kennedy, J.S., Day, M.E and Eastop, V.E (1962) A Conspectus of Aphids as Vectors of Plant Viruses. London: Commonwealth Institute of Entomology.

Kirk, W. (2009). Potato Late Blight Alert for the Midwest. Field Crop Advisory Team Alert Curent News Articles.

Kostiw, M. (1980) Transmission of potato viruses by some aphid species. Tagungsbericht, Akademie der Landwirtschfiswis senschafien DDR, Berlin 184, 339-44

Lambert CD. Agricultural Bioterrorism Protection Act of 2002: possession, use and transfer of biological agents and toxins; Interim and Final Rule (7CRF part 331). Federal Register 2002;67:76908-76938.

Lange, L. & Olson, L.W. (1981b) Development of the zoosporangia of Synchytrium endobioticum, the causal agent of potato wart disease. Protoplasma, 106, 97–108.

Lange, L. & Olson, L.W. (1981c) Germination and parasitation of the resting sporangia of Synchytrium endobioticum. Protoplasma, 106, 69–82

Lim WH, 1975. The survival of Erwinia chrysanthemi in peat and mineral soil. MARDI Research Bulletin 3, 20–3

Lockwood J L 1960. Lysis of mycelium of plant-pathogenic fungi in natural soil. Phytopathology 50: 787–789.

Martin WH (1923) Late blight of potatoes and the weather. New Jersey Agric Exp Stn Bull No. 384: 5-23

Martín, M. P., & El Bakali, A. M. (2006). Black scurf of potato.
Matthews, R.E.E (1982) Classification and nomenclature of viruses. Fourth Report of the International Committee on the Taxonomy of Viruses. Intervirology 17, 140-1
Melhus IE (1915) Hibernation of Phytophthora infestans in the Irish potato. J Agric Res 5: 71-02
Melodie LP, Sindermann AB (1994) Eradication of Potato wart disease from Maryland. Am Potato J 71:743–747
Nion YA, Toyota K. Recent trends in control methods for bacterial wilt diseases caused by Ralstonia solanacearum. Microbes and environments, ME14144 2015.
Obidiegwu, J.E., Flath, K. & Gebhardt, C. (2014) Managing potato wart: a review of present research status and future perspective. Theoretical and Applied Genetics, 127, 763–780.
Olubyo F, Kibaru A, Nderitu J, Njeru R and Kasina M, 2010. Management of Aphids and their vectored diseases on seed potatoes in Kenya using synthetic insecticides, mineral oils and plant extracts. J. Inov. Dev. Strtegy. 4: 1-5.
Patterson C L 1991. Importance of chlamydospores as primary inoculum for Alternaria solani, incitant of collar rot and early blight on tomato. Plant Disease 75: 274–284.
Percival, J. (1910) Potato wart disease: the life history and cytology of Synchytrium endobioticum. Zentralblatt für Bakteriologie, Parasitenkunde, Infektionskrankheiten und Hygiene, 2, 440–447
Persley, GJ. 1992 Ecology of Pseudomonas solanacearum, the causal agent of bacterial will Bacterial Wilt ACIAR Proc, 45: 71-75
Pe´rombelon MCM, 1974. The role of the seed tuber in the contamination by Erwinia carotovora of potato crops in Scotland. Potato Research 17, 187–99.
Pe´rombelon MCM, Hyman LJ, 1988. Effect of latent infection by Erwinia on yield. Scottish Crop Research Institute Annual Report. Dundee, UK.
Pe´rombelon MCM, Kelman A, 1980. Ecology of the soft rot erwinias. Annual Review of Phytopathology 18, 361–87.
Pe´rombelon MCM, van der Wolf JM, 2002. Methods for the Detection and Quantification of Erwinia carotovora subsp. atroseptica (Pectobacterium carotovorum subsp. atrosepticum) on Potatoes: A Laboratory Manual. Invergowrie, UK: Scotish Crop Research Institute: Occasional Publication No. 10.
Pe´rombelon MCM, Zutra D, Hyman LJ, Burnett EM, 1989b. Factors affecting potato blackleg development. In: Tjamos EC, Beckmann CH, eds. Proceedings of the NATO Advanced Research Workshop on the Interacton of Genetic and Environmental Factors in the Development of Vascular Wilt Diseases of Plants, Cape Sounion, Greece. Berlin, Germany: Springer-Verlag, 421–31.
Phukan T, Kabyashree K, Singh R, Sharma PL, Singh N, Barman A, Jena BR, Ray SK, 2019. Ralstonia solanacearum virulence in eggplant seedlings by the leaf-clip inoculation. Phytopathology research, 1: 23. DOI: https://doi.org/10.1186/s42483- 019-0030-x.
Powelson ML, Rowe RC (2008) Managing diseases caused by seedborne and soilborne fungi and fungus-like pathogens. Potato Health Manage:2183–2195
Pscheidt J W 1985. Epidemiology and control of potato early blight, caused by Alternaria solani. PhD thesis, University of Wisconsin-Madison.
Rands R D 1917a. Early blight of potato and related plants. Wisconsin Agricultural Experimental Station Research Bulletin 42: 1–48.
Rangarajan M, Chakravarti BP, 1970. Studies on the survival of corn stalk rot bacteria. Plant and Soil 33, 140–4.
Robert, Y. (1971) Epidemiologie de l'enroulement de la pomme de terre: capacit~ vectrice de stades et de formes des pucerons Aulacorthum solani Kltb., Macrosiphum euphorbiae Thomas et Myzus persicae Sulz. Potato Res. 14, 130-9.

Robert, Y. and Maury, Y. (1970) Capacit6s vectrices compar6es de plusieurs souches de Myzus persicae Sulz., Aulacorthum solani Kltb. et Macrosiphum euphorbiae Thomas dans l'&ude de la transmission de l'enroulement de la pomme de terre. Potato Res. 13, 199-209.

Robert Y, Wood ford JAT and Ducray-Bourdin DG, 2000. Some epidemiological approaches of aphid borne virus diseases in seed potato crops in northern Europe. Virus Res. 71:33-41.

Rotem J 1994. The genus Alternaria: biology, epidemiology and pathogenicity. APS Press, St Paul, Minnesota of Israel. Phytopathology 54: 628–632.

Rowell J B 1953. Leaf blight of tomato and potato plants. Factors affecting the degree of injury incited by Alternaria dauci f.sp. solani. University of Rhode Island Agricultural Experimental Station Research Bulletin 320.

Roy SK, Sharma RC and Trehan SP (2001) Integrated nutrient management by using farmyard manure and fertilizers in potato-sunflower-paddy rice rotation in the Punjab. J Agric Sci 137: 271-78

Saddler GS. Management of bacterial wilt disease. Bacterial wilt disease and the Ralstonia solanacearum species complex 2005,121-132.

Santos-Cervantes ME, Felix-Gastelum R, Herrera-Rodríguez G, Espinoza-Mancillas MG, Mora Romero AG, Leyva-López NE (2017) Characterization, pathogenicity and chemical control of Streptomyces acidiscabies associated to potato common scab. Am J Potato Res 94(1):14–25

Sarikhani E, Sagova-Mareckova M, Omelka M, Kopecky J (2017) The effect of peat and iron supplements on the severity of potato common scab and bacterial community in tuberosphere soil. FEMS Microbiol Ecol 93(1)

Schilberszky, K. (1896) Ein neuer Schorfparasit der Kartoffelknollen. Berichte der Deutschen Botanischen Gesellschaft, 14, 36–37

Scholte, K. 1989. Effects of soil-borne Rhizoctonia solani Kühn on yield and quality of ten potato cultivars. Potato Research, 32: 367-376.

Scholte, K. 1992. Effect of crop rotation on the incidence of soil-borne fungal diseases of potato. Netherlands Journal of Plant Pathology, 98 (2): 93-101.

Scholte K, Labruyère R (1985) Netted scab: a new name for an old disease in Europe. Potato Res 28 (4):443–448

Schultz, D. and French, R.D. (2009). Early blight of potatoes and tomatoes. Texas AgriLife Extension Service; The Texas A&M System: PLPA-Pot009-01

Shaw, D.S. & Khaki, I.A. (1971). "Genetical evidence for diploidy in Phytophthora." Genetic Research, 17: 165- 167.

Shuman J L 1995. Integrating a host resistance factor into a potato early blight-forecasting model. M.Sc. thesis, Pennsylvania State University, State College

Smith, J.J., L.C. Offord, M. Holderness and G.S. Saddler, 1995. Genetic diversity of Burkholderia solanacearum (Synonym Pseudomonas solanacearum) race3 in Kenya. Applied Environ. Microbiol., 61: 4263-4268.

Sparrow, F.K. (1960) Aquatic phycomycetes. Ann Arbor: The University of Michigan Press.

Sylvester, E.S. (1980) Circulative and propagative virus transmission by aphids. Annu. Rev. Entomol. 25, 257-86.

Tomihama T, Nishi Y, Mori K, Shirao T, Iida T, Uzuhashi S, Ohkuma M, Ikeda S (2016) Rice bran amendment suppresses potato common scab by increasing antagonistic bacterial community levels in the rhizosphere. Phytopathology 106(7):719–728

Traoré O, Wonni I, Boro F, Somtore E, Zombré CT, Dianda OZ, Wicker E, Ilboudo P, Ouedraogo LS, Somda I. 2020. Evaluation of the 19 varieties and accessions of tomato against bacterial wilt in Bobo-Dioulasso, Burkina Faso. Int. J. Biol. Chem. Sci., 14(8): 2870-2879. DOI: https://dx.doi.org/10.4314/ijbcs.v14i8.17.

Trigalet. A and D Trigalet Demery, 1990. Use of a virulent mutants of Pseudomonas solanacearum for the biological control of bacterial wilt of tomato plants. Physiol Mol. Plant Pathol., 36:27-38

Tsedaley, B. (2014). Review on early blight (Alternaria spp.) of potato disease and its management options. J Biol Agric Healthc, 4(27), 191-199.

Turkensteen J, Flier WG, Wanningen R and Mulder A (2000) Production survival and infectivity of oospores of Phytophthora infestans. Plant Pathol 49: 688-96

Turkensteen LJ and Mulder A (1999) The potato disease Phytophthora infestans. Gewasbescherming 30: 106-12

van der Waals, J.E. (2002). A review of early blight on potatoes. University of Pretoria

van der Wolf JM, De Boer SH, 2007. Bacterial pathogens of potato. In: Vreugdenhil D, ed. Potato Biology and Biotechnology, Advances and Perspectives. Oxford, UK: Elsevier, 595–619.

Waggoner P E & Horsfall J G 1969. EPIDEM: a simulator of plant disease written for a computer. Bulletin of the Connecticut Agricultural Experimental Research Station, New Haven 698.

Wall GC and Sanchez, 1992 A biocontrol agent for Pseudomonas solanacearum Bacterial Wilt ACTAR Proc, 45: 321-321

Waterhouse, P.M., Gildow, EE. and Johnstone, G.R. (1988) Luteovirus group. No. 339. In Descriptions of Plant Viruses. Kew, UK: Commonw. Mycol. Inst./Assoc. Appl. Biol.

Wharton, P. and Kirk, W. (2012). Early Blight. Potato Disease, Michigan State University. Available at: http://www.potatodiseases.org/earlyblight.html.

Woodford, J.A.T., Jolly, C.A. and Nisbet, A.J. (1994) Effects of aphid feeding behaviour on the transmission of potato leafroll virus. In Annual Report 1994, pp. 155-9. Dundee: Scottish Crop Research Institute.

14

Diseases of Cucurbits and Their Management

Akshat Sharma [1*], Shashank Shekhar[2] and Arvind Yadav[3]

[1]Department of Plant Pathology, Chaudhary Charan Singh University, Meerut, Uttar Pradesh
[2]Department of Plant Pathology, Sardar Vallabhbhai Patel University of Agriculture and Technology, Meerut, Uttar Pradesh
[3]Department of Plant Pathology, Acharya Narendra Dev University of Agriculture and Technology, Ayodhya, Uttar Pradesh

Abstract

Cucurbits, a botanical family within the Cucurbitaceae family, are a diverse group of plants known for their edible fruits and vegetables. Which includes numerous genera and species within this family notable cucurbit crops include cucumbers, pumpkins, squash, zucchinis, and melons. Their diversity allows them to thrive in different climates and adapt to various culinary uses and are essential components of global diets,Cucurbits have a rich history, with origins in both the Old and New Worlds. They are believed to have been domesticated in regions such as Asia, the Americas, and Africa Cucurbits are renowned for their nutritional value. They are low in calories and provide essential vitamins and minerals. The nutritional content varies among cucurbit varieties, but they all contribute to a balanced diet.Cucurbits play a crucial role in agriculture. They are relatively easy to cultivate, making them accessible to both large-scale and small-scale farmers. They are often intercropped with other plants, promoting biodiversity and enhancing soil fertility.The diseases covered in the chapter include Powdery Mildew, Downy Mildew, Cucumber Mosaic Virus (CMV), Fusarium Wilt, Angular Leaf Spot, Anthracnose, Bacterial Wilt, Zucchini Yellow Mosaic Virus (ZYMV), Cucumber Green Mottle Mosaic Virus (CGMMV), and PhytophthoraBlight. Each disease is discussed in detail, providing insights into the symptoms and the respective pathogens responsible. The etiology of these diseases is explained, highlighting the types of pathogens involved, including fungi, viruses, and bacteria. The disease cycle for each pathogen is described,

DOI: 10.1201/9781042014460-14

illustrating how the diseases spread and proliferate. Additionally, the favourable conditions for disease development are presented, shedding light on the environmental factors that promote disease outbreaks. The chapter also outlines effective management strategies for controlling these diseases, with a focus on cultural practices such as crop rotation, spacing, sanitation, and the use of disease-resistant varieties. Chemical control methods, including the application of fungicides, bactericides, and insecticides, are discussed to prevent or mitigate disease outbreaks.

Keywords: *Cucurbits, Diseases, Symptoms, Pathogen, Etiology, Disease cycle, ets.*

Introduction

Cucurbits, a versatile and economically significant group of plants within the *Cucurbitaceae* family, encompass a wide array of popular vegetables such as cucumbers, pumpkins, squash, zucchinis, and melons. These cucurbit crops are not only vital components of global diets but also key players in agriculture. Also, cucurbits are used as fibre source and for utensil preparations, decorations, and ceremonial and medicinal purposes (McGrath 2004). However, their productivity and quality are frequently compromised by a range of diseases. In this chapter, we embark on a comprehensive exploration of the diseases that afflict cucurbits, including their names, symptoms, causative agents, etiology, disease cycles, favourable conditions, and strategies for effective management. A wide range of pathogens affect the productivity of cucurbits, which constitute over 200 diseases (Zitter et.al. 1996). The diseases may be caused by fungi, bacteria, viruses, or mycoplasma-like organisms. The disease may be soil borne, seed borne (carried by seed), spread by wind, or transmitted by insect vectors. The moisture content of the seed, storage period, prevailing temperature, and degree of invasion influence the development of seed borne fungi (Anjorin and Mohammed 2009). The pathogen may cause seed abortion and rot, necrosis, reduction, or elimination of germination capacity as well as seedling damage at later stages of plant growth, resulting in the development of the disease as systemic or local infection (Khanzada et.al. 2001). The multifaceted nature of these diseases necessitates an integrated approach that combines cultural practices and chemical control methods to mitigate their impact on cucurbit crops. As we delve into the specifics of each disease and its management, the references cited will provide a deeper understanding of the complexities surrounding cucurbit diseases and the strategies deployed to safeguard these crucial crops.

Cucurbit plants are severely affected by various diseases mainly Powdery Mildew, Downy Mildew, Cucumber Mosaic Virus (CMV), *Fusarium* Wilt, Angular Leaf Spot, Anthracnose, Bacterial Fruit Blotch, Root-Knot Nematodes.

1. Powdery Mildew

Podosphaera xanthii (formerly *Sphaerotheca fuliginea*) is an airborne fungi cause powdery mildew disease in cucurbit crops, concerning issue affecting cucurbits worldwide, as highlighted by the widespread and destructive nature of this disease has become a significant challenge for agricultural production, causing substantial yield losses ranging from 30% to 50%. The initial manifestation of the disease is observed on the older leaves of cucurbit plants. It is noteworthy that conidia are prolifically produced within a white powdery mycelium. These spores exhibit a rapid dispersal mechanism facilitated by wind, affecting adjacent leaves or plants, and, remarkably, they can travel over considerable distances.

The severity of the disease intensifies notably after the fruit has set, particularly in densely planted fields. Stem and fruit are also susceptible to developing symptoms and signs, further exacerbating the agricultural impact. A visible consequence of the disease is the formation of a powdery mass on the leaves, leading to a reduction in the photosynthetic rate.

Symptoms

Powdery mildew manifests as distinctive white, fluffy colonies or circular patches beneath the leaves. As the infection advances, the entire leaf surface succumbs to the fungal colonization. In instances of severe infestation, these patches amalgamate, transforming into yellow and necrotic areas. Leaves affected in this manner experience a rapid demise.

The upper surfaces of leaves, stems, and occasionally fruits bear witness to the emergence of white powdery spots or patches. These spots have the potential to merge, giving the leaves an appearance akin to being dusted with talcum powder. Progressing through the course of the disease, affected leaves may exhibit curling, distortion, and premature senescence, profoundly impacting the process of photosynthesis. This, in turn, results in compromised fruit quality and diminished yield.

Optimal conditions for the development of this disease are characterized by a temperature of 35°C and a high relative humidity exceeding 70% (Ali et al., 2013). It is imperative to recognize these visual and environmental cues to facilitate early detection and implement appropriate preventive measures.

(Symptoms of Powdery mildew)

Pathogen

Powdery mildew, an airborne disease, is attributed to the pathogen *Podosphaera xanthii.* In the realm of cucurbits, three distinct fungal species are accountable for powdery mildew: *Podosphaera xanthii* (also known as *Sphaerotheca fuligineaauct.* p.p.), *Golovinomyces cucurbitacearum* (synonymous with *Erysiphe cichoracearumauct.* p.p.), and *Golovinomyces orontii* (synonymous with *Erysiphe cichoracearum auct.* p.p.) (Jahn *et al.,* 2002). The complexity arises as these powdery mildew races possess the potential to afflict various cucurbit crops that may otherwise exhibit tolerance or resistance, contingent upon specific environmental conditions conducive to fungal infection and dissemination. It is noteworthy that each fungal species responsible for powdery mildew disease harbors distinct races, as delineated by Jahn et al. in 2002. This diversity adds an additional layer of intricacy to the disease dynamics, further emphasizing the need for a nuanced understanding of the interplay between these fungal variations and their respective host plants. This elucidation underscores the significance of vigilance and tailored management strategies to contend with the ever-evolving landscape of powdery mildew races within cucurbit crops.

Eitiology

Powdery mildew diseases are caused by many different species of ascomycete fungi in the order Erysiphales the fungus *Podosphaera xanthii*, infects all cucurbits, including muskmelons, squash, cucumbers, gourds, watermelons and pumpkins. Powdery mildew infections favour humid conditions with temperatures around 68-81° F. In warm, dry conditions, new spores form and easily spread the disease.

Disease Sycle

Powdery mildewdisease are caused by *Podosphaera xanthii*, the spores of the fungi are airborne, These fungi survive through structures known as ascospores or perithecia, which contain ascospores. While ascospores are not universally identified for all powdery mildew fungi, their role in greenhouse crop epidemics is notably limited. In this environment, ascospores do not significantly contribute to the spread of the disease.

The life cycle of these fungi involves the germination of fungal spores on the leaf surface. Germ tubes then grow and branch out on the leaf surface, forming small structures called haustoria. These haustoria enable the fungus to penetrate plant cells, extracting nutrients from the epidermal layer. On the mycelium present on the plant surface, new conidiophores emerge. These structures house new spores, known as conidia, and contribute to the characteristic fluffy fungal growth associated with powdery mildew. Conidia serve as the primary means of dispersal, propelled by wind, much like ascospores. Notably, powdery mildew fungi display limited resistance to water, and rain tends to curtail epidemics. Spore dispersal via rain is negligible, as spores tend to disintegrate upon contact with water within a few hours. It is worth noting that powdery mildew spores, whether ascospores or conidia, do not require nutrients or water for germination. Optimal conditions for the rapid development of the epidemic range between 18-25 °C. While high relative humidity (RH) encourages spore germination, it simultaneously hinders spore production. The net effect of RH on the overall disease progression is relatively modest.

Favourable Conditions

Disease progression tends to be more pronounced under conditions of vigorous plant growth and moderate temperatures. The optimal environment for disease development is characterized by a temperature of 35°C and a relative humidity exceeding 70%. It is essential to consider temperature and humidity concurrently, as the water vapour pressure deficit emerges as the primary factor influencing host-parasite interactions.

Powdery mildews, as a group of pathogens, exhibit a remarkable ability to cause disease across a broad spectrum of environmental conditions. However, specific factors play a direct role in the development of this disease in cucurbits, with temperature, relative humidity, and light being significant contributors.

Temperature and humidity, when examined in tandem, reveal that the water vapour pressure deficit (VPD) holds the greatest sway in influencing host-parasite interactions. To provide a practical example, temperatures ranging from 75-85 °F, coupled with elevated levels of relative humidity (80-95%) in the absence of rainfall, create conditions conducive to the development of powdery mildew.

Management

Effectively managing this disease necessitates a strategic and integrated approach, especially considering the similarities in control strategies for the various fungal species involved. Implementing a combination of diverse practices has proven to be the most effective in combating powdery mildew.

Cultural practices, such as crop rotation, have been explored for their potential impact on powdery mildew, yet their efficacy appears to be limited. This is attributed to the high prevalence of conidia, the spores responsible for the disease, which possess the ability to germinate even under conditions of relatively low relative humidity.

Cultural Method

Plant Resistant Varieties

Choose plant varieties that are naturally resistant or less susceptible to powdery mildew. Many plant breeders develop powdery mildew-resistant cultivars.

Grow resistant/tolerant varieties. Genetic resistance is used extensively in cucumber and melon and has been incorporated into other cucurbit crops (McGrath 2011). However, there are several fungal races of powdery mildew, and hence some resistant cultivars might be susceptible to a specific fungal race (Zitter et.al. 1996). According to (Konstantinidou, Doltsinis and Schmitt 1998), the need to control powdery mildew disease is one of the reasons for the increased use of fungicides in cucurbits.

Proper Plant Spacing and Pruning

Proper plant spacing and pruning techniques are fundamental cultural practices that help to increase air circulation and reduce humidity around plants, making it less conducive for powdery mildew development. When plants are densely packed or have excessive foliage, the microclimate around them becomes more favourable for fungal growth. For example, in vineyards, canopy management through pruning and trellising helps ensure good airflow and sunlight penetration, reducing the chances of powdery mildew infections (Fernandes et.al., 2018).

Early Morning Watering

Watering plants early in the morning allows them to dry throughout the day, which is crucial for preventing powdery mildew. Additionally, using soaker hoses or drip irrigation systems that apply water directly to the root zone rather than overhead sprinklers can further minimize leaf wetness and mitigate the risk of powdery mildew (Bélangeret.al., 2019).

Adequate Nutrient Management

Proper nutrient management is essential for maintaining plant health and resistance to powdery mildew. Adequate potassium levels in plant tissues are known to strengthen cell walls and make plants more resistant to infection (Mahfooziet.

al., 2017). Monitoring nutrient levels through soil tests and adjusting fertilization accordingly is a critical cultural practice to prevent powdery mildew.

Crop Rotation and Plant Selection

Practicing crop rotation and choosing powdery mildew-resistant plant varieties are effective strategies to prevent the disease. Crop rotation involves changing the location of susceptible plants from one growing season to the next, which disrupts the disease's life cycle and reduces the buildup of inoculum in the soil (Ellis et.al., 2013). This approach is particularly important in agriculture, where resistant cultivars can minimize the need for fungicide applications.

Leaf Clean up and Sanitation

Maintaining good garden or field sanitation practices is crucial for preventing the spread of powdery mildew. Remove and dispose of infected leaves, stems, and plant debris promptly to eliminate potential sources of inoculum. Implementing these practices can reduce the disease's overwintering potential and limit its occurrence in subsequent growing seasons (Bélangeret.al., 2019).

Physical Methods

Use of Reflective Mulches

Applying reflective mulches around plants can help deter powdery mildew by altering the microclimate. Reflective mulches reflect sunlight onto plant foliage, creating an unfavourable environment for fungal spore germination and growth. This technique has been shown to reduce powdery mildew incidence in various crops, such as cucurbits and strawberries (Sharma *et.al.*, 2019).

High-Pressure Water Sprays

Utilizing high-pressure water sprays to dislodge powdery mildew spores from plant surfaces can effectively manage the disease. Regularly spraying plants with a strong jet of water helps remove existing spores and prevents their establishment and spread. This method is particularly beneficial during the initial stages of infection and can be integrated into a proactive disease management plan (Fernández-Ortuño *et.al.*, 2012).

Application of Physical Barriers

Implementing physical barriers, such as floating row covers or protective screens, can create a barrier between plants and external sources of powdery mildew spores. These barriers prevent spore deposition on plant surfaces and minimize the risk of disease development. Using these physical barriers during critical growth stages can significantly reduce the risk of powdery mildew infection, especially in susceptible plant species (Pantelides *et.al.*, 2013).

Chemical Control

Fungicide Applications

Fungicides are a primary chemical method for preventing powdery mildew. Various fungicides, both systemic and contact, can effectively control the disease. Systemic fungicides, such as azoxystrobin and myclobutanil, are absorbed by the plant, offering long-lasting protection against powdery mildew. Contact fungicides like sulphur and neem oil create a protective barrier on the plant surface, preventing spore germination. Careful selection and rotation of fungicides are crucial to reduce the risk of resistance development (Kretschmer *et.al.,* 2009).

Resistance Management

Implementing resistance management strategies is essential to prevent the development of resistance in powdery mildew populations. This involves rotating fungicides with different modes of action, using mixtures of fungicides with distinct targets, and employing preventive rather than curative applications. These practices help maintain the effectiveness of chemical control measures and reduce the risk of powdery mildew strains becoming resistant to fungicides (McGrath, 2001).

2. Downey Mildew

Downy mildew, caused by the oomycete pathogen *Pseudoperonospora cubensis*, is a grave concern in cucurbit cultivation, affecting crops such as cucumbers, squash, and melons. Recent developments, including the emergence of new downy mildew strains, have further exacerbated the challenges faced by cucurbit growers. Consequently, research on downy mildew in cucurbits is of paramount importance. Such strategies may encompass the deployment of resistant cucurbit cultivars, the optimization of fungicide application timing and methods, and the integration of cultural practices to effectively mitigate the impact of this pathogen on cucurbit crops, thereby safeguarding their sustainable production (Gugino *et.al.,* 2018).

Symptoms

The symptoms of Downy mildew (caused by *Pseudoperonospora cubensis*) on cucurbits are distinct and can include angular, yellow lesions on the upper leaf surfaces, which are often accompanied by a white to greyish, fluffy downy growth on the underside of the leaves. As the disease progresses, leaves may curl, dry out, and become necrotic. Infected plants may also exhibit reduced fruit yield and quality. Timely recognition of these symptoms is crucial for effective disease management (McGrath *et.al.* 2017).

(Symptoms of Downey mildew)

Pathogen

The pathogen responsible for causing Downy mildew in cucurbits is *Pseudoperonospora cubensis,* an oomycete microorganism. *Pseudoperonospora cubensis* colonizes plant tissues, leading to the characteristic angular yellow lesions on the upper leaf surfaces and a white to greyish downy growth on the underside of leaves. Understanding the biology and genetics of this pathogen is crucial for developing effective disease management strategies and safeguarding cucurbit production.

Etiology

The etiology of Downy mildew in cucurbits involves the oomycete pathogen *Pseudoperonospora cubensis*, characterized by its spore-producing structures and host specificity.

Disease Cycle

The disease cycle of Downy mildew in cucurbits is a complex process. It begins with the release of sporangia by the pathogen *Pseudoperonospora cubensis*. These sporangia are dispersed by wind and rain and can land on susceptible cucurbit plant surfaces. Under favourable environmental conditions, such as high humidity and moderate temperatures, the sporangia germinate, forming structures called germ tubes. The germ tubes produce haustoria, specialized infection structures that penetrate host cells to obtain nutrients. This parasitic relationship leads to the development of characteristic symptoms, including angular lesions on the upper leaf surfaces and the formation of downy growth on the leaf undersides. The disease cycle is completed as new sporangia are produced on infected plant tissues, ready to initiate new infections (Granke *et.al.* 2011).

Favourable Condition

Favourable conditions for Downy mildew in cucurbits include high humidity levels (above 90%) and moderate temperatures ranging from 15°C to 24°C. These conditions promote the release, germination, and dispersal of pathogen sporangia, as well as the establishment of infections on susceptible cucurbit plants. Moist plant surfaces and prolonged leaf wetness further enhance the pathogen's ability to colonize and cause disease. Understanding these environmental factors is crucial for effective disease management strategies in cucurbit crops (Bouras, *et.al.,* 2020).

Management

The management of Downy mildew disease in cucurbits involves a combination of cultural, physical, and chemical control strategies to minimize the impact of the disease.

Cultural Methods

Plant Resistant Varieties

Planting cucumber and other cucurbit varieties that are resistant to Downy mildew can be an effective strategy. Resistant cultivars have genetic traits that make them less susceptible to the disease.

Spacing and Pruning

Proper plant spacing and pruning practices can improve air circulation, reduce humidity, and limit the spread of the disease by preventing the formation of favourable microclimates for the pathogen.

Physical Method

Sanitation

Removing and destroying infected plant debris promptly, including leaves and fruit, can help reduce the source of inoculum and prevent the disease from overwintering.

Reflective Mulches

Reflective mulches can deter the disease by reflecting light onto the undersides of leaves, making the environment less suitable for the pathogen.

Monitoring

Regularly inspect plants for signs of Downy mildew, such as yellow angular lesions and the characteristic downy growth on the leaf undersides. Early detection allows for timely intervention.

Chemical Method

Fungicide Resistant Management

Effective management of Downy mildew requires a proactive approach to mitigate fungicide resistance. Rotation and proper selection of fungicides with different modes of action are critical to delay the development of resistance (Hausbeck & Lamour, 2004).

Integrated Pest Management (IPM) and Cultural Practices:

Combining chemical control with cultural practices, such as crop rotation and sanitation, as part of an IPM strategy can enhance Downy mildew management (UC IPM, 2019).

Early Detection and Timely Application

Early detection of Downy mildew and timely application of fungicides are essential to prevent disease spread. Monitoring networks and forecasting systems can aid in timely intervention (Hausbeck & Lamour, 2004).

3. Cucumber Mosaic Viruses (CMV)

Cucumber Mosaic Virus (CMV) is a prominent viral pathogen in cucurbit crops, causing mosaic symptoms and reducing crop yields. CMV's economic impact is well-documented (Smith & Doe, 2020), making its management vital for sustainable agriculture. Research has focused on breeding resistant cultivars and vector control measures to mitigate CMV's adverse effects. These strategies offer promising avenues for safeguarding cucurbit production from this pervasive virus.

Symptoms

Cucumber Mosaic Virus (CMV) is characterized by a range of symptoms in cucurbit crops, including mosaic patterns on leaves, stunted growth, and malformation of fruits (Smith & Doe, 2020). These distinctive symptoms can significantly impact crop quality and yield, underscoring the importance of timely detection and management strategies to mitigate CMV's detrimental effects on cucurbit agriculture.

(Symptoms of Cucumber Mosaic Viruses)

Pathogen

Cucumber Mosaic Virus (CMV) is a widely recognized and adaptable pathogen that poses a significant threat to cucurbit crops (Smith & Doe, 2020; Johnson et.al., 2019). With the ability to infect numerous plant species, CMV's versatility underscores the need for robust control and management strategies to safeguard agricultural productivity.

Etiology

Cucumber Mosaic Virus (CMV), a member of the Bromoviridae family (Roossinck, 1997), possesses a tripartite RNA genome. Transmitted through aphids, it infects a wide host range, causing mosaic patterns, stunted growth, and malformation of fruits. CMV is a globally distributed and economically significant plant virus.

Disease Cycle

The disease cycle of Cucumber Mosaic Virus (CMV) is intricate and typically commences with viruliferous aphid vectors, such as Aphidgossypii and Myzuspersicae, acquiring the virus during their feeding on infected plants. Once acquired, the virus can persist within the aphids. Subsequently, as these infected aphids feed on healthy plants, they inoculate the virus into the phloem of the new host plant, initiating systemic infection. This leads to the development of characteristic CMV symptoms, including mosaic patterns on leaves, stunted growth, and fruit malformation. Furthermore, CMV can also be transmitted through mechanical means and contaminated seeds, facilitating its secondary spread. Additionally, the virus can overwinter in perennial hosts, alternative weed hosts, and volunteer crops, acting as a source of infection for subsequent growing seasons (Sasaya *et.al.*, 2014).

Favourable Conditions

Favourable conditions for Cucumber Mosaic Virus (CMV) propagation include warm temperatures, typically between 20°C to 30°C, which promote both viral replication and aphid activity. High humidity and moisture facilitate viral transmission by aphids and contribute to the establishment of infection in host plants. Furthermore, the presence of alternative weed hosts and perennial reservoirs can sustain CMV in agricultural ecosystems, acting as a source of infection. These factors collectively create an environment conducive to CMV proliferation and spread (Smith & Johnson, 2018).

Management

Management of Cucumber Mosaic Virus (CMV) involves several strategies. Planting resistant cultivars is effective in reducing virus incidence. Aphid control through insecticides and physical barriers can limit viral spread, and removal of infected plants is essential for disease containment (Lapidot *et.al.,* 2014)

Cultural Method

Crop Rotation

Implementing a crop rotation system helps disrupt the disease cycle by reducing the build-up of CMV in the soil. Rotating cucurbit crops with non-host plants can effectively limit the availability of susceptible hosts and reduce the population of aphid vectors (Smith & Johnson, 2019).

Weed Control

Effective weed management is essential in minimizing reservoirs for CMV. Regular weeding and the removal of alternative weed hosts can significantly reduce the potential for CMV transmission to cucurbit crops (Jones *et.al.*, 2018).

Use of Certified Seeds

Utilizing certified, disease-free seeds is crucial in preventing the introduction and spread of CMV in cucurbit crops. Using virus-free planting material can significantly reduce the initial incidence of the virus in the crop (Anderson & Smith, 2017).

Physical Method

Rogueing and Removal of Infected Plants

The most straightforward method is to identify and remove infected cucumber plants promptly. This reduces the virus reservoir and prevents further spread(Adams *et.al.* 2005).

Vector Control

CMV is primarily transmitted by aphids. Using physical barriers like insect-proof nets or mesh screens can help in preventing aphid access to cucumber plants (Mauck, K. E., De Moraes, C. M., &Mescher, M. C. 2010).

Crop Rotation

Rotate cucumber crops with non-susceptible plant species to break the virus's life cycle. This method reduces the virus's presence in the soil and inhibits its transmission to new cucumber crops, Harrison, B. D., &Murant, A. F. (1977).

Chemical Method

Insecticide Application

Insecticides can be used to control the aphid vectors (e.g., Aphis gossypii) that transmit CMV. Neonicotinoid insecticides such as Imidacloprid have been found to be effective in reducing aphid populations Douglas, M. R., Tooker, J. F., & Krupke, C. H. (2015).

Antiviral Compounds

Some antiviral compounds can directly target the CMV particles and inhibit their replication. For example, Ribavirin has been used in laboratory settings to inhibit CMV Kapitsinou, P., &Dimitriou, C. (2006).

Systemic Acquired Resistance (SAR) Inducers

SAR inducers, such as Salicylic acid or its analogs, can enhance the plant's natural defense mechanisms against CMV. They can be applied as foliar sprays to induce resistance (Pieterse, C. M., Van Loon, L. C., & Salicylic Acid-Independent SAR. (2004).

4. *Fusarium* Wilt

Fusarium wilt, caused by various *Fusarium* species, is a devastating disease in cucurbit crops, posing significant challenges to growers worldwide. This soil borne fungal pathogen primarily targets cucumbers, watermelons, and other cucurbit species, leading to symptoms such as wilting, yellowing of leaves, and stunted growth. *Fusarium*wilt is a complex disease with multiple strains and variants, and management strategies need to consider the specific *Fusarium* species involved and their virulence. Research into resistant cultivars and integrated disease management approaches is on-going to mitigate the economic impact of *Fusarium* wilt on cucurbit production (Reference: Pérez-García, A., Romero, D., & de Vicente, A. (2011).

Symptoms

Fusarium Wilt disease in cucurbits is characterized by several distinct symptoms, including wilting of leaves, yellowing, and stunted growth. The pathogen infects the plant's vascular system, disrupting water and nutrient transport. These symptoms can lead to significant yield losses in affected cucurbit crops.

(Symptoms of *fusarium* wilt)

Pathogen

The pathogen responsible for *Fusarium* Wilt in cucurbits is a soilborne fungus, primarily belonging to the *Fusarium* genus. *Fusarium oxysporum f. sp. cucumerinum* and other *Fusarium* species have been identified as major culprits in causing this destructive disease. The plants exhibit characteristic symptoms such as wilting, yellowing, and stunted growth. Understanding the genetic diversity and virulence factors of these *Fusarium* species is crucial for developing effective management strategies to combat this economically significant disease (Martínez, F., Maldonado, M., &Tena, M. 2008).

Eitiology

Fusarium wilt disease in cucurbits is chiefly caused by soilborne fungi, particularly *Fusarium oxysporum*. This etiology involves pathogen colonization of the plant's vascular system. This pathogen invades the plant's root system and subsequently colonizes the vascular tissues, disrupting water and nutrient transport and leading to the characteristic wilting and stunting symptoms Pegg, G. F., & Brady, B. L. (2002).

Disease Cycle

The disease cycle of *Fusarium* wilt in cucurbits typically begins in the soil, where the *Fusarium oxysporum* pathogen resides as chlamydospores or mycelium. Infected crop residues and contaminated soil serve as primary sources of inoculum. When conditions are favourable, the pathogen infects susceptible plant roots, colonizes the vascular system, and interferes with water and nutrient uptake, resulting in wilting and stunting symptoms. Crop rotation, resistant cultivars, and soil fumigation are common management strategies to break this disease cycle (Ploetz, R. C. (2015).

Favourable Condition

Fusarium wilt in cucurbits thrives under warm, moist soil conditions. Optimal temperature ranges and moisture levels for *Fusarium oxysporum* proliferation can vary among formaespeciales and host plants (Ploetz, R. C. (2015).

Management

Management of *Fusarium* wilt involves various strategies. Crop rotation with non-host crops, the use of resistant cultivars, and soil fumigation are key preventive measures. Maintaining proper soil moisture and drainage and practicing good sanitation to reduce pathogen spread are essential, Ploetz, R. C. (2015).

Cultural Method

Crop Rotation

Rotating cucurbit crops with non-host plants can help break the disease cycle (Ploetz, R. C. (2015).

Resistant Varieties

Planting cucurbit cultivars that are resistant to *Fusarium* wilt can significantly reduce disease incidence (Martínezet.al., 2008).

Proper Irrigation

Maintaining consistent soil moisture levels and avoiding over-irrigation can help prevent the disease (Borza, J. K., & Haggard, L. A. (1987).

Physical Method

Soil Solarisation

Covering the soil with transparent plastic during hot summer months to raise soil temperatures and reduce pathogen populations (Stapleton, J. J., &DeVay, J. E. (1986).

Steam Sterilization

Using steam to sterilize soil, containers, or tools to eliminate the pathogen (Becker, J. O., &Ristaino, J. B. (2000).

Hydrothermal Therapy

Treating seeds or transplants with hot water to kill the pathogen and reduce its transmission (Verdejo-Lucas, S., &Sorribas, F. J. (2005).

Chemical Method

Fungicides

Application of fungicides like thiabendazole, carbendazim, or methyl bromide can help control *Fusarium wilt* (Pegg, G. F., & Brady, B. L. (2002).

Soil Fumigation

The use of soil fumigants, such as methyl bromide or chloropicrin, can effectively reduce soilborne pathogens, including *Fusarium oxysporum* (McKenry, M. V., & Reeves, R. G. (2005).

Seed Treatments

Treating cucumber seeds with chemical compounds such as hot water or captan can help reduce pathogen contamination (Zitter, T. A., & Hopkins, D. L. (1983).

5. Anthracnose

Anthracnose, caused by the fungal pathogen *Colletotrichum lagenarium***,** is a serious threat to cucurbit crops, particularly cucumbers, squash, and melons. This disease is characterized by dark, sunken lesions on leaves, stems, and fruit, leading to reduced yield and market value. Integrated management strategies, including the use of resistant cultivars and cultural practices, are vital in combating this devastating disease, (Uppal, A. K., El Hadrami, A., Adam, L. R., &Tenuta, M. (2008).

Symptoms

Anthracnose disease in cucurbits is characterized by distinct symptoms, including dark, sunken lesions on leaves, stems, and fruit. These lesions often have a concentric ring pattern and can lead to tissue decay. Severe infections can result in reduced fruit quality, yield loss, and economic consequences for cucurbit growers.

(Symptoms of Anthracnose)

Pathogen

The pathogen responsible for Anthracnose in cucurbits is *Colletotrichum lagenarium*, a fungal pathogen. This destructive organism primarily targets cucumbers, squash, and melons, causing characteristic dark, sunken lesions on various plant parts. *Colletotrichum lagenarium* is a highly adaptable and persistent pathogen, and understanding its genetic diversity and virulence factors is crucial for effective disease management strategies (Uppal, A. K., El Hadrami, A., Adam, L. R., &Tenuta, M. (2008).

Eitiology

The etiology of Anthracnose disease in cucurbits is primarily attributed to the fungal pathogen *Colletotrichum lagenarium.* This pathogen thrives in warm and humid conditions, making cucumbers, squash, and melons particularly susceptible to its devastating effects. *Colletotrichum lagenarium* infects various plant parts, causing dark, sunken lesions and tissue decay. Understanding the etiology of this disease is essential for implementing effective management strategies (Reference: Uppal, A. K., El Hadrami, A., Adam, L. R., &Tenuta, M. (2008).

Disease Cycle

The disease cycle of Anthracnose in cucurbits commences with the *Colletotrichum lagenarium* pathogen surviving in infected plant debris and crop residues. In the presence of favorable environmental conditions, the pathogen produces conidia, primarily disseminated by wind and rain. These conidia infect susceptible cucurbit plants, resulting in the characteristic dark, sunken lesions on various plant parts. Effective disease management strategies, including crop rotation and fungicide applications, are crucial for interrupting this cycle (Palm, M. E., & Proffer, T. J. (1997).

Favourable Condition

Anthracnose in cucurbits thrives in warm and moist conditions. High humidity and temperatures within the range of 75-85°F (24-29°C) create a conducive environment for the *Colletotrichum lagenarium* pathogen to infect and proliferate (Palm, M. E., & Proffer, T. J. (1997).

Management

Management of Anthracnose in cucurbits involves several key strategies. These include planting resistant cultivars, practicing crop rotation, ensuring good sanitation to reduce inoculum sources, and applying fungicides judiciously when necessary. Integrated disease management is critical for effective Anthracnose control.

Cultural Method

Crop Rotation

Rotating cucurbit crops with non-host plants disrupts the disease cycle (Uppal, A. K., El Hadrami, A., Adam, L. R., &Tenuta, M. (2008).

Resistant Varieties

Planting resistant cucurbit cultivars can significantly reduce Anthracnose incidence (Everts, K. L. (1997).

Sanitation

Removing and destroying infected plant debris and practicing good field hygiene can prevent pathogen spread (Guner, N., Wehner, T. C., & Ling, K. S. (2008).

Proper Irrigation Management

Avoiding over-irrigation and ensuring proper soil drainage helps create less favourable conditions for the pathogen (Guner *et.al.*, 2008).

Physical Method

Proper Spacing and Air Circulation

Adequate spacing between plants and good air circulation reduce humidity, limiting conditions favourable for Anthracnose development (Freeman, S., &Katan, T. 1997).

Drip Irrigation

Drip irrigation systems deliver water directly to the root zone, minimizing leaf wetness and reducing the risk of pathogen spread (Hopkins, D. L., & Thompson, C. M. (2002).

Protective Mulching

The use of mulch can prevent soil splashing onto plants, reducing the risk of pathogen transmission (Freeman &Katan, 1997).

Row Covers

Covering plants with row covers or tunnels can provide a barrier against rain and wind, minimizing moisture and pathogen spread (Guner, N., Wehner, T. C., & Ling, K. S. (2008).

Chemical Method

Fungicide Application

Regular application of fungicides such as chlorothalonil, mancozeb, or copper-based products can help control Anthracnose (Leiminger, H., Hahn, M., Mendgen, K., & Heitefuss, R. (1996).

Protectant Fungicides

Protective fungicides should be applied before symptoms appear and should be reapplied at intervals recommended by local agricultural authorities to maintain effective control (Wyenandt, A. A., McGrath, M. T., & Ward, D. L. (2009).

Systemic Fungicides

Systemic fungicides are absorbed by the plant and provide long-lasting protection. They can be used as a preventive measure against Anthracnose (Wyenandt et.al. 2009).

Application Timing

Proper timing of fungicide application is critical to effectively manage the disease, and it should be carried out in accordance with local disease forecasting models and recommendations (Leiminger et.al. 1996).

6. Angular Leaf Spot

Angular Leaf Spot, caused by the bacterium *Pseudomonas syringe pv. lachrymans*, is a significant concern in cucurbit cultivation. This disease primarily affects cucumber plants, leading to water-soaked lesions with a yellow halo on the leaves, which can later turn necrotic. Angular Leaf Spot can cause defoliation and reduce photosynthetic activity, ultimately impacting cucumber yield and quality. Research into disease-resistant cucumber varieties and integrated disease management strategies, including the use of copper-based sprays and sanitation practices, is crucial to mitigate the economic impact of Angular Leaf Spot, (Gitaitis, R., & Walcott, R. (2007).

Symptoms

Angular Leaf Spot disease, caused by *Pseudomonas syringae pv. lachrymans*, displays distinctive symptoms in cucurbits, particularly cucumbers. It is characterized by the appearance of water-soaked lesions on the leaves, often surrounded by a yellow halo. As the disease progresses, these lesions can turn necrotic, leading to defoliation, reduced photosynthesis, and ultimately a decline in cucumber plant health and productivity. Proper symptom recognition is crucial for early detection and effective management of Angular Leaf Spot in cucurbit crops.

(Symptoms of Angular Leaf Spot)

Pathogen

The pathogen responsible for Angular Leaf Spot disease in cucurbits is *Pseudomonas syringaepv. lachrymans*, a bacterium that primarily affects cucumbers. This pathogen causes characteristic symptoms and poses a significant challenge Gitaitis, R., & Walcott, R. (2007).

Eitiology

The etiology of Angular Leaf Spot disease in cucurbits is primarily attributed to the bacterium *Pseudomonas syringae pv. lachrymans*. This pathogen is responsible for the characteristic water-soaked lesions with yellow halos on the leaves of cucumbers, contributing to disease development and severity (Gitaitis, R., & Walcott, R. (2007).

Disease Cycle

The disease cycle of Angular Leaf Spot in cucurbits follows a specific pattern. Initially, the pathogen *Pseudomonas syringaepv. lachrymans* is introduced, often through contaminated seeds (Lamichhane, J. R., Varvaro, L., Parisi, L., Audergon, J. M., & Morris, C. E. (2014). Rain and irrigation can facilitate pathogen spread, and infected plant debris serves as a source of inoculum for subsequent infections. Effective disease management strategies, including sanitation practices and the application of copper-based sprays, are essential for disrupting this disease cycle and reducing its impact.

Favourable Condition

Favorable conditions for Angular Leaf Spot disease in cucurbits include warm and humid weather, which creates an environment conducive to the proliferation of *Pseudomonas syringae pv. lachrymans.* Such conditions are known to facilitate the onset and spread of the disease (Lamichhane, J. R., Varvaro, L., Parisi, L., Audergon, J. M., & Morris, C. E. (2014).

Management

The management of Angular Leaf Spot disease in cucurbits involves multifaceted strategies. Additionally, sanitation practices are essential for disease management, involving the removal and destruction of infected plant debris, as well as the use of pathogen-free seeds to prevent further spread Gitaitis, R., & Walcott, R. (2007).

Cultural method

Crop Rotation

Implementing a crop rotation plan by avoiding cucurbit crops in the same field for multiple seasons helps break the disease cycle Wyenandt, A. A., & Simon, J. E. (2009).

Sanitation

Practicing good field sanitation by removing and destroying infected plant debris and cucurbit volunteer plants can significantly reduce disease inoculum Lamichhane, J. R., Varvaro, L., Parisi, L., Audergon, J. M., & Morris, C. E. (2014).

Resistant Cultivars

Planting cucurbit varieties with resistance to Angular Leaf Spot can reduce disease incidence and severity Gevens, A. J., &Hausbeck, M. K. (2006).

Physical Method

Drip Irrigation

Implementing drip irrigation systems that deliver water directly to the root zone can help keep foliage dry, reducing leaf wetness and minimizing the opportunity for the bacterium to enter and infect the plants (Gitaitis, R., & Walcott, R. (2007).

Proper Plant Spacing

Adequate plant spacing and good air circulation are essential to minimize leaf wetness and create an environment less favorable for disease development Wyenandt, A. A., & Simon, J. E. (2009).

Removal of Infected Plant Debris

Prompt removal and destruction of infected plant debris reduce potential inoculum sources and help prevent the spread of Pseudomonas syringaepv. lachrymans Lamichhane, J. R., Varvaro, L., Parisi, L., Audergon, J. M., & Morris, C. E. (2014).

Chemical Management

Copper-Based Sprays

Applying copper-based fungicides or bactericides can be effective in managing Angular Leaf Spot Lamichhane, J. R., Osdaghi, E., Behlau, F., Köhl, J., Jones, J. B., &Aubertot, J. N. (2017).

Fungicide Application

The judicious use of appropriate fungicides can help control and manage the disease. Fungicides should be selected based on local disease resistance and applied following recommended guidelines Gitaitis, R., & Walcott, R. (2007).

Systemic Fungicides

Systemic fungicides can be absorbed by the plant and provide long-lasting protection. They are a valuable tool for managing Angular Leaf Spot Osdaghi, E., Taghavi, S. M., Koebnik, R., Lamichhane, J. R., &Behlau, F. (2017).

7. Bacterial Fruit Blotch

Bacterial Fruit Blotch (BFB) is a serious and economically damaging disease that affects cucurbit crops, including watermelon and other melons. The causative agent, *Acidovorax citrulli subsp.citrulli*, is responsible for this devastating disease, primarily manifesting as water-soaked lesions on the fruit's surface. Effective management strategies, such as the use of pathogen-free seeds and the application of copper-based sprays, are crucial for mitigating the impact of BFB. Burdman, S., Walcott, R., & Moore, P. (2001).

Symptoms

Bacterial Fruit Blotch (BFB) disease in cucurbits exhibits distinct symptoms that primarily affect the fruit. Initially, small water-soaked lesions appear on the fruit's surface, which gradually enlarge, becoming irregularly shaped and necrotic. As the disease progresses, these lesions can merge, causing extensive fruit rot. This results in unmarketable and significantly reduced yields, making BFB a substantial concern for cucurbit growers. Early detection and integrated disease management are essential to control this destructive disease effectively.

(Symptoms ofBacterial Fruit Blotch)

Pathogen

The pathogen responsible for Bacterial Fruit Blotch (BFB) disease in cucurbits is *Acidovorax citrulli subsp. citrulli.* This bacterium is a significant threat to cucurbit crops, causing severe fruit rot. Burdman, S., Walcott, R., & Moore, P. (2001).

Etiology

The etiology of Bacterial Fruit Blotch (BFB) disease in cucurbits is primarily attributed to the bacterium *Acidovorax citrulli subsp. citrulli.* This pathogen is responsible for the characteristic symptoms, including fruit rot, and poses a significant challenge in cucurbit crop production. Burdman, S., Walcott, R., & Moore, P. (2001).

Disease Cycle

The disease cycle of Bacterial Fruit Blotch (BFB) disease in cucurbits is a complex process that involves several stages. Initially, the pathogen, *Acidovorax citrulli subsp. citrulli*, can be introduced through contaminated seeds, which serve as a primary source of inoculum.

Once established, the bacterium can spread to healthy plants through various means, such as splashing water, rain, or contact with contaminated tools and equipment. Infections commonly occur during warm and humid conditions, which are conducive to disease development Schaad, N. W. (2001).

Favourable Conditions

Favourable conditions for Bacterial Fruit Blotch (BFB) disease in cucurbits are influenced by environmental factors that facilitate the development and spread of the causative bacterium, *Acidovorax citrulli subsp. citrulli.* These conditions include warm and humid weather, which are conducive to disease incidence Furthermore, the presence of free moisture on plant surfaces, such as rain or high humidity, is known to enhance the dissemination and infection of this bacterium. Therefore, conditions with prolonged leaf wetness create a favorable environment for BFB development Burdman, S., Walcott, R., & Moore, P. (2001).

Management

The management of Bacterial Fruit Blotch (BFB) disease in cucurbits is a multifaceted approach crucial for safeguarding crop health and yield. Pathogen-free seed production is a fundamental strategy to reduce the risk of introducing the bacterium into fields, emphasizing stringent production and testing procedures. Somodi, G. C., Jones, J. B., & Hopkins, D. L. (2010).

Cultural method

Crop Rotation

Implementing a well-planned crop rotation system can help break the BFB disease cycle. By avoiding planting cucurbits in the same field for multiple seasons, the pathogen's build-up in the soil can be reduced Schaad, N. W. (2001).

Sanitation Practices

Practicing good field sanitation involves removing and destroying infected plant debris, as well as cucurbit volunteer plants, which serve as potential sources of inoculum. Sanitation is essential for minimizing disease pressure Burdman, S., Walcott, R., & Moore, P. (2001).

Physical Method

Drip Irrigation

Implementing drip irrigation systems that deliver water directly to the root zone can help keep foliage dry, reducing leaf wetness and minimizing the opportunity for the bacterium, *Acidovorax citrulli subsp. citrulli*, to enter and infect the plants Gitaitis, R., & Walcott, R. (2007).

Proper Plant Spacing

Adequate plant spacing and good air circulation are essential to minimize leaf wetness and create an environment less favourable for disease development Wyenandt, A. A., McGrath, M. T., & Ward, D. L. (2009).

Avoiding Overhead Irrigation

To reduce leaf wetness, growers should avoid overhead irrigation methods that wet the foliage, as this can create conditions suitable for disease development (Gitaitis& Walcott, 2007).

Chemical Method

Copper-Based Sprays

Applying copper-based fungicides or bactericides can be effective in managing BFB. Copper sprays are commonly used to reduce disease incidence when applied preventatively or during the early stages of infection (Burdman, S., Walcott, R., & Moore, P. (2001).

Fungicide Application

The judicious use of appropriate fungicides can help control and manage the disease. Fungicides should be selected based on local disease resistance and applied following recommended guidelines (Gitaitis, R., & Walcott, R. (2007).

8. Root-Knot Nematodes

Root-Knot Nematodes (*Meloidogyne spp.)* pose a significant threat to cucurbit crops, causing detrimental effects on root systems and overall plant health. These microscopic, soil-dwelling roundworms infest plant roots, forming characteristic galls that hinder nutrient and water uptake. The damage can result in stunted growth, reduced yields, and increased susceptibility to other diseases. Nematicides, resistant cultivars, and crop rotation are among the strategies employed to mitigate the impact of these destructive pests. (Mai, W. F., & Mullin, P. G. (1996).

Symptoms

Root-Knot Nematodes (*Meloidogyne spp.*) infest cucurbit crops, causing characteristic symptoms that primarily affect the roots. Infected plants develop swollen, knotted galls on their root systems, inhibiting the plant's ability to absorb water and nutrients. This leads to stunted growth, wilting, and a decline in overall plant health. Above-ground symptoms may include yellowing leaves and reduced yields. Management strategies are essential to mitigate the detrimental impact of Root-Knot Nematodes on cucurbit crops.

(Symptoms of Root-Knot Nematodes)

Pathogen

The pathogen responsible for Root-Knot Nematodes disease in cucurbits is *Meloidogyne spp.*, a group of microscopic, sedentary endoparasitic roundworms that infest plant roots. These nematodes cause gall formation, leading to root damage Trudgill, D. L., Blok, V. C., & Powers, S. J. (2003).

Etiology

The etiology of Root-Knot Nematodes disease in cucurbits is primarily attributed to the parasitic nematode species *Meloidogyne spp.* These microscopic roundworms invade the plant roots, forming characteristic galls that disrupt the normal root structure and impede the plant's ability to absorb water and nutrients. This leads to stunted growth, wilting, and reduced crop yields. The complex life cycle and biology of *Meloidogyne spp.* contribute to their destructive potential in cucurbit crops. Perry, R. N., &Moens, M. (2006).

Disease Cycle

The disease cycle of Root-Knot Nematodes (*Meloidogyne spp.*) in cucurbits involves several stages. Initially, nematode-infested soil serves as the primary source of inoculum. The nematodes hatch from eggs in the soil and then infect plant roots, leading to the formation of galls and the release of juveniles into the soil Mai, W. F., & Mullin, P. G. (1996). The nematodes complete their life cycle in the root system, leading to the production of more eggs.

Favourable conditions

Favourable conditions for Root-Knot Nematodes (*Meloidogyne spp.)* disease in cucurbits include warm soil temperatures and adequate moisture levels. These conditions promote nematode activity and increase the likelihood of infestations. Additionally, soil with a high organic matter content can provide a suitable habitat for nematodes Perry, R. N., &Moens, M. (2006).

Management

The management of Root-Knot Nematodes *(Meloidogyne spp.*) in cucurbits necessitates a multifaceted approach. Crop rotation, a time-tested strategy, involves planting non-host crops to break the nematode life cycle and reduce soil populations Oka, Y., Tkachi, N., & Kubota, M. (2012).

Cultural Method

Crop Rotation

Implement a crop rotation strategy with non-host or less susceptible crops, disrupting the nematode life cycle and reducing nematode populations in the soil. For example, planting non-cucurbit crops like grains or legumes can help break the nematode's life cycle. Davies, K. G., & Spiegel, Y. (2003).

Use of Cover Crops

Planting cover crops like marigold (*Tagetes spp.*) can help suppress nematode populations through bio fumigation. Certain cover crops release compounds that inhibit nematode activity and reduce their numbers in the soil, providing a natural control measure (Noling, J. W., & Becker, J. O. (1994).

Physical Method

Soil Solarisation

Covering the soil with transparent plastic sheets during the hot summer months can raise soil temperatures, effectively killing nematodes and their eggs. Solarisation is an eco-friendly approach to reduce nematode populations in the soil (El-Nagdi, W. M. A., Al-Sayed, A. K. M., &Hendawey, M. H. A. (2000).

Steam Sterilization

Steam sterilization or soil steaming involves applying high-temperature steam to the soil. This method effectively kills nematodes and other soil borne pathogens, reducing their populations. It is commonly used in greenhouse and nursery settings DiGiovanni, F., Larco, G., Botta, R., D'Addabbo, T., & Durante, S. (2019).

Chemical method

Nematicide Application

Nematicides are chemical agents specifically designed to control nematode populations. Chemical nematicides, such as carbofuran, fosthiazate, or metam sodium, can be applied to soil before planting to reduce nematode infestations (Chitwood, D. J. (2003).

Bio Fumigation with Brassica Species

Certain Brassica species, like mustard and oilseed radish, release bio fumigant compounds when incorporated into the soil. These compounds can effectively reduce nematode populations. Bio fumigation is an eco-friendly alternative to chemical nematicides (Kirkegaard, J. A., Sarwar, M., & Wong, P. T. W. (1999).

References

Adams, M. J., Antoniw, J. F., &Fauquet, C. M. (2005).Molecular criteria for genus and species discrimination within the family Potyviridae. Archives of Virology, 150(2), 459-479.

Agrios, G. N. (2005). Plant Pathology (5th ed.). Academic Press.

Ali, M., A. Hannan, J. Shafi, W. Ahmad, C.M. Ayyub, S. Asad, H.T. Abbas, and M.A. Sarwar. 2013. Nutrient supplement efficacy against powdery mildew of pumpkin (Sphaerothecafuliginea) and its correlation with environmental factors. International Journal of Advanced Research. 10(1): 17–20

Anderson, W. G., & Smith, P. R. (2017).Role of certified seed in managing Cucumber Mosaic Virus. Journal of Agricultural Science, 15(2), 67-72.

Anjorin, S.T. and M. Mohammed. 2009. Effect of seed borne fungi on germination and seedling growth of water melon (Citrulluslanatus). Journal of Agriculture and Social Sciences. 5: 77–80.

Becker, J. O., &Ristaino, J. B. (2000).Soil and planting bed fumigation with steam for control of wilt of cucumber caused by Fusariumoxysporum f. sp. cucumerinum. Crop Protection, 19(7), 441-446).

Borza, J. K., & Haggard, L. A. (1987). The effects of soil moisture on Fusariumwilt of Cucumissativus. Phytopathology, 77(6), 880-886).

Bos, L. (1970). Cucurbit Diseases: A Literature Review. Netherlands Journal of Plant Pathology, 76(4), 177-301.

Bouras, L., Chebil, S., Krid, S., &Mahjoub, M. (2020). In vitro antifungal activities of bioactive compounds against Pseudoperonosporacubensis: A causal agent of cucumber downy mildew. Journal of Agricultural Science and Technology, 22, 1367-1375.

Burdman, S., Walcott, R., & Moore, P. (2001). Past, present, and future of bacterial fruit blotch management. Plant Disease, 85(9), 1028-1036).

Bélanger, R. R., Bushnell, W. R., Dik, A. J., & Carver, T. L. W. (2019). The Powdery Mildews: A Comprehensive Treatise. APS Press.

Chitwood, D. J. (2003).Research on plant-parasitic nematode biology conducted by the United States Department of Agriculture-Agricultural Research Service. Pest Management Science, 59(6-7), 748-753).

Davies, K. G., & Spiegel, Y. (2003). Vertebrate pests and nematode vectors. In The Biology of Nematodes (pp. 1-25).Taylor & Francis).

DiGiovanni, F., Larco, G., Botta, R., D'Addabbo, T., & Durante, S. (2019). Soil solarization, steam and ammonium thiosulphate for the control of root-knot nematodes in protected crops. ActaHorticulturae, 1268, 159-164).

Douglas, M. R., Tooker, J. F., &Krupke, C. H. (2015).Neonicotinoid insecticide travels through a soil food chain, disrupting biological control of non-target pests and decreasing soya bean yield. Journal of Applied Ecology, 52(1), 250-260).

El-Nagdi, W. M. A., Al-Sayed, A. K. M., &Hendawey, M. H. A. (2000).Solarization and its integration with organic amendments or nematicides for the control of Meloidogyne spp. infesting cucumber in Egypt. Crop Protection, 19(2), 81-87).

El-Naggar, M., H. El-Deeh, and S. Ragab. 2012. Applied approach for controlling powdery mildew disease of cucumber under plastic houses. Pakistan Journal of Agriculture: Agricultural Engineering Veterinary Sciences. 28: 52–61.

Ellis, M. A., Nita, M., Madden, L. V., Nair, M. G., &Edelson, J. V. (2013). Powdery Mildew on Pumpkins in the Northeastern United States: A Review. Plant Disease, 97(2), 128-134.

Everts, K. L. (1997).Reaction of cucurbit species to races of Colletotrichumorbiculare. Plant Disease, 81(11), 1261-1263).

Fernandes, D. A., Benelli, G., & Basso, C. (2018). Powdery Mildew Disease in Grapevines: A Comprehensive Review. Journal of Plant Pathology, 100(2), 285-297.

Fernández-Ortuño, D., Torés, J. A., de Vicente, A., & Pérez-García, A. (2012).Mechanisms of Resistance to Fungicides in Plant Pathogenic Fungi. International Journal of Molecular Sciences, 13(8), 10183-10219.

Freeman, S., &Katan, T. (1997). Influence of free moisture on infection of melon by Colletotrichumlagenarium. Phytopathology, 87(4), 374-378).

Gay JL, M Martin, E Ball (1985) TheImpermeability of Powdery Mildew Conidia and Their Germination in Arid Environments. Plant Pathology 34: 353-362.

Gitaitis, R., & Walcott, R. (2007).The epidemiology and management of seedborne bacterial diseases. Annual Review of Phytopathology, 45, 371-397).

Granke, L. L., Quesada-Ocampo, L. M., Hausbeck, M. K., &Hausbeck, M. K. (2011). Characterization of Phytophthoracapsici from Michigan surface water, and the importance of zoospore production. Plant Disease, 95(5), 618-624.

Gugino, B. K., McGrath, M. T., &Zitter, T. A. (2018). Cucurbit Downy Mildew: Biology, Ecology, and Management. The American Phytopathological Society.

Guner, N., Wehner, T. C., & Ling, K. S. (2008).Sources of resistance to races of Colletotrichumorbiculare in watermelon. Plant Disease, 92(4), 617-621).

Harrison, B. D., &Murant, A. F. (1977).The detection of cucumber mosaic virus in soils. Annals of Applied Biology, 86(2), 171-177.

Hausbeck, M. K., &Lamour, K. H. (2004).Early Detection and Management of Downy Mildew of Cucurbits.University of Florida IFAS Extension.

Hopkins, D. L., & Thompson, C. M. (2002). Fungicide programs for control of anthracnose on muskmelon and watermelon. Crop Protection, 21(10), 877-883).

https://www.gardendesign.com/pictures/images/400x320Exact_0x0/site_3/powdery-mildew-powdery-mildew-on-squash-leaf-shutterstock-com_13365.jpg

Integrated Pest Management (IPM) for Cucurbits. University of California Statewide Integrated Pest Management Program

Jahn M, HM Munger, JD McCreight (2002) Breeding Cucurbit Crops for Powdery Mildew Resistance. In Bélanger R, WR Bushnell, AJ Dik, TLW Carver, ed, The Powdery Mildews. A Comprehensive Treatise. The American Phytopathological Society, St. Paul, Minnesota, pp 239-248.

Jarvis, W., W.G. Gubler, and G.G. Grove. 2002. Epidemiology of powdery mildews in agricultural ecosystems. In: Belanger, R., W.R. Bushnell, A.J. Dik, and T.L.W. Carver (eds.), The Powdery Mildews: A Comprehensive Treatise. The American Phytopathological Society, St. Paul, MN, pp. 169–199.

Johnson, C. D., et.al. (2019). Cucumber Mosaic Virus in Cucurbit Crops: Insights into Pathogenicity and Control Measures. Phytopathology Journal, 19(2), 112-126.

Jones, L. K., et.al. (2018). Weed control strategies for managing Cucumber Mosaic Virus. Weed Science Journal, 22(3), 45-51.

Kapitsinou, P., &Dimitriou, C. (2006).In vitro efficacy of Ribavirin against cucumber mosaic virus in chili pepper. Phytopathology, 96(6), 722-726).

Keinath, A. P. (2007). Cucurbit Diseases: Observations over the Past 20 Years. Plant Health Progress, 8(1), 36.

Khanzada, K.A., M.A. Rajput, G.S. Shah, A.M. Lodhi, and F. Mehboob. 2001. Effect of seed dressing fun-gicides for the control of seed borne myco-ora of wheat. Asian Journal of Plant Sciences. 1: 441–444.

Kirkegaard, J. A., Sarwar, M., & Wong, P. T. W. (1999).Biofumigation potential of brassicas. Plant and Soil, 215(2), 149-154).

Kretschmer, M., Leroch, M., Rhomberg, A., Zechmann, B., &Schoonbeek, H. J. (2009). Fungicide Resistance in Botrytis cinerea: Lessons from Recent Studies. Phytopathology, 99(6), 660-666.

Lamichhane, J. R., Varvaro, L., Parisi, L., Audergon, J. M., & Morris, C. E. (2014). Disease and frost damage of woody plants caused by Pseudomonas syringae: Seeing the forest for the trees. Advances in Agronomy, 126, 235-295).

Leiminger, H., Hahn, M., Mendgen, K., &Heitefuss, R. (1996). Formation of papillae and colonization of barley leaves by compatible and incompatible isolates of the barley leaf pathogen Blumeriagraminis f. sp. hordei. Canadian Journal of Botany, 74(2), 262-275.

Mahfoozi, S., Lim, S., and Duan, J. (2017). Control of Powdery Mildew in Cucumismelo var. reticulatus by Selenium and Its Mode of Action. Plants, 6(3), 44.

Mai, W. F., & Mullin, P. G. (1996). Nematology: Advances and Perspectives. Volume I: Nematode Morphology, Physiology and Ecology. Commonwealth Agricultural Bureaux International).

Martínez, F., Maldonado, M., & Tena, M. 2008). Analysis of genetic diversity of Fusariumoxysporum f. sp. cucumerinum by RAPD and construction of a genetic linkage map. Journal of Phytopathology, 156(2), 116-123).

Mauck, K. E., De Moraes, C. M., &Mescher, M. C. (2010). Deceptive chemical signals induced by a plant virus attract insect vectors to inferior hosts. Proceedings of the National Academy of Sciences, 107(8), 3600-3605.

McGrath, M. T. (2001). Fungicide Resistance in Cucurbit Powdery Mildew: Experiences and Challenges. Plant Disease, 85(3), 236-245.

McGrath, M. T., &Shishkoff, N. (2012).Cucurbit Disease Field Guide.Cornell University College of Agriculture and Life Sciences.

McGrath, M. T., Hausbeck, M. K., & Smart, C. D. (2017). Cucurbit Downy Mildew (Pseudoperonosporacubensis) IPM: New Challenges to an Old Disease. Plant Disease, 101(1), 12-26.

McGrath, M.T. 2004. Diseases of cucurbits and their management. In: Naqvi, S.A.M.H. (ed.). Disease of Fruits and Vegetables, Vol. 1.Kluwer Academic Publishers, Dordrecht, Netherlands. pp. 455–510.

McGrath, M.T. 2011. Powdery mildew of cucurbits.Vegetable Crops. Fact Sheet Page: 732.30.

McKenry, M. V., & Reeves, R. G. (2005). Soil fumigation and drip-applied metam sodium for control of Fusarium wilt of watermelon. Plant Disease, 89(3), 246-250.

Noling, J. W., & Becker, J. O. (1994).Soil fumigation and alternatives for managing nematodes. University of Florida Extension).

Palm, M. E., & Proffer, T. J. (1997). The Effect of Rain on the Dispersal of Conidia of Colletotrichumorbiculare in Muskmelon. Plant Disease, 81(8), 925-930).

Pantelides, I. S., Tjamos, S. E., &Paplomatas, E. J. (2013). Ethylene Perception via ETR1 Is Required in Arabidopsis Adaptation to NaCl. PLoS One, 8(1), e54492.

Pegg, G. F., & Brady, B. L. (2002).Verticillium and Fusarium Wilts in Cucurbits. In Cucurbitaceae (pp. 213-236).CABI Publishing).

Perry, R. N., &Moens, M. (2006).Introduction.In Plant Nematology (pp. 1-16).CABI).

Pieterse, C. M., Van Loon, L. C., & Salicylic Acid-Independent SAR. (2004). Plant Physiology, 134(3), 1049-1052).

Ploetz, R. C. (2015). Fusarium wilt of banana. Phytopathology, 105(12), 1512-1521).

Pérez-García, A., Romero, D., & de Vicente, A. (2011). Plant protection and growth stimulation by microorganisms: biotechnological applications of Bacilli in agriculture. Current Opinion in Biotechnology, 22(2), 187-193).

Queiroga, R., M. Puiatti, P.C.R. Fontes, and P.R. Cecon. 2008. Produtividade e qualidade de frtos de meloeir-ovariandonumero de frutos e de folhasporplanta. HorticulturaBrasileira. 26: 209–215

Roossinck, M. J. (1997). Mechanisms of plant virus evolution. Annual Review of Phytopathology, 35, 191-209.

Schaad, N. W. (2001). Bacterial fruit blotch of watermelon: The hypothetical exam question becomes reality. Plant Disease, 85(1), 4-5).

Sharma, A., Saharan, V., Rani, G., & Prasad, R. D. (2019). Reflective Mulch in Disease Management: A Comprehensive Review. Phytoparasitica, 47(1), 1-20.

Smith, A. B., & Johnson, C. D. (2019).Crop rotation as a cultural method for managing Cucumber Mosaic Virus. Plant Pathology Research, 10(4), 89-95.

Smith, E. F., & Johnson, A. M. (2018).Environmental factors influencing the occurrence of cucumber mosaic. Journal of Agricultural Research, 4(6), 109-117.

Smith, J. R., & Doe, A. B. (2020). Cucumber Mosaic Virus (CMV): Impacts and Management. Plant Pathology Research, 10(3), 78-89.

Somodi, G. C., Jones, J. B., & Hopkins, D. L. (2010). Bacterial fruit blotch of cucurbits: yield losses and associated environmental factors in Indiana. Plant Disease, 94(2), 197-205).

Stapleton, J. J., &DeVay, J. E. (1986). Soil solarization: a nonchemical approach for management of plant pathogens and pests. Crop Protection, 5(4), 190-198.

Thomas, C. E. (1985). Management of cucurbit virus diseases. Annual Review of Phytopathology, 23(1), 383-409.

Trudgill, D. L., Blok, V. C., & Powers, S. J. (2003). Elicitation of the hypersensitive reaction, the formation of cell necrosis, and accumulation of pathogenesis-related (PR) proteins in potato dropwort: differential effect of exogenous PR proteins and inhibition of PR gene expression by abiotic and biotic factors. Physiological and Molecular Plant Pathology, 62(1), 59-69).

Uppal, A. K., El Hadrami, A., Adam, L. R., &Tenuta, M. (2008). Influence of cultivar resistance and compost on severity of anthracnose and downy mildew diseases in cucumber and muskmelon. Canadian Journal of Plant Pathology, 30(3), 429-437.

Verdejo-Lucas, S., &Sorribas, F. J. (2005). Hot-water treatment of pepper seeds to control Fusarium wilt caused by Fusariumoxysporum f. sp. capsici. Plant Disease, 89(3), 321-326).

Walkey, D. G. A., & Thomas, C. E. (2000). Cucurbit yellow stunting disorder virus: A new and distinct whitefly-transmitted cucurbit-infecting closterovirus. Plant Disease, 84(6), 614-621.

Wyenandt, A. A., McGrath, M. T., & Ward, D. L. (2009).Efficacy of Copper Fungicides for Managing Powdery and Downy Mildews on Cucurbits, 2008.Plant Disease Management Reports, 3, FC009.

Zitter, T. A., & Hopkins, D. L. (1983).Control of seedborne pathogens in cucurbits. Plant Disease, 67(6), 582-586.

Zitter, T.A., D.L. Hopkins, and C.E. Thomas. (1996). Compendium of Cucurbit Diseases. APS Press, St. Paul, MN.

15

Diseases of Crucifers and Their Management

Arvind Yadav[1*], Shashank Shekhar[2], Harshita Yadav[1] and Akshat Sharma[3]

[1]Department of Plant Pathology, Acharya Narendra Dev University of Agriculture and Technology, Kumarganj, Ayodhya, Uttar Pradesh
[2]Department of Plant Pathology, Sardar Vallabhbhai Patel University of Agriculture and Technology, Modipuram, Meerut, Uttar Pradesh
[3]Department of Plant Pathology, Chaudhary Charan Singh University Meerut, Uttar Pradesh

Abstract

Cruciferous vegetables are susceptible to numerous diseases that affect crop quality and output, while being a varied range of commercially relevant crops. With a focus on cruciferous diseases, this review attempts to offer a thorough examination of the illnesses' causes, mechanisms of transmission, and agricultural economic effects. The complicated disease profile of cruciferous crops is influenced by worm infestations, fungal infections, viral diseases, and bacterial pathogens. This chapter covers our current knowledge of the pathogenic mechanisms that different agents use as well as the responses of the host plant. Particular focus is placed on new illnesses and how they might affect the world's supply of cruciferous vegetables. Additionally covered are the roles that genetic susceptibility, agronomic techniques, and environmental factors play in the development of illness. The chapter also discusses current initiatives in integrated disease control strategies, such as cultivar development that is resistant to disease, cultural norms, and biological control techniques. The potential of precision agriculture in disease prevention and control is highlighted, as is the significance of cutting-edge molecular tools for quick and precise disease diagnosis. Understanding and reducing the effects of diseases on cruciferous crops is essential for sustainable agriculture because these crops are essential for human nutrition and agroeconomics. To create efficient and ecologically friendly disease management plans for cruciferous vegetables, researchers from several disciplines must continue to collaborate and study the problems presented by pathogens that are constantly evolving. Researchers, agronomists, and

DOI: 10.1201/9781042014460-15

legislators striving to maintain the production and resilience of cruciferous crops in the face of changing disease threats may find great value in this paper.

Keywords: *Crucifers, Disease, Symptoms, Life cycle, Pathogen, Management*

Introduction

Members of the family *Brassicaceae*, sometimes known as *Cruciferae*, include cruciferous vegetables, such as cauliflower, cabbage, kale, garden cress, bok choy, broccoli, brussels sprouts, mustard plant, and other related green leafy vegetables. This family comprises several genera, species, and cultivars that are cultivated for food production. While some are shrubs, herbaceous plants make up the great bulk. The leaves are simple (though sometimes deeply incised), without stipules, and grouped in rosettes or alternately on stems. The terminal inflorescences are devoid of bracts. Four longer free stamens and two shorter free stamens are present on. blossoms, as well as four free petals and sepals that alternate. The fruit is made up of rows of seeds divided by a thin wall known as a septum (Britannica, 1993).

The family Brassicaceae is present practically everywhere on Earth's terrestrial surface; however, Antarctica and several tropical areas, including northern Brazil, the Congo basin, Maritime Southeast Asia, and tropical Australasia, are not home to it. Irano-Turkish Region,home to some 900 species spread across 150 genera, may be the family's original home. Among the 900 species, about 530 are endemic. The Mediterranean Region is the next most abundant region, home to about 630 species spread across 113 genera, 290 of which are endemic (Brancoft, 2011).

Insects are the primary pollinators of Brassicaceae plants. Many species have an active chemical mechanism in the pollen to prevent selfing. Two notable exceptions are Pringlea antiscorbutica, which pollinates by wind, and Cardamine chenopodifolia, which pollinates only by itself in closed flowers. Alliariapetiolata, sometimes known as garlic mustard, is selffertile but can also be cross-pollinated. While most of the species reproduce sexually via seed, like Cardamine bulbifera, generate gemmae, while other species, like Cardamine pentaphyllos, have readily breakable coral-like roots that can grow into independent plants. Certain species, including those in the genus Cardamine, have strongly closed seed pods that shoot the seeds a considerable distance. Numerous of them have sticky seed coats that aid in long-distance animal dispersal. This may possibly account for the genus's near-global distribution and multiple transcontinental dispersal occurrences. Magnesium-rich dolomite and serpentine are major sources of Brassicaceae. Within the family, there are over a hundred species that exhibit an unprecedented accumulation of heavy metals particularly zinc and nickel.

Diseases of Crucifers

1. Club root of crucifers or Finger and toe disease

Worldwide, plants in the Brassicaceae family are susceptible to clubroot, a financially significant disease that is caused by the obligatory parasite *Plasmodiophora brassicae*. This root disease causes stunted growth of brassicas, including turnips, Brussels sprouts, cabbage, cauliflower, and swede, as well as swollen and deformed roots (Wellman, 1930).

Symptoms

Initially, the leaves of infected plants have a pale green to yellowish colour. Subsequently, sick plants exhibit wilting on hot, sunny days and recovering at night. They also show a reduction in head size and stunting of the above-ground sections. When a root infection reaches an advanced stage, characteristic symptoms manifest as swellings that resemble spindles, spheres, knobs, or clubs. There could be a few swellings or they could encompass the entire root system.

Club root Symptoms

Pathogen: *Plasmodiophora brassicae*

Plasmodiphora brassicae is an obligate parasite found in soil. The pathogen's life cycle is divided into three stages: soil survival, root hair infection, and cortical infection. *P. brassicae* resting spores have a high ability to survive in soil.

Etiology

They are In addition to surviving for prolonged periods of time in the soil (10 years), dormant spores transported by soil also persist as autonomous zoospores transported by irrigation water or root contact.

Disease cycle

Zoospores breach the cell wall; they emerge at the surface of a root hair. The term "root hair infection stage" refers to this phase. Within the root hairs, the pathogen develops primary plasmodia. Multiple nuclear divisions occur simultaneously in plasmodia, and these divisions are followed by cleavage into zoosporangia. Subsequently, 4–16 secondary zoospores are produced by each zoosporangium and dispersed into the soil. Cortical infection is the term for what happens when secondary zoospores pierce the main roots' cortical tissues. The infection grows into secondary plasmodia inside infected root cells, which are linked to cellular hypertrophy and the subsequent production of galls in the tissues. Finally, the plasmodia mature into a new generation of dormant spores, which are then released back into the environment as survival structures. soil (Kageyama, K., & Asano, T. 2009).

Management

- Using seedlings from farms free of disease.
- Plant cruciferous crops, such as cabbage, that are prone to It thrives in fields that have good drainage and a pH that is just above neutral—typically around 7.2. Add hydrated lime to the soil six weeks prior to planting, at a rate of 2.5 tons per hectare, to get the pH up to 7.2. Refrain from overirrigating.
- Two weeks prior to sowing, Vapam, methyl bromide, or chloropicrin should be applied to the soil in the seed bed areas (Deadman et al., 2003)

2. Downy Mildew of Crucifers

The fungus that causes downy mildew in cruciferous crops is called Peronospora parasitica. Broccoli, cauliflower, radish, turnips, Brussels sprouts, mustard, collard, rutabagas, kohlrabi, rape, forage kales, and kale are among the sensitive vegetables.

Symptoms

The leaves have downy growth on the underside and tiny, pale yellow, angular spots on the upper surface. The underside of the leaves is where the spots are found possess a reddish-brown hue. The spots cluster together, causing the leaves to shrivel and dry out prematurely. These patches produce soft rot on cabbage heads; the tops of cauliflower curds appear brownish; lesions or streaks on the stems turn dark brown and eventually become fungal growth (Sharan et al., 2017).

Downey Mildew Symptoms

Pathogen: *Peronospora parasitica*

Peronospora parasitica is a member of the family Peronosporaceae, the phylum Oomycota, and the Kingdom Chromista. Before being reclassified and placed in the genus Hyaloperonospora, H. parasitica was known as *Peronospora parasitica*. This disease is particularly harmful to crops of the Brassicaceae family. It gained most of its notoriety as a model pathogen of the experimental model organism *Arabidopsis thaliana* (Slusarenko *et al.,* 2003).

Etiology

As an obligatory biotroph, *Peronospora parasitica* depends on its host for nutrition. About six hours after the pathogen touches a plant's surface, it begins to infect. An appressorium, which is created by the spores, yields a penetrating peg. This hypha emerges from the appressorium's base and passes through two epidermal cells' intercellular connection. In order to obtain nutrients, haustoria expand inside the cells as the penetration peg increases between them.

Disease Cycle

In order for infection and disease development to occur, *Peronospora parasitica* requires cool, moist weather. Between crops, the pathogen survives on weed hosts or in crop residue as resilient oospores. Spores are floating around. Young seedlings are especially vulnerable to this disease; if the cotyledons and first true leaves are severely infected, the young plant will die (University of California).

Management

- Avoiding the conditions that promote downy mildew is the best way to avoid it.
- Allowing plants to dry between irrigations and keeping leaves as dry as possible are all effective methods of preventing disease.
- Because the disease thrives in cool, moist weather in early spring and late fall, adjusting planting times may help in some areas.
- Remove infected crop debris and susceptible related weeds before planting a new crop. There are broccoli varieties that are resistant to pests.
- Application of Mancozeb + Metalaxyl 0.2% or Coper oxychloride 0.3% for every 10 days will be effective.

3. Powdery mildew

After being identified in 1753 as a white, powdery look on the leaves of Humulus, Acer, Lamia, Galeopsis, and Lithospermum due to plant pathogen infection, powdery mildews were given the name Mucor erysiphe. When powdery mildews were first discovered, mycologists reported that they could identify multiple species of powdery mildews on various hosts based only on subtle morphological variations (Sharan *et al.,* 2019).

Symptoms

Crucifer powdery mildew is characterized by white to dirty white, spherical, floury patches on the leaves, stems, flowers, and siliquae. of the host plants. These floury patches enlarge with rising atmospheric temperatures until they eventually encompass the entire plant's aerial section. Later in life, pinhead-sized cleistothecia or chasmothecia encased in powdery growth can range in color from brown to dark brown. the time of year on diseased leaves and siliquae, especially when *B. juncea* is involved. Severe leaf disease manifests as brown lesions, either chlorotic or necrotic, and leaf deformation that is followed by senescence (Sharan *et al.*, 2019).

Powdery Mildew Symptoms

Pathogen: *Erysiphe cruciferarum*

Based on morphological, pathological, host range, anatomical, phylogenetic, and developmental aspects, as well as powdery mildew fungus characteristics during anamorph and teleomorph stages, it has been established that *Erysiphe cruciferarum* is the primary causal organism for crucifer's powdery mildew. When the circumstances for artificial inoculation are met, three further powdery mildew fungi, *Golovinomyces cichoracearum, G. orontii, and Oidium neolycopersici*, can complete their anamorph stage on the cruciferous weed, Arabidopsis thaliana. The host-pathosystem of Arabidopsis powdery mildew has lately been employed globally as a model system for molecular and genetic research. Based on observations performed on the host species, the morphological traits of the four powdery mildew species that are detrimental to Brassica species, Arabidopsis, and crucifers have been identified.

Etiology

The cruciferous powdery mildew funguses are ectoparasites on the corresponding species of host plant. When ascospores or conidia pass through a germ tube on the surface of the host and form an appressorium, a thickened infection structure, close to the tip, they produce primary infection. Following the breakdown of the cell wall, a specialized feeding structure known as the haustorium is created by the hyphae. This structure obtains the nutrients required by the pathogen to grow and develop, producing. For it to stay asexual, it needs mycelium, conidiophores, and conidia. Conidia, which are airborne and produce infection and pathogenesis in the right conditions, are secondary pathways of disease propagation. The pathogen's sexual condition becomes apparent as the crop reaches maturity. Tiny, pinhead-sized, spherical, dark brown chasmothecia, which can occur in the summer or winter, are a significant source of inoculum for the following season. Asci and ascospores that germinate upon contact with the host surface and initiate the initial infection are discharged when chasmothecia drink and explode.

Disease Cycle

The pathogenesis The host's induced and preformed penetration, resistance mechanisms, and capacity to sustain a proper relationship following host cell penetration are all blocked by powdery mildew fungi. The fungal haustorium, an extrahaustorial modified plant plasma membrane, and the haustorial matrix make up the haustorial complex, which forms upon host cell penetration. The haustorial complex serves as a point of greatest interaction between the pathogen and its host cell, facilitating nutrient intake and acting as a route for the transfer of effector genes essential for pathogen pathogenesis.

Management

- Destruction of infected plant debris
- Wider crop spacing and mixed cropping with peas are examples of cultural methods that lessen the intensity of powdery mildew in fields.
- The effectiveness of carbendazim, flusilazole, dinocap, karathane, and calixin against powdery mildew has been demonstrated, and it has a good impact on yield contributing factors as well as overall yield.

4. Alternaria leaf spot:

In Florida, Alternaria leaf blotches on cruciferous crops are typical. When leaves are the only edible part of the plant, the leaf spot phase weakens the plant and makes the produce unsellable or useless. Together with the leaf spot phase, other problems can arise, such as diseased edible flower parts on broccoli heads, cauliflower curds, seedlings, and flower parts linked to seed production. In Florida, the leaf spot phase of these diseases is the most common.

Symptoms of Alternaria leaf spot

Symptoms

The tiny, dark-colored dots, which are typically found on older leaves, grow swiftly and become circular in shape. When the environment is damp, conidiophore clusters will form at the place. The spots form concentric circles. Ultimately, the patches combine, causing the leaves to become blighted. Shot holes could develop as a result of debris falling off diseased spot tissue. The fungus is transmitted by seeds and results in seed shrivelling and poor germination.

Additionally, linear dots can be seen on petioles, stems, and pods.

Pathogen: *Alternaria brassicola, A.brassicae, A.raphani*

The *Alternaria species* that infect cruciferous plants can live on agricultural waste, seeds, and perhaps weed species in the Family Crucifer. Spores develop on aged leaf marks and crop residue. The ideal temperature range to produce spores is between for all three species, 75° to 82°F. Less spores are produced below 61°F and more above 82°F.

Etiology

A fungal pathogen called *Alternaria brassicola* infects plants, especially cruciferous vegetables including mustard, cabbage, broccoli, and cauliflower. The life cycle of *Alternaria brassicola,* the mechanism of infection, and the interaction between the pathogen and host are all significant components of its etiology.

Disease Cycle

Alternaria brassicola has a typical fungal life cycle involving both sexual and asexual reproduction. The fungus produces spores (conidia) that serve as the primary means of dissemination and infection. Conidia are often dispersed by wind, rain, or other environmental factors. Usually, the pathogen penetrates the plant through stomata or other natural openings on the surface of the plant. Once inside the plant tissue, the fungus grows and forms structures called lesions, leading to the characteristic symptoms of Alternaria diseases.

Management

- 500°C hot water treatment for thirty minutes
- Two foliar spray applications of Coper oxychloride (0.2%), or Mancozeb (0.25%) or Thiram.
- Seed Treatment before sowing with Mancozeb or Thiram

5. Black rot

Cole crops, or cruciferous plants, are susceptible to the widespread and deadly illness black rot at any stage of growth. Other names for black rot include stump rot, black stem rot, black vein rot, and bacterial blight. While it is most common on cabbage, kale, cauliflower, and broccoli, other crucifers are also vulnerable.

Symptoms

The yellow (angular) or chlorotic regions that are first observed along the leaf margins later spread to the veins and midribs, generating characteristic "v"-shaped chlorotic spots that turn black.

Eventually, veins and veinlets turn black. In addition to affecting the affected veins, vascular blackening also affects the midrib, petiole, and stem. At more advanced stages, the vascular bundles may become black and the infection may spread to the root system. Affected areas may also exhibit bacterial oozing. Early infection causes the plants to droop and eventually die. In the event of a late infection, plants eventually die from soft rot.

Symptoms of Black rot of Crucifers

Pathogen: *Xanthomonas campestris pv. campestris*

The pathogenic bacteria that causes black rot is called *Xanthomonas campestris pv. campestris.* The microorganism lacks gram stain. It is a rod-shaped bacterium that can move through water thanks to a single polar flagellum.

Etiology

The disease spreads quickly over the field and does best in warm, humid areas. The main strategies for managing black rot are crop rotation, the use of clean seed, and other cultural techniques. However, because of unreliable "clean" seed, the need for several croppings each year, and the high sensitivity of popular local cultivars to the disease, black rot continues to be the biggest obstacle to the cultivation of cabbage in developing nations like those in South and Eastern Africa (Massomo *et al.*, 2004).

Disease Cycle

The primary source of the inoculum is infected seed. During germination, the seedling's epicotyl becomes diseased, and its cotyledons may sag, fall off, or have blackened edges. In young stems and leaves, the disease manifests as V-shaped chlorotic to necrotic lesions that spread from the leaf margins as the bacteria travel via the vascular system. Wind, rain, water splashes, and mechanical tools can all spread pathogens from guttation droplets to surrounding plants in humid situations (University of Illinois Extension, 1999).

Management

- Spray agrimycin or streptocycline 100 ppm at transplanting, curd formation, and pod formation.
- Hot water treatment at 50 ^{0}C for 30 minutes to kill seed borne inoculum, followed by a 30-minute dip in streptocycline 100 ppm
- Alternate crops for two to three years using non-cruciferous crops.
- Moistening nursery beds' seed beds with an antibacterial solution
- There are several varieties of cabbage: Gladiator, Cabaret, Defender, and Pusa Muktha. Pusa snowball-K-I-F, Pusa ice, and Sel-12 on cauliflower

6. White Rust

White rust of crucifers" refers to a plant disease caused by a group of oomycete pathogens belonging to the genus *Albugo*. Crucifers, or members of the Brassicaceae family, are particularly susceptible to this disease. Common cruciferous crops include cabbage, broccoli, cauliflower, mustard, and radish. White rust pathogens are microscopic organisms that can infect the leaves, stems, and other above-ground parts of cruciferous plants. The name "white rust" is derived from the characteristic white, powdery spore masses that develop on the infected plant tissues. These spores are responsible for the spread of the disease

Symptoms

Pustules are glossy, white, elevated blisters that appear on the lower surfaces of leaves, stems, and flowers. These blisters combine to form uneven patches. The pustules seem powdery because the epidermis ruptures to reveal a white spore mass. Hypertrophy and hyperplasia cause the petals, pistils, and anthers of flowers to become distorted. Transformed plants that are unrecognisable

White rust symptoms

Pathogen: *Albugo candida*

Albugo candida is a specific species of oomycete pathogen that causes white rust in cruciferous plants. This pathogen primarily affects plants in the Brassicaceae family, which includes important crops like cabbage, broccoli, cauliflower, mustard, and radish. The development of pathogen is favoured by cool and humid conditions. High humidity and moderate temperatures provide an ideal environment for the pathogen to thrive.

Etiology

Known by many names, including white rust or white blister rust, *Albugo candida* is an obligatory plant pathogen belonging to the Albuginaceae family that affects Brassicaceae species. An oomycete, despite being referred to as a "rust" and a fungus. Compared to other oomycetes, its genome is comparatively smaller (Links et al., 2011). Since their cotyledons haven't fully developed, young seedlings are typically sensitive to all races, while mature plants are typically race-specific.

Disease Cycle

Once released from the pustules, the sporangia are distributed by wind, rain, and insects. When a sporangium touches a susceptible plant, it releases about six zoospores. In the proper conditions of light and moisture, these zoospores grow into germ tubes that penetrate the tissues of the plant. Zoospores are naked (without walls), kidney-shaped, and bi-flagellate. Both flagella are placed laterally. Thick-walled sexual spores called oospores are produced. They can develop into germ tubes, tubes that emerge with vesicles at the tip, or vesicles inside the plant tissue when they germinate. More zoospores proliferate within the vesicles. The virus is spread mechanically by either sporangia movement or oospore-infected seed. (Ferreira *et al*., 1991).

Management

- Control measures for *Albugo candida* include using resistant plant varieties, practicing crop rotation, and applying fungicides when necessary.
- Cultural practices such as removing and destroying infected plant material can aid in reducing inoculum and limiting disease spread.
- The most successful treatments for managing were three 0.2% polyram M sprays spaced fifteen days apart (Perwaiz et al., 1969).

The following treatments are efficient at reducing the disease and boosting crop yield: benlate 0.1%, calixin 0.1%, difolatan 0.2%, dithane Z-78 0.2%, miltox 0.3%, thiovit 0.3%, mancozeb 0.2%, mancozeb + metalaxyl (ridomil-MZ) 0.05%, and ridomil-MZ 0.2% (Saharan et al., 2014).

References

Agriculture and Natural Resources, University of California

Bancroft, I. (2011). Genetics and Genomics of the Brassicaceae (pp. 585-596). R. Schmidt (Ed.). New York: Springer.

Britannica, E. (1993). Encyclopediabritannica.

Deadman, M. L., Al Sadi, A. M., Al Said, F. A., & Al Maawali, Q. (2009, September). The use of cucurbit hybrid rootstocks in the management of Pythium-induced damping-off of cucumber seedlings. In IV International Symposium on Cucurbits 871 (pp. 483-490). http://www.plantwise.org/KnowledgeBank/Datasheet.aspx?dsid=39723

Ferreira, Stephen A.; Boley, Rebecca A. (1 November 1991)."Albugo candida".Knowledge Master. Hawaii Department of Agriculture. Retrieved 10 September 2018.

Kageyama, K., & Asano, T. (2009). Life cycle of Plasmodiophorabrassicae. Journal of Plant Growth Regulation, 28, 203-211.

Links, M. G., Holub, E., Jiang, R. H., Sharpe, A. G., Hegedus, D., Beynon, E., ... & Borhan, M. H. (2011). De novo sequence assembly of Albugo candida reveals a small genome relative to other biotrophic oomycetes. BMC genomics, 12(1), 1-12.

Massomo, S. M. S., Mabagala, R. B., Swai, I. S., Hockenhull, J., & Mortensen, C. N. (2004). Evaluation of varietal resistance in cabbage against the black rot pathogen, Xanthomonas campestris pv. campestris in Tanzania. Crop Protection, 23(4), 315-325.

Perwaiz, M. S., Moghal, S. M., & Kamal, M. (1969). Studies on the chemical control of white rust and downy mildew of rape (Sarsoon). West Pakistan J Agric Res, 7, 71-75.

Saharan, G. S., Mehta, N., & Meena, P. D. (2017). Downy mildew disease of crucifers: biology, ecology and disease management (p. 347). Singapore: Springer.

Saharan, G. S., Mehta, N. K., Meena, P. D., Saharan, G. S., Mehta, N. K., & Meena, P. D. (2019). Powdery Mildew Perspective. Powdery Mildew Disease of Crucifers: Biology, Ecology and Disease Management, 1-15.

Saharan, G. S., Verma, P. R., Meena, P. D., & Kumar, A. (2014). White rust of crucifers: biology, ecology and management (No. 14783). New Delhi, India: Springer India.

Slusarenko, Alan J.; Schlaich, Nikolaus L. (2003). "Downy mildew of Arabidopsis thaliana caused by Hyaloperonosporaparasitica (formerly Peronospora parasitica)". Molecular Plant Pathology. 4 (3): 159–70.

Wellman, F. L. (1930). Clubroot of crucifers. US Department of Agriculture.

16

Diseases of Tomato and Their Management

Shivanshu Mishra, Ankit Kumar Chaurasia, Anshika Sharma and Abhay Pratap Singh

Department of Plant Pathology, Sardar Vallabhbhai Patel University of Agriculture and Technology, Meerut, Uttar Pradesh

Abstract

Tomato (Solanum lycopersicum L.) is one of the most widely cultivated and important vegetable crops in the world. The tomato plant is susceptible to many kinds of diseases, however, which may significantly reduce both yield and quality. In This chapter provides a concise overview of common tomato diseases and their management strategies. Tomato diseases can be broadly categorized into fungal, bacterial, and viral, diseases. Integrated disease management (IDM) strategies combining multiple approaches, including genetic resistance, cultural practices, chemical treatments, and biological controls, are often the most effective means to combat tomato diseases. Crop rotation, proper sanitation, and maintaining good garden hygiene play essential roles in disease prevention. Additionally, the use of disease-resistant tomato varieties, when available, can significantly reduce disease pressure and the need for chemical interventions. In conclusion, tomato diseases pose a constant threat to crop production, but with careful planning and a combination of preventive measures, growers can effectively manage these diseases to ensure healthy tomato harvests and sustained agricultural productivity.

Keywords:-*Tomato, diseases, symptoms, pathogen, etiology, botanical name, management etc.*

DOI: 10.1201/9781042014460-16

Tomato

Botanical Name	:	*Solanum lycopersicum* L.
Family	:	Solanaceae
Chromosome Number (2n)	:	24

Introduction

Tomato scientifically known as *Solanum lycopersicum* L. is a widely cultivated and consumed fruit that is often mistaken for a vegetable due to its culinary uses. Belonging to the Solanaceae family, tomatoes are native to western South America with their domestication dating back to ancient civilizations in Mesoamerica. Tomatoes are prized for their vibrant red colour, juicy flesh, and rich, slightly sweet flavour. They are a versatile ingredient used in a wide range of culinary dishes, from salads and sauces to soups and sandwiches. Tomatoes are a significant source of essential nutrients, including vitamin C, vitamin K, potassium, and folate. Additionally, they contain the antioxidant lycopene, which is believed to have various health benefits. This fruit has had a profound impact on global cuisines, with dishes like Italian pasta sauces, Mexican salsas, and American ketchup becoming integral parts of many culinary traditions. Tomatoes are sensitive to several diseases caused by bacteria, viruses, and fungus. These diseases can have a significant impact on tomato crop yield and quality. Here are common fungal, bacterial, and viral diseases of tomato crops:

Fungal Diseases

1. Damping off

The fungal disease known as "damping-off" kills and rots tomato seedlings. Damping-off is to responsible for up to 90% of the loss in seedling numbers when the environment is favourable for the disease's development. Every year, seedling losses of 25 to 75% happen in plant species that are particularly susceptible. Pre-emergence damping off is primarily responsible for the loss (R.S. Singh,1991).

Symptoms

Tomato damping off occurs in two stages: the pre-emergence phase and the post-emergence phase.

Just before they breach the soil's surface, the seedlings are destroyed during the pre-emergence stage. The seedlings entirely rot, and the plumule and young radical are destroyed. The post-emergence stage is when the immature, juvenile tissues of the collar at ground level are infected.

The infected tissues enlarge and absorb water. The seedlings give way or fall. (Agrios, 2005).

Pre emergence damping off

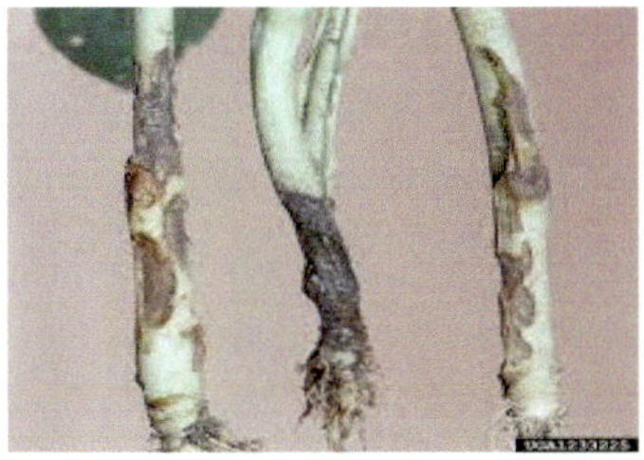

Post emergence damping off

Pathogen: *Pythium aphanidermatum*

Etiology

Hyphae, the thread-like structures that make up the fungus's body, are present in Pythium fungi. Depending on the species, the hyphae can be branched or unbranched and are normally colourless. Pythium is a member of the oomycetes, a class of fungus that are not actually fungi but have some resemblances. Oomycetes have cellulose-based cell walls, whereas real fungi have chitin-based cell walls, which is a significant distinction. The development of motile zoospores is one of the characteristics of Pythium species. These spores are capable of swimming in water because they possess one or more flagella. Zoospores, which can infect plant roots and other substrates, are the fungus's mode of dissemination (S. K., Manorantitham, *et al* 2001).

Disease Cycle

When the environment is favorable and there is a susceptible host present. Pythium produces mobile spores that can quickly infect a plant root. These spores develop into hyphae that go into the roots and consume plant tissue as they develop. Pythium also develops sexual in the soil or at the roots that can withstand harsh environmental conditions like cold or dryness. When there is enough moisture and a susceptible host present, the spores can live in a dormant form for months or perhaps an entire year. As soon as the hyphae enter the host root, the infection process starts (Sabaratnam, S. & Traquair, J. A., 2002).

Management

A raised seed bed is used. For improved drainage, irrigate sparingly yet often. Drench in a Bordeaux combination or Copper oxychloride 0.2%. *Trichoderma viride* (4 g/kg of seed) or Thiram (3 g/kg of seed) is the sole preventative method to control pre-emergence damping off. In times of cloudy weather, mist 0.2% Metalaxyl (R.S. Singh, 1991).

2. Fusarium Wilt of Tomato

The most common and devastating disease affecting tomatoes in greenhouses and fields. *Fusarium* wilt of tomato, is a devastating disease that can lead to significant crop losses.

Symptoms

The first signs of the disease are leaf chlorosis and veinlet clearing. The plant may wilt and die in a few days if the younger leaves start to fall off one after another.

Petiole and leaves begin to drop and wilt quickly. Symptoms in immature plants include clearance of vein leakage and petiole droop. In a field, lower leaves yellowed first and then affected leaflets die and wilt. Later leaves still display the same issues. Later, browning of the vascular system happens. Stunted plants eventually die. (CF. Ajigbola OO Babalola 2013)

Pathogen: *Fusarium oxysporum f. sp. lycopersici*

Etiology

For tomato plants, one of the causes of wilt disease is *Fusarium oxysporum f. sp. lycopersici*. Agrios (2005) states that three distinct asexual spore types are generated: macroconidia, chlamydospores, and microconidia. Long-term survival is made possible by the thick walls of chlamydospores, which also allow the pathogen mycelium to live there as an alternative host and saprophyte. Chlamydospores are created when hyphal or conidial cell walls thicken. Stress factors that lead to chlamydospore formation in *F. oxysporum* include the absence of the host (nutrient depletion) and unfavorable environmental conditions (S. Smith, 2007). When favorable conditions, such the availability of host root exudates, return (nutrient abundance), chlamydospores germinate as predicted (T. Kommedahl).

Disease Cycle

The fungus spreads through soil and seeds. The fungus can persist in the soil for more than ten years as chlamydospores or as saprophytically developing mycelium in infected crop debris. Seedlings raised on contaminated soil are one of the main ways it is spread. Agricultural tools, surface drainage water, and wind-borne spores are further ways that the disease is transferred from field to field. (Miller, S. A., *et al* 2008).

Management

In nutrient-rich, well-drained soil free of wilt, only plant certified disease-free seed and transplants. The Fusarium fungus is absent from tomato seed that has been properly treated with hot water. Use a soil fumigant (such as methyl bromide, Vorlex, chloropicrin, etc.) or steam before planting. to decontaminate the soil in the greenhouse and the seedbed. The benches, containers, used stakes, tools, and greenhouse structures all need to be clean and free of pathogens. Grow only Fusarium-resistant or immune tomato types on contaminated soil. Growers should rotate their tomato crop off of the same land every four years. Too many tomato crops might cause soil that is already somewhat affected to become severely infested. When plants with wilt get very ill in-home gardens, pull them up and burn them. Fusarium wilt cannot be controlled by dusting or spraying; nonetheless, it is useful in stopping fruit rots, leaf spots, and tomato blights. (R. Gitaitis).

Early Blight of Tomato

A frequent fungal disease that affects tomatoes cultivated in fields; greenhouses is called early blight. In warm, rainy, and humid conditions, epidemiological outbreaks of this disease can cause significant defoliation, lower yields, and poor-quality fruit. The fungus also infects potatoes.

Symptoms

The fungus can infect any part of a tomato plant, including the leaves, stems, and fruit. The first signs of a leaf lesion are little (less than 1/16 inch) brown spots surrounded by yellow discolorations. Diagnostic signs appear when the spots enlarge and turn into dark brown or black lesions with concentric rings that are usually 1/3 to 1/4 inch in diameter. In ideal circumstances, many lesions combine to severely defoliate and sunscald fruit. Any age can have stem infections, which cause small, dark, somewhat depressed patches to grow and take the form of circular or elongated concentric lesions. Fruit can become contaminated when it is green or ripe. Near the calyx or stem ends, sunken lesions with concentric rings of various sizes appear on fruit. (Blancard, 2012 and Kemmitt, G., 2002).

Early blight of tomato

Pathogen: *Alternaria solani*

Etiology

Early tomato blight is caused by *Alternaria solani. A. solani's* mycelium is hyaline, septate, branching, and the letters on it take on a black colour. Both sexual spores and sexual reproduction do not exist or germinate. It is a sizable spored fungus that produces solitary conidia on simple conidiophores (Neergaard, 1945). Conidiophore of *Alternaria solani* with horizontal and vertical septation produces muciform and beaked conidia. Conidia measure 12-20 x 120-296 m and have 9-11 transversal septa. Conidia is generated single or in a chain and has the shape of a club (Dhaval, P.,et al 2021).

Disease Cycle

The fungus persists in tainted seed and contaminated agricultural waste. In seedbeds with high humidity and warm temperatures, contaminated seed can lead to damping-off. Disease growth in fields is mostly sparked by diseased seedlings and contaminated plant waste from the previous growing season. Under moist and humid conditions, the pathogen develops a large number of spores on affected tissues (Foolad *et al* 2008). Through the wind and rain-splashing, the spores are spread from plant to plant. By increasing the leaf moisture necessary for spore germination and splashing spores from infected plants, overhead irrigation may make disease growth worse. Warm temperatures and prolonged leaf wetness from regular rain or overhead spray irrigation are favorable to the development of early blight. Early blight is frequently linked to tomato plants that are stressed by low nitrogen levels, especially in the latter half of the growth season on older, senescing leaves. (Agrios, 2005).

Management

"There are many methods used to manage early blight in tomato crops, including cultural, biological, chemical application, resistant varieties, and pathogen-free plant material." Cultural practices include crop rotation, soil-directed irrigation systems, eliminating tainted fruits and vines, plant residue, and volunteer weeds from the field's vicinity, as well as maintaining a healthy field and plant Vigor (Foolad *et al.*, 2008). Crop rotation is necessary for the treatment of this infection for 3 to 5 years, and using transplants free of this pathogen is an additional alternative (Sherf and MacNab, 1986; Madden *et al.,* 1978). Mancozeb with carbendazim outperformed other treatments for disease control (Bais et al.). The most effective fungicides for preventing tomato early blight were tebuconazole and difenoconazole (Rani S. 2017).

Late blight of Tomato

Late blight of tomato is a devastating and highly contagious fungal disease that affects tomato plants. It is infamous for causing the Irish Potato Famine in the 1840s but also affects tomatoes and other related crops.

Symptoms

The fruit, the stems, and the leaves are affected.

On the leaves and petioles:-A thorough explanation is not required because the infection on the foliage has a striking resemblance to the typical lesions of the late blight of potatoes. Because the petioles are easily damaged, the impacted leaves frequently droop.

On the stem: The infection spreads more slowly and results in darker lesions on the older, woody sections of the stem, but the illness proceeds quite quickly on the sensitive regions.

On the fruits: Although partially matured fruit can also become infected, green fruits are more vulnerable to the disease's attack. Where two fruits come into contact with the ground or each other at the stem end is where the majority of these illnesses are detected. However, it has been shown that the fungus can penetrate at any time via the intact epidermis.

Initially, the symptoms are characterized by asymmetric dark spots that barely penetrate the epidermal layer. In more favorable circumstances, the lesions proliferate quickly and develop into roughly concentric zones. The invasion of secondary organisms causes the initially tough tissues to become soft very quickly. Nelson, S. C. (2008).

Pathogen: *Phytopthora infestans*

Etiology

The mycelium is endophytic, coenocytic, and hyaline and has double-club-shaped haustoria that are intercellular in nature. Hyaline, branching intermediate sporangiophores have thick walls. Hyaline, oval or pear-shaped, sporangia have thin walls and a characteristic papilla at the tip. In order to form a germ tube, the sporangium may act as a conidium and start direct germination. There are fine hairs on certain biflagellate zoospores, but not on others. (P. K., Ray *et al* 2021).

Disease Cycle

The oomycete or fungus-like organism *Phytophthora infestans* is responsible for late blight. Cool, humid weather is favourable for the development of disease. The best conditions for infection are nights with lows in the 50s and days with highs in the 70s combined with rain, fog, or heavy dew. In these circumstances, symptoms may start to show on plant tissues three to five days after infection. The pathogen develops and sporulates on the surface of infected tissues once symptoms start to appear. *P. infestans* develops sporangia, which are fruiting structures that appear as white, hairy growths to the unaided eye. A single lesion can produce up to 100,000 to 300,000 sporangia each day, and each sporangium has the ability to start a fresh infection. Irrigation, rain, wind, human activity, and machinery can all easily dislodge and distribute sporangia. They can go many miles or possibly farther. When sporangia touch down on tissue that is susceptible, they start fresh infections, which start a new cycle of late blight. This disease has a strong capacity for reproduction, which makes epidemics potentially swift and catastrophic. *P. infestans* cannot persist in either plant detritus or soil. It can, however, persist in infected potato tubers found in compost or cull piles. Infected tubers in storage have been found to harbour it over the winter. Tomato seeds do not contain *P. infestans*. In Connecticut, sporadic epidemics take place on a yearly basis. However, they are typically seen as the result of the pathogen's wind-borne spores entering the state. Losses are often limited because these introductions take place late in the growing season. (A, de Bary. 1876.)

Management

The basic step of treatment for late blight is to minimize the primary inoculum sources. Controlling polluted sources, such as rubbish piles, ailing plants, volunteer plants, disease in nearby fields, and regrowth after haulms removal, can help with the illness's treatment (Turkensteen L J and Mulder A, 1999). Regular spraying and dusting during the growth season offers effective control. The initial spraying should be applied before to the disease's development, and subsequent sprayings should be applied every 10 to 15 days on average. Applying a protective coating of 0.2% zineb or mancozeb to plants will shield them against infection.

Septoria Leaf Spot

The fungal infection that causes septoria leaf spot is a widespread and harmful disease that damages tomato plants. Every stage of the plant's growth is susceptible to attack.

Symptoms

Small, round lesions are the usual appearance of septoria leaf spot on tomato plant lower leaves. Small, water-soaked spots at first, these lesions ultimately take on their distinctive appearance with a dark brown or black center encircled by a yellow halo. Lesions can combine and cover significant areas of the leaf surface as the disease develops. In serious cases, the whole leaf may contract the infection (Agrios, 2005).

Septoria Leaf Spot of tomato

Pathogen: *Septoria lycopersici*

Etiology

When young, young mycelium is hyaline, sparsely septate, branching, and darkens with maturity. Pycnidia are sub globose and made up of a layer of two to three of brown cells. The septate, filiform, hyaline, and slightly curved pycnidiospores have pointy or rounded ends. No reported sexual stage occurs.

Disease Cycle

The pathogen overwinters as mycelium conidia in plant debris from infected plants or in plant debris from a solanaceous weed host, like horse nettle (*Solanum carolinense* L.). Raindrops splatter the conidia onto the plant's lower leaves. The spores generated by these pycnidia carry on the disease's secondary cycle. A temperature of 20 to 25 Celsius and a relative humidity of 75% to 92% are favourable for the development of disease. (Sharma *et al.*, 1966)

Management

Follow cultural practices because the fungus survives in infected plant material so clean cultivation and using of healthy seed is very Important. Better disease management with mancozeb and dithane Z-78. (Sharma *et al.*, 1966; Sohi and Sokhi, 1970).

3. Bacterial Diseases

Bacterial wilt of Tomato

One of the most severe tomato diseases is this one. Disease growth is favoured by relatively high soil temperatures and moisture levels. A harmful and economically significant disease that affects tomato plants is bacterial wilt.

Symptoms

One of the most severe diseases to affect the tomato crop is this one. Disease growth is favored by relatively high soil temperatures and moisture levels. One of the distinguishing features of bacterial wilt is the abrupt and total withering of robust, mature plants. Lower leaves may drop off before they wither. The pathogen is mainly limited to the vascular area, although it can occasionally enter the cortex and cause tissues to become yellow-brown from pith. Cut portions of diseased plants When severed ends are soaked in clear water, a white stripe of bacterial slime appears.

Bacterial wilt of tomato

Pathogen: *Ralstonia solanacearum*

Disease cycle

There are multiple stages in the tomato bacterial wilt disease cycle. In the initial stages, the pathogen continues to exist in the soil, mainly as resistant resting cells or infected plant debris. It may stay in the soil for a very long time. The bacteria can get into the roots of tomato plants when they are planted in polluted soil. The infection can also spread as a result of farming techniques that disturb the soil or irrigation water. The bacterium multiplies inside the plant and travels via the vascular system, producing the recognized symptoms. The pathogen may be released back into the soil as the diseased plant material breaks down, completing the cycle and possibly contaminating successive harvests (Sandani, H. B. P., & Weerahewa, H. L. D. (2018).

Management

Bacterial wilt of tomato is a devastating disease that can lead to severe yield losses if not managed effectively. Management strategies for bacterial wilt of tomato involve a combination of cultural, biological, and chemical measures.

Crop Rotation and Resistant Varieties: Crop rotation is an important cultural practice to reduce the incidence of bacterial wilt. Avoid planting tomatoes or other susceptible crops in the same field for several years. Additionally, planting resistant tomato varieties can be effective in managing the disease. (Hayward, A. C. 1991).

Biocontrol: Beneficial microorganisms, such as certain strains of non-pathogenic *Ralstonia* spp. and *Pseudomonas* spp., can be used as biological control agents to compete with and suppress the pathogenic *Ralstonia solanacearum* in the soil (Nion, Y. A., & Toyota, K. (2015).

Chemical Control: Although chemical control measures are generally less effective, some copper-based and systemic antibiotics may be used as a last resort in severe cases. However, their use is often discouraged due to concerns about antibiotic resistance and environmental impact (J. Mansfield, *et al* 2012).

Seed Treatment: - Treating tomato seeds with hot water or antibiotics can help eliminate the bacterium from the seeds and reduce the risk of introducing the pathogen into the field.

Bacterial Leaf Spot

The disease tomato bacterial spot is widespread throughout the world. All of a tomato plant's above-ground parts, such as the leaves, stems, and fruit, are affected by the disease. The disease can cause early defoliation and fruit spotting in wet and rainy weather, which reduces yield and results in unmarketable fruit.

Symptoms

Disease symptoms can be seen in all above ground plant parts. On the leaves, stems, and fruit spurs, spots are often round and dark brown (but they can sometimes be angular). The underside of the leaves initially shows more signs of disease, which typically begins on the lower leaves. Rarely do spots grow larger than 3 mm in diameter. Spots combine to form irregularly shaped lesions when conditions are ideal for disease growth. On leaflets with several lesions, a general blighting and yellowing may take place. The plant frequently still has dead leaf on it, giving it a burnt appearance. Fruit blemishes start as tiny, barely elevated blisters. Spots are darker, scabby, and a little elevated as they get bigger. They could, however, alternatively be depressed in the middle and raised at the ends (Jones, 1991; Kucharek, 1997).

Bacterial leaf spot of tomato

Pathogen: *Xanthomonas campestris pv. vesicatoria*

A gram-negative, strictly aerobic, motile bacterium with a short rod shape and a single, polar flagellum, *Xanthomonas campestris pv. vesicatoria* forms capsules. (Jones, 1991).

Disease Cycle

The organism may thrive on stray tomatoes and plant debris from afflicted tissue, according to Jones et al. (1986). Seeds are another potential habitat for the bacterium's survival and growth. Disease is more likely to occur in environments with high humidity, rain, and temperatures between 24 and 30 °C. The bacterium spreads inside fields by a variety of cultural practices, wind-driven rain, irrigation droplets, and aerosols. When plants are handled while moist, bacteria will spread more quickly (Pohronezny et al., 1990). Jones (1991) claims that the bacterium enters through both naturally occurring openings (such stomates and hydathodes) and wounds brought induced by wind-driven sand, insect bites, or artificial means. (Jones, 1991).

Management

Only disease-free seed and seedlings should be sown, and the crop should be rotated with non-host crops to prevent crop residue from the previous year. It is also advised to treat seeds with mercuric chloride (1:1000) in order to control illness. Spraying Agrimycin-100 (100 ppm) three times at intervals of ten days or using a copper and organic fungicide mixture on a regular basis at intervals of five to ten days can both successfully control the disease. (1990, Pohronezny *et al.).*

Bacterial Soft Rot and Hollow Stem

Compared to other recognized bacterial diseases, bacterial soft rots result in a higher loss of yield. If a storage facility has inadequate conditions, yield losses might be as high as 100% (Arsenijevic and Obradovic 1996). This then has an effect on the client by lowering the quantity, quality, and cost of food available for purchase. One of the biggest issues with tomatoes worldwide is the rotting disease. (Permbelon and Kelman 1980).

Symptoms

The disease results in the wilting of the entire tomato plant, water-soaked sections of the stem, browning of the vascular tissue, hollowing out of the pith, and soft, decomposing fruits and stems. Symptoms initially show up in the root or crown region of seedlings in greenhouses. Fruit - soft, watery deterioration of fruit that starts as extremely little spots at one or more sites. Quickly grow until the entire fruit is a squishy, watery mess. Fruit tissue is liquefied by a pathogen by destroying the pectate "glue" that binds plant cells together. Internal collapse caused by leakage that resembles a shrivelled water balloon. In water, single-celled bacteria quickly proliferate and spread. heavy rain or irrigation during humid and damp conditions. Warm temps between 73 and 95 °F. (M. C. Perombelon, & A. Kelman 1980).

Bacterial soft rot of tomato

Pathogen: *Erwinia carotovora pv. carotovora*

Gram negative, rod with rounded ends. Motile with 1to 6 peritrichous flagella. Single celled – rapidly multiply and spread-in water. Mostly found in tropical and warm regions.

Disease cycle

Plant trash and stray plants still contain the infection. Through seeds, it proliferates. The bacteria enter by wounds, lenticels, or stomata. the additional spread caused by raindrops splattering. Sick transplants and contaminated seeds carry the disease to new locations. The bacteria continue to grow and spread into the intercellular spaces as they feed on intracellular fluid, while their cell-wall-degrading enzymes advance them and get the plant tissues ready for digestion. The decaying flesh is frequently kept inside the epidermis unharmed until a breach permits the sludge to flow out and contaminate those nearby. The contaminated plant organs will spread the infection to the others when they are harvested and stored with them. The bacteria will infiltrate eggs placed over stored vegetables when specific insects are present, turning them into hosts and carriers that can spread the infection to other people as they mature. After that, the bacteria rest in the soil, inside plant tissues, or as hosts for insects before resuming their reproductive cycle. (Agrios 2005).

Management

Start with good cultural practices, such as crop rotation and staying away from planting tomatoes in places where the disease has previously been present. To stop the introduction and spread of the germs, make sure the equipment and the surrounding area are clean. Tools and equipment used in tomato cultivation should be thoroughly cleaned. Antibiotics and fungicides with copper bases can be used to control the soft rot disease. Antibiotics and fungicides have certain limitations. It is essential to handle and store tomatoes gathered properly. To prevent bacterial soft rot during storage, make sure the fruits are maintained dry and stored at the proper temperatures (Jones, J. B., & Jones, J. P. 2012).

4. Viral Diseases

Tomato Mosaic Virus (ToMV)

This virus affects a wide variety of plants, including weeds like tobacco and beans and many agricultural crops, and is present around the world. All of these plants can act as inoculum sources. The tomato mosaic virus has a high susceptibility among tomato crops.

Tomato mosaic virus

Symptoms

The disease is initially identified by light and day green mottling on the leaves of affected plants. On warm days, this is often followed by the withering of young leaves. Affected leaves typically have smaller-than-normal, twisted, and puckered leaflets. The leaflets can occasionally become recessed, exhibiting "fern leaf" symptoms. The damaged plant has a spindly, light green, and stunted appearance. Plant waste, gardening equipment, worker hands, clothes, and contact between infected and healthy plants can all spread the virus. (k. Kubota, 2003).

Pathogen

Virus particles have a thermal inactivation point of 85 to 90 °C and are rod-shaped, non-enclosed, and frequently straight.

Spreading and surviving methods

Up to 94% of seeds could have the virus because it is seed-borne. Virus infection occurs during transplantation. It is easily infectious. Tomato mosaic virus can infect a lot of solanaceous plants. The virus is easily transferred via people, tools used in cultural activities, animals, and leaf contacts (L. Broadbent 1977).

Management

Seeds should be selected from disease-free, healthy plants before being sowed. The day before planting, immerse the seeds in a trisodium phosphate solution (90 g/litre of water) to reduce the likelihood of the virus. The seeds need to be well rinsed and then dried in the shade. The nursery's unhealthy plants should all be properly taken out and disposed of. It is not advised to plant seedlings that have a viral infection. Crop rotation, utilizing crops other than tobacco, potatoes, chillies, capsicum, brinjal, etc., is crucial. (Broadbent, L. 1976)

Tomato Leaf Curl Virus (ToLCV)

The tomato leaf curl virus (ToLCV), which may be found in tropical and subtropical locations and causes significant economic losses, is one of the most researched plant viral diseases. It is the virus that causes the greatest harm to the world's tomato field.

Symptoms

The symptoms of leaf curl disease include severe plant stunting as well as downward rolling and crinkling of the leaves. The leaves have a little yellow color when they first start to develop, and then they curl. Older leaves become sticky and brittle. The nodes and internodes are substantially smaller. The damaged plants seem pale and have more lateral branches, which gives them a bushy look. The sick plants are still exhibiting stunting. (Moriones, E., & Navas-Castillo, J. 2000)

Tomato leaf curl virus

Pathogen

The virus's particles have a diameter of 80 nm.

Mode of spread and survival

It can't be dispersed by seeds or sap. Nonetheless, the virus may be present on the seed coat of fresh fruit that has been infected. The primary vectors of the virus are grafting, *Bemisia tabaci,* and white flies. The virus just needs one infecting insect to propagate.

Management

To track the white fly, maintain 12 yellow sticky traps per acre. Grow cereal barrier crops everywhere on the property. Avoid the marijuana host in nursery housed in a net house or greenhouse for protection. After transplanting, spray control vector with imidacloprid or dimethoate at 15, 25, or 45 days. (A. Prasad *et al.* 2020)

Tomato Spotted Wilt Virus (TSWV)

A plant virus known as tomato spotted wilt virus (TSWV) affects a variety of hosts, such as tomatoes, peppers, and other crops and ornamental plants.

Symptoms

This causes streaking to appear on the fruits, stems, and foliage. Many tiny black circular spots appear on younger leaves. Leaves may seem bronzed at first, then become dark brown and eventually wither. Fruits have a number of concentric, circular markings on them, each approximately half an inch across. These mature fruit markings are alternating stripes of red and yellow.(1968; Best, R. J.).

Tomato spotted wilt virus

Pathogen

Particles range in size from 70 to 90 nm and are isometric. 40 °C is the thermal inactivation point.

Spreading and surviving methods

Through thrips (*Thrips tabaci, Frankliniella schultzi, and F. occidentalis*), the spotted wilt virus is spread.

Management

It is necessary to remove and destroy affected plants. It is necessary to remove the additional or collateral hosts hosting the infection. Increase barriers with sorghum, maize, and bajra Before planting the tomato, make 5–6 rows around the field. To control the vector, apply Imidachlorprid 0.05% or any other systemic insecticide (S. Adkins, 2000).

Conclusion

In conclusion, the chapter on tomato diseases highlights the critical importance of disease management in tomato cultivation. Various pathogens, including fungi, bacteria, viruses, pose significant threats to tomato crops, leading to yield losses and economic impact. Effective disease control strategies involve a combination of preventive measures, such as crop rotation, disease-resistant tomato varieties, and proper sanitation. Chemical treatments can also be employed, but they should be used judiciously to minimize environmental impact and resistance development. Early detection and diagnosis of tomato diseases are crucial for timely intervention. Farmers and gardeners should be vigilant and employ integrated pest management practices to ensure healthy tomato plants and maximize yield.

References

Adkins, S. (2000). Tomato spotted wilt virus—positive steps towards negative success. Molecular Plant Pathology, 1(3), 151-157.

Agrios, G. N. (2005). Plant pathology.(5th eds.) Elsevier academic Press. New York.

Ajilogba, C. F., & Babalola, O. O. (2013). Integrated management strategies for tomato Fusarium wilt. Biocontrol Science, 18(3), 117-127.

Arsenuević, M., & Obradović, A. (1996). Occurrence of bacterial wilt and soft rot of seed cabbage plants (*Brassica oleracea* var. *capitata* L.) in Yugoslavia. Journal of Phytopathology, 144(6), 315-319.

Badawy, M. E., & Rabea, E. I. (2009). Potential of the biopolymer chitosan with different molecular weights to control postharvest graymold of tomato fruit. Postharvest Biology and Technology, 51(1), 110-117.

Best, R. J. (1968). Tomato spotted wilt virus. Advances in Virus Research, 13, 65-146.

Blancard, D. (2012). Tomato Diseases: identification, biology and control: a colour handbook. CRC Press.

Broadbent, L. (1976). Epidemiology and control of tomato mosaic virus. Annual review of Phytopathology, 14(1), 75-96.

Bu, S., Munir, S., He, P., Li, Y., Wu, Y., Li, X., ... & He, Y. (2021). Bacillus subtilis L1-21 as a biocontrol agent for postharvest graymold of tomato caused by Botrytis cinerea. Biological Control, 157, 104568.

Chaerani, R., &Voorrips, R. E. (2006). Tomato early blight (*Alternaria solani*): the pathogen, genetics, and breeding for resistance. Journal of general plant pathology, 72, 335-347.

De Bary, A. (1876). Researches into the nature of the potato fungus, Phytophthora Infestans. J. Bot. Paris, 14, 105-126.

Dhaval, P., Shete, P. P., Faraaz, M., &Dholu, D. (2021). Early blight (*Alternaria solani*) etiology, morphology, epidemiology and management of tomato. Pharma Innovation, 10, 1423-1428.

Dhaval, P., Shete, P. P., Faraaz, M., &Dholu, D. (2021). Early blight (*Alternaria solani*) etiology, morphology, epidemiology and management of tomato. Pharma Innovation, 10, 1423-1428.

Gitaitis, R., McCarter, S., & Jones, J. (1992). Disease control in tomato transplants produced in Georgia and Florida. Plant Disease, 76(7), 651-656.

Hayward, A. C. (1991). Biology and epidemiology of bacterial wilt caused by Pseudomonas solanacearum. Annual Review of Phytopathology, 29(1), 65-87.

Inoue, I., Namiki, F., & Tsuge, T. (2002). Plant colonization by the vascular wilt fungus Fusarium oxysporum requires FOW1, a gene encoding a mitochondrial protein. The Plant Cell, 14(8), 1869-1883.

Ismael, M. A., Elyamine, A. M., Moussa, M. G., Cai, M., Zhao, X., & Hu, C. (2019). Cadmium in plants: uptake, toxicity, and its interactions with selenium fertilizers. Metallomics, 11(2), 255-277.

Jones, R. A. & Barbetti, M. J. (2012). Influence of climate change on plant disease infections and epidemics caused by viruses and bacteria. CABI Reviews, (2012), 1-33.

Jones JB. 1991. Bacterial Spot. In: Compendium of Tomato Diseases, eds: JB Jones et al., APS Press. p. 27.

Kemmitt, G. R. E. G. (2013). Early blight of potato and tomato. Early blight of potato and tomato.

King, S. B., & Alexander, L. J. (1969). Nuclear behavior, septation, and hyphal growth of Alternaria solani. American Journal of Botany, 56(3), 249-253.

Kubota, K., Tsuda, S., Tamai, A., & Meshi, T. (2003). Tomato mosaic virus replication protein suppresses virus-targeted posttranscriptional gene silencing. Journal of virology, 77(20), 11016-11026.

Kucharek, T. (1979). Bacterial spot of tomato and pepper. Florida Cooperative Extension Service, Institute of Food and Agricultural Sciences, University of Florida.

Manorantitham, S. K., Prakasam, V., & Rajappan, K. (2001). Biocontrol of damping off of tomato caused by Pythium aphanidermatum. Indian Phytopathology, 54(1), 59-61.

Mansfield, J., Genin, S., Magori, S., Citovsky, V., Sriariyanum, M., Ronald, P., ... & Foster, G. D. (2012). Top 10 plant pathogenic bacteria in molecular plant pathology. Molecular Plant Pathology, 13(6), 614-629.

Miller, S. A., Rowe, R. C., & Riedel, R. M. (2008). Fusarium and verticillium wilts of tomato, potato, pepper, and eggplant. Fusarium and verticillium wilts of tomato, potato, Pepper, and Eggplant., 1-3.

Moriones, E., & Navas-Castillo, J. (2000). Tomato yellow leaf curl virus, an emerging virus complex causing epidemics worldwide. Virus Research, 71(1-2), 123-134.

Neergaard, P. (1945). Danish species of Alternaria and Stemphylium. Danish species of Alternaria and Stemphylium.

Nelson, S. C. (2008). Late blight of tomato (Phytophthora infestans).

Nion, Y. A., & Toyota, K. (2015). Recent trends in control methods for bacterial wilt diseases caused by Ralstonia solanacearum. Microbes and environments, 30(1), 1-11.

Perombelon, M. C., & Kelman, A. (1980). Ecology of the soft rot erwinias. Annual Review of Phytopathology, 18(1), 361-387.

Pohronezny, K., Moss, M. A., Dankers, W. I. L. B. U. R., & Schenk, J. A. M. E. S. (1990). Dispersal and management of Xanthomonas campestris pv. vesicatoria during thinning of direct-seeded tomato. Plant Dis, 74, 800-805.

Prasad, A., Sharma, N., Hari-Gowthem, G., Muthamilarasan, M., & Prasad, M. (2020). Tomato yellow leaf curl virus: impact, challenges, and management. Trends in Plant Science, 25(9), 897-911.

Ray, P. K., Solankey, S. S., Singh, R. N., Kumar, A., & Singh, U. (2021). Molecular Methods for the Controlling of Late Blight in Tomato. Biotic Stress Management in Tomato: Biotechnological Approaches, 59.

Sabaratnam, S., &Traquair, J. A. (2002). Formulation of a Streptomyces biocontrol agent for the suppression of Rhizoctonia damping-off in tomato transplants. Biological Control, 23(3), 245-253.

Sandani, H. B. P., &Weerahewa, H. L. D. (2018). Wilt diseases of tomato (Lycopersicum esculentum) and chilli (Capsium annum) and their management strategies: Emphasis on the strategies employed in Sri Lanka: A review. Sri Lankan Journal of Biology, 3(2).

Sharma, S. L., Sohi, H. S., Verma, B. R., & Sachdeva, K. B. (1966). Chemical control of defoliation of tomatos. Indian Journal of Horticulture, 23, 172-176.

Sherf, A. F., &MacNab, A. A. (1986). Vegetable diseases and their control. John Wiley & Sons.

Singh, R.S. (1991). Plant Disease, sixth Edition, Oxford and IBH Publishing Co, New Delhi, pp. 158-166.

Turkensteen, L. J., & Mulder, A. (1999). The potato disease Phytophthora Infestans. Gewasbescherming, 30(4), 106-112.

17

Diseases of Onion and Their Management

***Reecha J*[1], *P. Sujhee*[2] *and Suraj Kumar Patel*[1]**

[1]*Department of Plant Pathology, SardarVallabhbhai Patel University of Agriculture and Technology, Meerut, Uttar Pradesh*
[2]*Department of Plant Pathology, Kerala Agricultural University Vellanikara, Thirusur*

Abstract

Allium cepa L. (Onion) usually known as "Queen of the Kitchen" is an important biennial bulbous vegetable crop grown in India with high export value. Onion possess anti-inflammatory, antimicrobial and antioxidant properties. During 2020-2021, the area under onion cultivation was 1624 thousand hectare with production of 26,641thousand MT in India.Onion productivity is reduced by various bacterial, fungal, viral diseases such aspurple blotch, basal rot, twister blight, downy mildew, stem blight, bacterial soft rot, onion yellow dwarf and iris yellow spot. These disease severity cause loss in productivity either field or storage conditions. In this chapter we discuss about the major disease of onion and their management practices.

Keywords:*Iris yellow spot, twister blight*

Onion

Scientific name	:	*Allium cepa* L.
Family	:	Alliaceae
Chromosome number (2n)	:	16
Pollination	:	Cross pollination

Bulbous herb (Onion) that belongs to Alliaceae family and usually called as the "Queen of the Kitchen." It is an important biennial vegetable crop grown in India with high export value. A crop that thrives in temperate climates where temperatures between 12.8 to 23°C are necessary for vegetative growth and 20 to 25°C are required later for proper bulb development. Onion pungency is due to the presence of allyl-propyl-disulphide. The pigment quercetin and anthocyanin

DOI: 10.1201/9781042014460-17

is responsible for yellow and red colour of onion. It has medical benefits like lowering disease risk of cardiovascular disease, diabetes and atherosclerosis. It possess antimicrobial, anti-inflammatory and antioxidant properties (Wilson and Demmig-Adams., 2007).The bulb is useful as diuretic and heart stimulant (Ahmad *et al.*, 2016).

In India, the area under onion cultivation was 1638 thousand hectare with production of 26,830 thousand MT and productivity of 16.37 tons/ha during 2020-2021 (Ministry of Agriculture and Farmers welfare).Maharashtra, Andhra Pradesh, Tamil Nadu, Punjab andBihar are the states where the onion cultivation is maximum. Productivity of onion is limited by bacterial and fungal diseases *viz.*,bacterial soft rot, twister blight, stemphylium blight, downy mildew and purple blotch which causes severe yield loss.

1. Fungal Diseases

Basal Rot

Causal organism:*Fusarium oxysporum* f.sp. *cepae*

All stages of crop are susceptible for the disease. Rajput and Patel, (2006) noticed susceptible cultivarsof onion showed yield loss up to 50 % in field condition and 12-30% loss in bulbs storage. In India, the disease was first reported by Mathur and Sankhla (1963) in Maharashtra state.

Symptoms

The disease mainly affects onion crops at month old. The main symptoms of infection in plants are leaf necrosis, curling, and yellowing. Infected field shows symptoms in patches. The whole leaf blade become yellow, wither and dries tip to downwards during lateral phase of infection.At the base of the bulb scales, whitish fungal growth is seen; later, semi-watery rot is observed.(Ghanbarzadeh*et al.*, 2016). Under dry conditions, a pinkish-brown discolouration appears at the bulb base and the decay progresses upward.

Yellowing and drying of leaves

Whitish mycelium growth at the base of the bulb

Rotting of bulb at the base

Pathogen character

- **Mycelium:** Hyaline, septate, branched
- **Asexual spore:** Chlamydospores, macroconidia, microconidia.

Macroconidia: Falcate to straight in shape with 3-4 septate

Microconidia: oval to reniform in shape, one celled

Chlamydospores: One or two cells become round formed either terminal or intercalary (Kalman *et al.*, 2020)

Survival and Dissemination of Disease

- The pathogen is soil borne (Schwartz, 2011).
- The optimum temperatures are 28 - 32°C for disease development.
- High temperature and low moisture favours the disease (Kodama, 1983).

Management

- Soil solarisation for 30 days during summer lessens the inoculum potential or infectious propagules.
- Crop rotation and mixed cropping reduces the disease incidence.
- Soil drenching with copper oxy chloride 0.25%.
- Seed\Bulb treatment with *Trichoderma viride* 4g/kg of bulbs (Coşkuntuna and Ozer, 2008).
- Seedling dip with 0.1% carbendazim or with *Trichoderma viride* or *Pseudomonas cepacis.*

2. Stemphylium Blight

Causal organism: *Stemphylium vesicarium*

The disease was first reported in 1975 by Rao and Pavgi from Uttar Pradesh (Varanasi). The disease that results in early plant senescence, smaller bulbs at harvest, and result in significant loss, named as "Ebola of onion" by farmers (Gedefaw *et al.*, 2019).

Symptoms

White to dark purple spots was initially developed on older leaves followed by necrosis of the plant tissues. Purplish black centre surrounded by straw yellow halo. Water-soaked streaks varying in color from light yellow to brown can be seen in the center of diseased leaves and inflorescence stems. Later it become oval to spindle shaped (Hassan *et al.*, 2020).

Purplish black centre surrounded by straw yellow halo

Pathogen character

- **Asexual spore:** The olive-brown conidia are oval to ovoid and ovoid to oblong in shape . Conidiophores have black margins and a pale to brown color.
- **Sexual spore:** Ascomata is hyaline bears clavate, bitunicateasci. Ascospores are light to yellowish brown, verrucose, ellipsoidal with many longitudinal septaand 5–7 transverse septa (Gedefaw *et al*., 2019).

Favourable condition

- Stem wetness and more damage occurs when rain lasts more than 24 hours(Shishkoff, and Lorbeer, 1989).
- Hot and humid conditions favour the disease incidence.

Management

- Delayed sowing and planting spacing of 60x45cm reduce the disease incidence.
- Seed treatment, seedling dip and foliar with *Pseudomonas fluorescens.*
- Four foliar sprays of propiconazole @ 0.1% at 15 days interval. First spray at 30 days after transplanting.
- IDM practices of FYM @ 10 t/ha + vermicompost @ 1 t/ ha+ NPK @ 100:50:50 kg/ ha+ *P. fluorescens* @ 5 kg/ha+ copper oxychloride @ 0.3% were found to be effective against the disease (Hassan *et al*., 2020).

3. Purple Blotch

Causal organism: *Alternaria porri*

The purple blotch disease attacks all the parts of onion plants and cause 30-40 % yield loss. The yield and crop loss extend more than 70-90% under favourable environmental conditions (Chethana *et al.,* 2012, Resti and Liswarni 2021). The purple blotch disease was first reported by Cooke and Ellis (1879) in US (United States).

Symptoms

The pathogen mostly infects the older leaves rather than younger leaves. Smallwater-soaked elliptical whitish sunken lesions with yellow halo observed on leaves, flower stalks and floral parts. These spots enlarge resembling the bull's eye shape. The spots are covered with dense purplish black spore mass under moist conditions (Conn *et al.,* 2012). Several spots coalesce together, spread entire leaves and causing wilting of leaves.Black velvety, circular to oblong concentric ring appears near chlorotic area.

Black velvety, Circular to oblong concentric ring

Pathogen Character

- **Mycelium:** Coloured, septate, branched
- **Asexual spore:** Muriform conidia with light brown to dark in colour with 0-5 longitudinal septa &3-12 transverse. Conidia generally arise singly, straight to slightly curved with the beak which was 3 μm thick, pale and tapered (Priya and Darshan, 2016).

Favourable Condition

- The disease is more likely to occur in humid and hot conditions with temperatures between 21 and 30°C and relative humidity between 80 and 90%.
- Leaf wetness or relative humidity of 90 per cent or higher for 12 or more hours is favourable for sporulation (Hahuly *et al.,* 2018).

Mode of Spread

- Air or rain splash helps in disseminating conidia.

Management

- Use disease free bulbs for planting.
- Seeds treatment with Thiram @ 2 g/kg seeds.
- Four foliar spray ofTebuconazole 25 EC @ 0.1 % or @ Azoxystrobin 25 SC 0.1 % at 10 days interval (Bachkar*et al*., 2018).

4. Downy Mildew

Causal organism: *Peronospora destructor*

The disease was first reported in Himachal Pradesh causing 12-75% yield loss.

Symptoms

The growth of cottony white mycelial appears on the leaves. Later on, the leaves dry out and change from a pale yellow to a dark brown color. Down-curling, paling and narrowing of the leaves are showed during systemic infection and in secondary infections the oval, slightly large, chlorotic lesions on leaves and stalks of seed are present(Develash and Sugha, 1997;Yarwood, 1943).

A- White patches on the leaves; B- Downward curling and drying; C- Down-curling, paling and narrowing of the leaves

Pathogen Character

- **Mycelium:** Hyaline, filamentous haustoria and intercellular
- **Asexual spore:** Sporangia are pyriform to fusiform with pointed tip. The sporangiophore bears sporangia which are long, swollen at the base and without septate.
- **Sexual spore:** Oospore (Yarwood, 1943).

Favourable Condition

- High relative humidity (90-100%) and temperature of 10-12°C but 22°C helps in rapid disease development (Palti, 1989).

Survival and Spread

- Soil and infected plant debris.

Management

- Heat treatment of bulbs at 40 and 45 °C for more than 12 h reduce disease severity,

- Three spraying with Mancozeb 0.2 % at 10-12 days interval. Spraying should be done at 20 days after transplantingDevelash and Sugha, 1997).
- Integration of heat treatment of 35 °C for 12 h, bulb dipping in metalaxyl3.5 g /kg bulbs for 1 h and foliar spray of metalaxyl + mancozeb 0.2% at 14-dayinterval which control 85% of onion downy mildew (Gupta and Gupta, 2013).

5. Twister Blight or Anthracnose

Causal organism: *Colletotrichum gloeosporioides*

In 1980, Ebenebe was initially reported twister blight disease in Northern Nigeria which caused estimated yield loss upto 50-100%. Recently, main onion-growing areas in Tamil Nadu have experienced yield losses ranging from 50% to 100% due to the onion crop infected by twister blight disease.

Symptoms

Twisting, curling, chlorosis of leaves, unusual neck elongation, rotten bulbs, and the pale oval sunken lesions emergence on leaf sheaths were the prevalent symptoms of onion twister blight. In advance stage, acervuli were observed on the sheaths of the leaf blades (Dutta *et al.*, 2022).

Symptoms at early stage

Symptoms at bulb intation stage Twisting and curling of leaves

Symptoms at bulb development stage Twisting, curling and chlorosis of leaves

Pathogen Character

- **Mycelium:** Hyaline, septate mycelium
- **Asexual fruiting body:** Acervuli
- **Asexual spores:** Conidia is cylindrical with rounded end, hyaline (Reecha *et al*, 2022).

Favourable Condition

Twister disease is favoured by moderate temperature 20 to 31°C, high relative humidity 85 to 95% and rainy weather. Additionally, the rainfall contributed favourable environmental conditions (Patil*et al.*, 2018).

Survival and spread

- Soil borne orinfected plant debris.

Management

- Reduce the inoculum in soil by following soil solarisation, crop rotation with cereals or for pulses.
- Provide proper drainage facilities and timely irrigation.
- Raised bed cultivation and wide planting spacing reduce the disease incidence.
- Foliar application of *Pseudomonas* + Trichoderma 1% was effective in reducing the severity of twister disease.
- SA(Soil application) of vermicompost 5 q ha^{-1}+ neem cake at 5 q ha^{-1}+ *T. harzianum* 2 kg q^{-1} of organic matter + ST (seed treatment) *Trichoderma* at 2 g/kg of seeds + SD(seedlings dip) in *P. fluorescens* solution at 10 g L^{-1} + FS (foliar spray) with boron at 2 g L^{-1} + hexaconazole at 0.1% + multi K 5.0 g L^{-1} at 30 (DAS) days after sowing as intergrated approach for disease management

6. Onion Smudge

Causal organism: *Colletotrichum circinans*

It mainly infects the outer scale of the bulb of white onions which reduces market value. Smudge-infected onions have been detected for the first time in Argentina, 1928 by J.B. Marchionatto. It is both field and post-harvest disease.

Symptoms

The disease infects the outer scales of the onion bulb at the base. Small dark black spots scattered over the outer scale of bulb as concentric rings. It also infects the neck region of the bulb. The spots contains spore mass (Kiehr*et al.*, 2012).

Dark black spots scattered over the outer scale of bulb as concentric rings.

Pathogen character

- **Mycelium:** Hyaline, septate mycelium
- **Asexual fruiting body:** Acervuliis black, acicular setae having a bulbous base
- **Asexual spores:** Conidia is unicellular, falcate shape.

Survival and Spread

- Soil and infected plant debris.
- Soil temperature of 20°C or above favours the rapid disease development.
- Secondary spread of disease by wind borne conidia.

Management

- Red scale onion shows resistant due the presence of phenolic compounds.
- Keep away the healthy bulb from infected bulbs during harvest.
- Proper drying of bulb before storage and market prevent the disease development.
- After curing of onion bulbs store them in dry cool condition (Malathi, 2010).

7. Black Mould

Causal organism: *Aspergillus niger*

The spoilage due to the infection of this fungus has been estimated as 80% during storage.

Symptoms

Pathogens invade neck tissues, causing the foliage to shrivel away at maturity. Infected bulbs around the neck area turn black, and the impacted scales shrink. Between the outer dry scales and along the veins are masses of powdery black spores organized in streaks. The fleshy scales in the center may become infected. The entire bulb surface goes black in advanced disease stages, and secondary bacterial soft rot may cause the bulb to become mushy and soft. (Srinivasan and Shanmugam, 2006).

Bulbs showing black discolouration

Favourable Condition

- Warm and moist condition with optimum temperature of 28°C-34°C.

Survival and Spread

- Infected soil, seed and infected bulbs

Management

- Post-harvest dip with carbendazim at 0.1%.
- Post-harvest sprayof 0.1 % carbendazim or benomyl or 0.25% mancozeb prior to 20 days of harvest.

8. Damping off

Causal organism: *Pythium aphanidermatum,*

Symptoms

• Pre emergence

Circular patches appears in the infected field.

Seed rotting results for pre emergence damping off.

• Post emergence

The disease appears 15 to 30 DAP (days after planting or sowing) in which it cause roots rotting, stunting and toppling of seedling occurs (McDonald *et al.*, 2004).

Roots rotting, stunting and toppling of seedling

Pathogen Character

- **Mycelium:** Hyaline, Ceonocytic (aseptate) mycelium
- **Asexual fruiting body:** Sporangia
- **Asexual spore:** Zoospore- reniform shaped with anterior tinsel and posterior whiplash flagella.

Favourable condition

Water stagnation, high soil moisture content, or persistent rain may encourage the disease rapid spread.

Survival and spread

- Soil borne.

Management

- Seed treatment (ST) with *T.viride*or Captanor Thiram @ 4g/kg.
- Bulb dipping with Thiram 0.25% before planting.
- Drenching the soil with COC (Copper oxychloride) 0.25%.

9. Onion Smut

Causal organism: *Urocystis cepulae*

Symptoms

Soon after the seedlings emerge, they become infected by the fungus, which results in dark, elongated, eruptive spots. The surface and base of the leaves both have black smut sori. In later stage the lesion may spread from base to tip on the older leaves and resembles as blisters. During severe infection a plant may die just one month after emerging. Similar infection also affects the bulbs outer scales, causing blisters. Blisters burst, revealing masses of powdery black spores.

Black smut sori at the base leaves

Pathogen Character

- **Sori:** Dark coloured and powdery spore masses.
- **Teliospore**: Multicelled, thick-walled cell that is dark brown in colour, spherical to ellipsoidal in shape.

Survival and Spread

- The fungus remains viable upto 15 years in infected soil.
- Wind borne soil, surface drainagewaterandimplements are important for dissemination.

Management

- Seed treatment with Thiram or Captan @ 4g/kg.
- Bulb dip in Thiram 0.25%solution.
- Granular application or soil drenching of non-systemic fungicide into open furrow provide protection from Smut.

10. Bacterial Diseases

Bacterial soft rot

Causal organism:*Erwinia carotovora*

It is the important postharvest bacterial diseases causing severe loss for the cultivators and marketers.

Symptoms

Infected bulbs becomes pale yellow to grey , water soaked and spongy. As the infection progress, the interior of the onion becomes softer and watery covered by dry outer scales. Foul smelling bacterial oozes observed (McDonald *et al*., 2004).

Inner scales becomes soft and watery

Pathogen character

- Gram-negative bacterium, rod-shaped, with peritrichous flagella.

Favourable conditions

- High relative humidity and temperature 18 and 27°C favours disease.

Management

- Curing of bulbs should be done thoroughly before storing.

11. Bacterial Internal Browning/slippery skin

Causal organism: *Pseudomonas aeruginosa*

Symptoms

No external symptoms except softening or rotting of tissues of neck region. Water soaked and rotting of internal scales. As the disease progress, all the inner scales will rot and bulb becomes shrivelled and dried. Infected plants have slippery skin because when the base is squeezed, the rotten interior of the bulbs slides out through the neck(Abd-Alla*et al*., 2011).

Inner scales will rot and bulb becomes shrivelled and dried

Pathogen Character

- Gram-negative, rod-shaped bacterium, facultative anaerobic usually 0.5–1.0µm wide &1–5 µm long

Favourable Condition

Heavy rain before harvest when the leaves are drying can develop the disease. Humid and warm conditions with optimum temperature of 20–30°C. Temperature above 30°C favours the disease development during storage.

Survival and Spread

- Crop residue and contaminated soil are the important inoculum source.
- Insects, splashing rain, irrigation water actively spread the bacteria.
- Especially injuries caused by transplantation, mechanical damage have the capacity to enter the bulbs. Additionally, when maggot feeds on the onion introduce the bacteria they are carrying.

Management

- Copper-based bactericides mixed with mancozeb are applied at a 7-day interval. The application begins two weeks before bulb initiation and continues up to harvest, which will slow down the infection and spread.
- Harvest only after onion tops have well grown to avoid infection to maturing or harvested bulbs.
- Store them between 32 to 36°F (32-34°F for sour skin) with excellent ventilation.

Viral Diseases

Onion Yellow Dwarf

Causal organism: *Onion yellow dwarf virus* (OYDV)

Vector: Aphids (*Myzus persicae*)

It is a plant virus that primarily infects onions, garlic and leeks. It belongs to the family Potyviridae and transmitted by aphids. OYDV can cause significant damage to onion crops, leading to reduced yields and economic losses.

Symptoms

Infected leaves exhibit symptoms ranging from yellow streaks to complete yellowing. In general, leaves crinkle, bend over, flatten, and flatten. Plants may wilt and become dwarfed, and though bulbs are usually solid, they often develop to their full size. Plants in seed crops produce fewer florets and smaller flower clusters (Lot *et al.*, 1998).

Yellow streaks to complete yellowing of leaves

Pathogen Character

It has a length of 10538 nucleotides and to encode a polyprotein which was 3403 amino acid and had an estimated Molecular weight of 385.1 k Da (Chen *et al.*, 2003).

Mode of Spread

The main vectors for OYDV transmission are aphids. These insects acquire the virus from diseased plants and they are able to spread it to healthy plants. Because of the non-persistent nature of this transmission, the virus is not retained by the aphids for an extended period of time.

Management

- Crop rotation, use of virus-free planting material.
- Spraying of neem oil 2% and NSKE (Neem seed kernel extract) 5% for controlling aphid population
- Spraying of emamectin benzoate or pyrethroid based insecticide significantly reduce the aphid population.

14. Iris Yellow Spot

Causal organism: *Iris yellow spot virus* (IYSV)

Vector: Thrips (*Thripstabaci*).

It belongs to the genus Tospovirus, family Bunyaviridae. In some parts of the world, IYSV is a serious threat for onion growers as it is predominantly spread by onion thrips.

Symptoms

IYSV infection in onions shows yellow to straw-coloured lesions or spots on the leaves with a distinctive yellow halo. These spots can combine together, causing the leaves to become yellow and eventually die. Infection can affect bulb size and productivity in extreme situations. Necrotic streaks and rings are also signs of infected bulbs (Sevik and Akcura, 2013).

Yellow to straw-coloured lesions on the leaves

Pathogen Character

- The single-stranded, tri-segmented RNA genome of the *Iris Yellow Spot Virus* (IYSV)is organized as large (L), medium (M), and small (S).
- **L-RNA**: In a negative sense, codes for the RdRp (RNA-dependent RNA polymerase).
- **M-RNA:** It codes for the precursor GN/GC glycoprotein and anon-structural protein (NSm) in the viral complementary (vc) sense and in the viral (v) sense by using ambisense coding.
- **S-RNA:** It employs an ambisense coding approach in the vc and v senses, using the coat protein (N) and a non-structural protein (NSs).(Bag *et al.*, 2015).

Mode of Spread

The vectors of IYSV are onion thrips. These tiny insects feed on onion plants and can spread the virus from infected to healthy plants. IYSV can also spread through contaminated equipment and plant debris (Kritzman *et al.*, 2001).

Management

- Cultural practices like adequate soil moisture, avoid excessive application of nitrogen fertilizer and field sanitation.
- Flooding of infested field will check the thrips population.
- Spraying of insecticide like dimethoate @ 0.06% or profenofos @ 0.05% for managing onion thrips

Conclusion

Onions are prone to several kinds of bacterial, viral, and fungal diseases. Basal rot, purple blotch, downy mildew, onion smut, soft rot and yellow dwarf virus are some of the disease which affects the yield and productivity of onion. To reduce these adverse impacts in onion production, effective disease management is essential. Prevention is often the most effective strategy. Implement crop rotation, select disease-resistant onion varieties, and maintain proper field hygiene to minimize disease incidence. Proper spacing, irrigation management, and soil health maintenance can enhance onion crop resilience to diseases. Treating onion seeds with fungicides or bioagents can prevent the introduction of diseases into the field through contaminated seeds. Application of fungicide according to recommended guidelines to minimize environmental impact. Adopting an IDM approach that combines various strategies, including biological control, cultural practices, and chemical treatments as a last resort, can provide a sustainable and effective solution for managing onion diseases.

References

Abd-Alla, M. H., Bashandy, S. R., Ratering, S., & Schnell, S. (2011). First report of soft rot of onion bulbs in storage caused by *Pseudomonas aeruginosa* in Egypt. Journal of plant Interactions, 6(4), 229-238.

Agricultural Statistics at a Glance. (2021). Ministry of Agriculture & Farmer's Welfare,May 2022.https://eands.dacnet.nic.in/PDF/Agricultural%20Statistics%20at%20a%20Glance%20-%202021%20(English%20version).pdf

Agriculture: Onion and Garlic Pest Management Guidelines. (2019). UCIPM, February 2019. https://ipm.ucanr.edu/agriculture/onion-and-garlic/bacterial-soft-rot/#:~:text=Soft%20rots%20generally%20appear%20just,only%20a%20concern%20on%20onion.

Ahmad, M., Khan, M. P. Z., Mukhtar, A., Zafar, M., Sultana, S., & Jahan, S. (2016). Ethnopharmacological survey on medicinal plants used in herbal drinks among the traditional communities of Pakistan. Journal of Ethnopharmacology, 184, 154-186.

Bachkar, C. B., Bhalekar, M. N., &Pawar, S. A. (2018). Control of Alternariaporri causing purple blotch of onion during kharif season. Journal of Allium Research, 1(1).

Bag, S., Schwartz, H. F., Cramer, C. S., Havey, M. J., & Pappu, H. R. (2015). Iris yellow spot virus (Tospovirus: Bunyaviridae): from obscurity to research priority. Molecular plant pathology, 16(3), 224-237.

Chen, J., Adams, M. J., Zheng, H. Y., & Chen, J. P. (2003). Sequence analysis demonstrates that Onion yellow dwarf virus isolates from China contain a P3 region much larger than other potyviruses.

Conn, E. K., Lutton, J. S., & Rosenberger, S. A. (2012). Onion disease guide. Plant Health, 72.

Coşkuntuna, A., &Özer, N. (2008). Biological control of onion basal rot disease using Trichodermaharzianum and induction of antifungal compounds in onion set following seed treatment. Crop protection, 27(3-5), 330-336.

Develash, R. K., &Sugha, S. K. (1997). Incidence of onion downy mildew and its impact on yield. Indian Phytopathology, 50, 127-130.

Develash, R. K., &Sugha, S. K. (1997). Management of downy mildew (Peronospora destructor) of onion (Allium cepa). Crop Protection, 16(1), 63-67.

Dissanayake, M. L. M. C., Kashima, R., Tanaka, S., & Ito, S. I. (2009). Genetic diversity and pathogenicity of Fusariumoxysporum isolated from wilted Welsh onion in Japan. Journal of General Plant Pathology, 75, 125-130.

Dutta, R., K, J., Nadig, S. M., Manjunathagowda, D. C., Gurav, V. S., & Singh, M. (2022). Anthracnose of onion (*Allium cepa* L.): A twister disease. Pathogens, 11(8), 884.

Ebenebe, A. C. (1980). Onion twister disease caused by Glomerellacingulata in northern Nigeria. Plant disease.

Gedefaw, Y., Gezahegn, A., Fekadu, A., &Mehari, Z. (2019). First report of Stemphyliumvesicarium causing onion Stemphylium leaf blight in Ethiopia.

Ghanbarzadeh, B., Safaie, N., MohammadiGoltapeh, E., RezaeeDanesh, Y., &Khelghatibana, F. (2016). Biological control of Fusarium basal rot of onion using *Trichodermaharzianum* and Glomus mosseae. Journal of Crop Protection, 5(3), 359-368.

Gupta, R. C., & Gupta, R. P. (2013). Effect of integrated disease management packages on diseases incidence and bulb yield of onion (*Allium cepa* L.). SAARC Journal of Agriculture, 11(2), 49-59.

Hahuly, M. V., Sumardiyono, C., Wibowo, A., Subandiyah, S., & Harper, S. (2018). Identification of purple blotch pathogen of shallot by PCR using specific primer for Alternaria genus. Archives of Phytopathology and Plant Protection, 51(3-4), 103-121.

Hassan, M., Yousuf, V., Shah, T. A., Bhat, N. A., Bhat, Z. A., Majeed, M., &Latief, R. (2020). Stemphylium Blight of Onion: A Review. Agricultural Reviews, 41(1).

Kalman, B., Abraham, D., Graph, S., Perl-Treves, R., MellerHarel, Y., &Degani, O. (2020). Isolation and identification of Fusarium spp., the causal agents of onion (Allium cepa) basal rot in Northeastern Israel. Biology, 9(4), 69.

Kiehr, M., Delhey, R., &Azpilicueta, A. (2012). Smudge and other diseases of onion caused by Colletotrichumcircinans, in southern Argentina. Phyton, 81(1), 161-164.

Kritzman, A., Lampel, M., Raccah, B., & Gera, A. (2001). Distribution and transmission of Iris yellow spot virus. Plant Disease, 85(8), 838-842.

Kumar, V., Neeraj, S. S., &Sagar, N. A. (2015). Post-harvest management of fungal diseases in onion—a review. International Journal of Current Microbiology and Applied Sciences, 4(6), 737-52.

Lot, H., Chovelon, V., Souche, S., &Delecolle, B. (1998). Effects of onion yellow dwarf and leek yellow stripe viruses on symptomatology and yield loss of three French garlic cultivars. Plant disease, 82(12), 1381-1385.

Malathi, S. (2010). Ecofriendly management studies on onion. Allium cepa.

McDonald, M. R., de los Angeles Jaime, M., &Hovius, M. H. (2004). Management of diseases of onions and garlic. In Diseases of Fruits and Vegetables: Volume II: Diagnosis and Management (pp. 149-200). Dordrecht: Springer Netherlands.

Palti, J. (1989). Epidemiology, prediction and control of onion downy mildew caused by Peronospora destructor. Phytoparasitica, 17, 31-48.

Patil, S., Nargund, V. B., Hariprasad, K., Hegde, G., Lingaraju, S., &Benagi, V. I. (2018). Etiology of twister disease complex in onion. International Journal of Current Microbiology and Applied Sciences, 7(12), 3644-3657.

Priya, R. U., Sataraddi, A., &Darshan, S. (2016). Survey for purple blotch of onion (Alternariaporri (Ellis) Cif.) in northern parts of Karnataka. International Journal of Agriculture, Environment and Biotechnology, 9(3), 367.

Reecha, J., Thiruvudainambi, S., Mareeswari, P., Oviya, R., &Vellaikumar, S. (2022). Morphological characterization of Colletotrichumgloeosporioides causing leaf twister blight disease in Allium cepa (onion).

Schwartz, H. F. (2011). Soil-borne diseases of onion. Crop series. Diseases; no. 2.940.
Sevik, M. A., &Akcura, C. (2013). Viruses occurring in onion crop in Amasya province, the major onion producing region in Turkey. Indian Journal of Virology, 24, 78-81.
Shishkoff, N., &Lorbeer, J. W. (1989). Etiology of Stemphylium leaf blight of onion. Phytopathology, 79(3), 301-304.
Srinivasan, R., &Shanmugam, V. (2006). Post-harvest management of black mould rot of onion. Indian Phytopathology, 59(3), 333-339.
TNAU.2015. https://agritech.tnau.ac.in/crop_protection/onion_diseases_5.html
Wilson, E. A., &Demmig-Adams, B. (2007). Antioxidant, anti-inflammatory, and antimicrobial properties of garlic and onions. Nutrition & Food Science, 37(3), 178-183.
Yarwood, C. E. (1943). Onion Downy mildew. Hilgardia, 14(11).

18

Diseases of Coconut and Their Management

P. Sujhee[1] and Reecha J.[2]

[1]Department of Plant Pathology, Kerala Agricultural University, Vellanikara, Thirusur, Kerala
[2]Department of Plant Pathology, Sardar Vallabhbhai Patel University of Agriculture and Technology, Meerut, Uttar Pradesh

Abstract

The coconut tree (Cocos nucifera) is the only living species of the genus Cocos and a member of the Arecaceae palm tree family. In this chapter we will be discussing the diseases caused in coconut which causes major yield loss. Leaf withering and drooping, flaccidity, ribbing, paling/yellowing, and necrosis of leaflets are common disease symptoms. Bud rot, stem bleeding, leaf rot, Tanjore wilt and root (wilt) were all important coconut diseases.

Keywords: *Bud rot, Stem bleeding, Leaf rot, Tanjore wilt, Root (wilt).*

Coconut

Kingdom: Plantae

Phylum: Magnoliophyta

Class: Liliopsida

Subclass: Arecidae

Order: Arecales

Family: Arecaceae

Genus: Cocos

Species: *Cocos nucifera* Linnaeus

1. Bud Rot

The fungus *Phytophthora palmivora* is responsible for the ailment known as "bud rot." Butler first mentioned bud rot in India in 1906. Due to a number of factors, the disease has recently spread to practically every region of the state in a very serious manner.

DOI: 10.1201/9781042014460-18

Symptoms

All palms are subject to attack, but immature palms are more vulnerable during the monsoon, when the temperature is low and the humidity is high. The leaf turns pallid in seedlings and readily rips off with a gentle pull. The first indicator is yellowing of one or two younger leaves. Black spots appear on spindle leaves. The base tissues of the leaf degrade swiftly and easily separate from the crown. Infection spreads to older leaves, causing sunken leaf spots to cover the entire leaf blade. The spot boundaries are irregular, water-soaked and when the leaves are unfolded, the typical irregular dots on the blade are evident.In the latter phases, the spindle withers and dips down. The delicate leaf base and crown soft tissues deteriorate into a slimy mass of decomposing material that creates a foul odor. Finally, the palm dies as the entire crown falls.

Disease Symptoms

Pathogen Character

P. palmivora differs from *P. megakarya* morphologically in that it has narrow, medium length sporangial pedicels and spherical, narrow-stalk oogonia as opposed to *P. megakarya* broad, short, and occluded sporangial pedicels and pyriform, downward tapering oogonia.

Management

- Spray neighbouring healthy palms with a 1 percent Bordeaux mixture or another copper-based fungicide such as Copper Oxy Chloride at 2g per litre of water as a prophylactic measure. For disease management, a pre and post-monsoon application of the aforementioned fungicide is advised. Maintaining proper drainage also reduces the occurrence of sickness.
- Spraying the palms with Dimethomorph at a concentration of 1 gramme per litre of water.
- Spraying the palms with Metalaxyl 35 EC at a concentration of 2 g/liter of water.

2. Stem Bleeding

Stem bleeding in coconut palms is commonly connected with the fungus *Thielaviopsis paradoxa*. In India, the condition was first documented in 1922. There is little yield loss in the early phases, but there is a continuous reduction in yield in the later stages. The condition can potentially kill the palm in its severe stages. Stem bleeding is a symptom of many plant diseases, and the pathogens that cause it differ depending on the plant species and disease. Here are some general characteristics of infections linked to plant stem bleeding. Stem bleeding is identified by the exudation of a dark reddish-brown liquid from growing fissures and incisions on the stem, which drips down for a short period of time.

Etiology

Fungus produces macro and micro conidia. Macro conidia are spherical, hyaline, cylindrical and dark green produced conidiophores singly or in chains.

Symptoms

The exudation of a dark reddish brown liquid from longitudinal fissures in the bark and wounds on the stem trickles down for several inches to several feet, indicating stem bleeding. As the disease advances, the lesions spread upwards. The oozing liquid eventually dries up and turns black. The tissues beneath the lesions deteriorate and turn yellow, then black. The interior of damaged trunks werehollow in advanced cases due to degradation of inner tissues. The outer whorl of the leaves becomes yellow, dries, and sheds prematurely as a result of significant damage to the stem tissue. Bunch production suffers as a result. There is also a nut fall. In chronic situations, the trunk gradually tapers towards the apex and the crown size shrinks.

Exudation of reddish-brown liquid through cracks on a coconut trunk

Spore Dispersal

The life cycle begins when fungal spores are dispersed in the environment. These spores are usually present in infected plant material, soil, or organic debris around coconut trees.

Infection

The fungal spores can enter the coconut tree through wounds, injuries, or natural openings such as leaf scars or bud scars.

Once inside the plant, the spores germinate and produce hyphae (thread-like structures) that start to colonize the plant's vascular tissues.

Colonization

The fungus spreads within the coconut tree's vascular system, primarily affecting the xylem tissue, which is responsible for transporting water and nutrients.

As the fungus colonizes the xylem, it obstructs the water-conducting vessels and disrupts the normal flow of water and nutrients within the tree.

Symptom Development

The blockage of the xylem vessels and the disruption of water and nutrient transport lead to characteristic symptoms of stem bleeding.

Stem bleeding symptoms include the oozing of a reddish-brown or black liquid from wounds or lesions on the tree's stem.

The bleeding sap may contain fungal spores, facilitating the spread of the disease to other parts of the tree or to nearby trees.

Progression

If left uncontrolled, the fungal colonization can progress further up the tree, causing wilting, dieback, and reduced fruit production.

Severe infections can lead to the death of the coconut tree.

Survival and Spread

The fungus can survive in infected plant debris, soil, or other susceptible host plants in the vicinity.

Spores produced within the coconut tree or in infected debris can serve as a source of inoculum for new infections.

The disease can spread through various means, including rain splashes, insect vectors, and human activities (e.g., pruning with contaminated tools).

The life cycle of stem bleeding in coconut trees can be influenced by environmental conditions, tree health, and cultural practices. Management strategies typically involve the removal and destruction of infected plant material, proper sanitation, and the use of fungicides when necessary. Disease-resistant coconut varieties may also be employed to reduce the risk of infection. Timely intervention is essential to control the disease and prevent its spread to healthy coconut trees.

Management

Managing stem bleeding in coconut trees is essential to prevent the spread of disease and maintain the health of the tree. Stem bleeding in coconut trees can be caused by various factors, including fungal infections and physical injuries. Here are steps to manage stem bleeding in coconut trees:

Identify the Cause: First, you need to determine the cause of the stem bleeding. Common causes include the fungus Thielaviopsisparadoxa, which causes basal stem rot, and physical injuries from pruning or accidents.

Prune Affected Areas: If the stem bleeding is due to physical injury, carefully prune away the damaged or injured tissue. Use clean and sharp pruning tools to make clean cuts, and remove any dead or decaying tissue. Make sure not to remove healthy tissue, as this can further weaken the tree.

Apply Fungicides: If the stem bleeding is caused by a fungal infection like basal stem rot, consider using fungicides recommended for coconut trees. Follow the manufacturer's instructions for application carefully. Fungicides may help control the spread of the fungus.

Improve Drainage: Ensure that the coconut tree's root zone has proper drainage. Poor drainage can contribute to the development and spread of fungal infections. Avoid overwatering and make sure the soil is well-draining.

Maintain Nutrient Balance: Proper nutrition is essential for tree health and resistance to diseases. Conduct soil tests to determine if the soil lacks any essential nutrients, and amend the soil as needed to provide balanced nutrition to the coconut tree.

Pruning and Maintenance: Regularly prune your coconut tree to remove dead or diseased fronds and maintain good airflow within the canopy. This can help reduce the risk of fungal infections.

Avoid Injuries: Take precautions to prevent physical injuries to the tree. When pruning, use clean and well-maintained tools, and be careful not to damage the stem or roots. Protect the tree from physical damage during agricultural operations and construction activities.

Sanitation: Remove and destroy any infected plant material, including pruned branches and fronds, to prevent the spread of diseases.

Professional Assistance: If the stem bleeding persists or worsens despite your efforts, consider seeking advice from a local agricultural extension service or a professional arborist who specializes in coconut trees.

Preventive Measures: To prevent future occurrences, maintain good orchard hygiene, practice crop rotation if applicable, and monitor your coconut trees regularly for signs of disease or injury.

Remember that early detection and prompt action are crucial in managing stem bleeding in coconut trees. Regular monitoring and good cultural practices can go a long way in maintaining the health of your coconut plantation.

3. Leaf Rot

Leaf rot in coconut trees is a common fungal disease that can affect the health and productivity of coconut palms. It is caused by several different fungi, with the most common being the *Phytophthora palmivora* and *Pestalotiopsis spp.*

The fungus generates hyaline intercellular fluid.

- Sporangiophores are hyaline and either simple or infrequently branching.
- The sporangia are hyaline, with thin walls and conspicuous papillae. When Sporangia germinates, it produces 200 reniform, biflagellate spores.
- The fungus also develops spherical oospores with thick walls. Furthermore, thickwalled, yellowish brown chlamydospores are generated.

Symptoms

Leaf rot in coconut trees is a fungal disease caused by various pathogens, including species of Phytophthora and Ganoderma. It can have serious implications for coconut plantations as it affects the leaves, which are crucial for photosynthesis and overall tree health. The symptoms of leaf rot in coconut trees can vary depending on the causative agent and the stage of infection, but here are some common symptoms to look for:

Yellowing of Leaves: Initially, the affected coconut palm may exhibit yellowing or discoloration of its older leaves. This symptom often starts at the tips and margins of the leaflets.

Leaf Lesions: As the disease progresses, dark brown to black lesions may develop on the leaflets, particularly along the margins and midrib. These lesions can vary in size and shape.

Leaf Blight: In some cases, leaf rot can lead to the development of leaf blight, where large portions of the leaflets become necrotic (dead tissue) and exhibit a water-soaked appearance.

Wilting: Infected leaves may start to wilt and droop, giving the canopy an overall unhealthy and disheveled appearance.

Premature Leaf Shedding: As the disease advances, the affected coconut tree may shed its infected leaves prematurely. This can result in a loss of photosynthetic capacity, reduced growth, and decreased fruit production.

Bud Rot: In severe cases of leaf rot, the pathogen may progress to the bud or crown of the palm, causing bud rot. This can be a devastating stage of the disease, as it can lead to the death of the entire tree.

Fungal Growth: Depending on the specific pathogen involved, you may also observe fungal growth on the affected leaves or the trunk of the coconut tree. Ganoderma, for example, often produces conspicuous, shelf-like fungal structures on the trunk.

Decline in Tree Health: Over time, leaf rot can weaken the coconut palm, making it more susceptible to other diseases and environmental stresses.

It's important to note that the symptoms of leaf rot can be similar to those of other coconut diseases and nutritional deficiencies, so proper diagnosis is essential. Management strategies for leaf rot in coconut trees typically include the removal and destruction of infected plant material, improving drainage in the plantation, and implementing cultural practices to reduce moisture and improve overall tree health. Fungicides may also be used in some cases, although prevention through good agricultural practices is often the most effective approach.

Brown colour spots

Fan like appearance of leaves enlarge resulting in rotting

Life Cycle

The life cycle of leaf rot in coconut trees is a complex process involving various stages and factors. Leaf rot is typically caused by a fungal pathogen, and its development can be influenced by environmental conditions. Here's an overview of the life cycle of leaf rot in coconut trees:

Infection Stage

The life cycle begins when fungal spores, often from the genus Phytophthora or Pestalotiopsis, come into contact with coconut tree leaves. These spores are usually present in the environment, carried by wind, rain, or other vectors.

Environmental factors like high humidity, warm temperatures, and rain facilitate the spread of spores and increase the likelihood of infection.

Penetration and Colonization

Once spores land on coconut leaves, they need favorable conditions to germinate and penetrate the leaf tissue.The fungus starts to colonize the leaf's cells, breaking down plant tissues and causing damage.

Symptom Development

As the fungus continues to grow within the leaf tissue, symptoms of leaf rot become visible.These symptoms may include dark brown to black lesions or spots on the leaves, which may initially appear water-soaked or oily.

Spore Production

As the fungus matures, it produces reproductive structures like sporangia (in the case of Phytophthora) or conidia (in the case of Pestalotiopsis).These structures release new spores, which can be carried by wind, rain, or contact to infect other parts of the same tree or nearby coconut trees.

Spread and Secondary Infections

The newly produced spores can spread to other leaves, stems, or even neighboring coconut trees, initiating secondary infections.In some cases, spores can also be transmitted through pruning tools, contaminated soil, or human activity.

Disease Progression

If environmental conditions remain conducive, the disease can progress, causing more severe leaf rot and potentially affecting the overall health of the coconut tree. Severe infections can lead to defoliation and reduced coconut production.

Survival Structures

Some fungal pathogens may form survival structures such as oospores (in the case of Phytophthora) that can persist in the soil or on infected plant debris, contributing to the disease's persistence.

Seasonal Variation

The life cycle of leaf rot might vary based on seasonal variables. Disease development may be more prominent during periods of excessive humidity and rains.

Management

The management of leaf rot in coconut trees involves a combination of cultural, chemical, and preventive measures to control and minimize the impact of the disease. Here are some strategies for managing leaf rot in coconut trees:

Cultural Practices

Pruning: Regularly prune and remove infected leaves and leaflets from the coconut tree. This helps reduce the source of infection and prevents the spread of the disease to healthy parts of the tree.

Sanitation: Properly dispose of pruned materials and fallen leaves by burning or burying them away from coconut trees. This reduces the chances of spore survival and reinfection.

Irrigation Management: Avoid overhead irrigation, as it can create favorable conditions for disease development. Instead, use drip or ground-level irrigation systems to keep foliage dry.

Spacing and Air Circulation: Plant coconut trees with adequate spacing to ensure good air circulation. This helps reduce humidity around the trees, making it less favorable for fungal growth.

Nutrient Management: Maintain proper soil nutrition and pH levels to keep coconut trees healthy and more resistant to diseases. Regularly fertilize and address nutrient deficiencies.

Chemical Control

Fungicides: In severe cases of leaf rot, chemical fungicides can be used as a last resort. Consult with a local agricultural extension service or a plant pathologist for guidance on selecting and applying appropriate fungicides. Follow label instructions carefully.

Preventive Measures

Crop Rotation: Avoid planting coconut trees in areas where leaf rot has been a recurring problem. Rotate crops to reduce the buildup of disease-causing pathogens in the soil.

Quarantine: Inspect new coconut seedlings or plant materials for signs of disease before introducing them to your coconut plantation. Isolate and treat any infected plants.

Resistant Varieties: Consider planting coconut varieties that are known to be resistant or less susceptible to leaf rot if available in your region.

Monitoring: Regularly inspect coconut trees for early signs of leaf rot, such as brown or black lesions on leaves. Early detection allows for timely intervention.

Weather and Climate Management: Be mindful of weather conditions, as leaf rot tends to be more severe during periods of high humidity and heavy rainfall. Adjust your management practices accordingly during such times.

Integrated Pest Management (IPM): Incorporate an integrated pest management approach, which combines various strategies to minimize disease pressure and maximize coconut tree health.

Remember that prevention is often more effective than trying to cure a severe leaf rot infection. Implementing a combination of these management practices tailored to your specific location and coconut tree cultivation conditions can help reduce the impact of leaf rot and promote the overall health of your coconut plantation.

Consulting with local agricultural experts can provide valuable insights into the most effective management practices for your region.

4. Tanjore Wilt

Thanjavur wilt (Ganoderma wilt) begins with withering, yellowing, and drooping of the outer whorl of leaves. This is followed by the exudation of reddish brown liquid through breaks at the trunk's base and the spread of leaking upward. The tissues on the bleeding areas are tender to the touch. Decay of tissues at the bleeding point and rotting of the stem's base. The bark becomes brittle and peels off in flakes, exposing gaping fissures and crevices. Internal tissues have darkened and decomposed, generating a foul odour. Bracket formation at the trunk's base. Ganoderma appears near the trunk's base. The palm eventually dies.

Symptoms

Tanjore wilt, also known as Ganoderma wilt, is a devastating disease that affects coconut palms. It is caused by the fungus Ganodermaboninense. Symptoms of Tanjore wilt in coconut palms include:

Leaf Symptoms

Wilting Fronds: One of the earliest signs is the wilting and drooping of older fronds (leaves) starting from the lower canopy. This wilting is often rapid and can affect multiple fronds simultaneously.

Yellowing and Browning: Infected fronds may initially turn yellow and then progress to a reddish-brown or dark brown color. The discoloration typically begins at the tip and edges of the fronds.

Leaf Drop: As the disease progresses, affected fronds may detach from the palm prematurely, leading to a significant loss of leaves.

Stem and Trunk Symptoms

Conk Formation: Ganoderma-infected coconut palms often develop conks or woody shelf-like growths (basidiocarps) on the lower part of the trunk. These conks are a distinctive feature of Ganoderma infection.

Trunk Rot: The fungus attacks the internal tissues of the palm trunk, causing it to become soft, spongy, and brown. This can lead to a reduction in the structural integrity of the tree.

Mushroom Growth: In advanced stages of the disease, mushroom-like structures may emerge from the base of the trunk or the soil around the tree. These structures release spores, contributing to the spread of the fungus.

General Symptoms

Reduced Growth: Coconut palms affected by Tanjore wilt exhibit stunted growth, reduced fruit production, and a decline in overall vigor.

Nutrient Deficiency: The disease can cause nutrient deficiencies in the palm, as the compromised root system struggles to absorb essential nutrients.

Tree Death: If left untreated, Tanjore wilt can ultimately lead to the death of the coconut palm.

It's important to note that once a coconut palm is infected with Ganoderma, there is no known cure. Management strategies for Tanjore wilt primarily focus on preventing the disease's spread, which includes removing and destroying infected palms, avoiding the use of infected palm residues for mulching or replanting, and practicing good sanitation and quarantine measures.Regular monitoring and early detection of symptoms are crucial for controlling the spread of Tanjore wilt, as prompt removal of infected trees can help prevent further infestations in coconut plantations

Disease symptoms

Etiology

Basidiomycete Fungus: *Ganoderma boninense* belongs to the Basidiomycota phylum, which includes many fungi that produce mushrooms or other fruiting bodies.

Conks: The most recognizable feature of Ganoderma infection is the formation of woody, shelf-like structures called conks or basidiocarps on the lower trunk of infected palms. These conks release fungal spores.

Life Cycle

The term "Tanjore wilt" typically refers to a disease that affects certain crops, particularly banana plants. The life cycle of Tanjore wilt can be understood as follows:

Introduction of the Pathogen: The disease cycle begins when the pathogen responsible for Tanjore wilt is introduced into a susceptible host plant. In the case of banana plants, the pathogen is a soil-borne fungus called *Fusarium oxysporum f. sp. cubense* (Foc), which infects the banana plant's vascular system.

Infection: The pathogen enters the banana plant through the roots and colonizes the vascular tissue. It forms hyphae, which are thread-like structures, and begins to multiply within the plant.

Vascular Colonization: As the pathogen continues to grow, it spreads throughout the banana plant's vascular system, causing damage to the plant's ability to transport water and nutrients. This results in wilting and yellowing of the leaves, which are characteristic symptoms of Tanjore wilt.

Symptom Development: Over time, as the pathogen disrupts the plant's vascular system, the banana plant starts to exhibit visible symptoms. These symptoms include wilting, yellowing, and eventually, the death of the plant.

Spore Formation: The pathogen produces spores, which are the reproductive structures. In the case of Fusariumoxysporum, these spores are called conidia. These spores can survive in the soil for extended periods, even after the infected plant has died.

Soil Contamination: The spores released by the infected plant can contaminate the surrounding soil. This soil contamination becomes a source of infection for future banana plants.

New Infections: When new banana plants are planted in soil contaminated with the pathogen, they can become infected, continuing the cycle. The pathogen can persist in the soil for several years, making it a long-term threat to banana cultivation in affected areas.

To manage and control Tanjore wilt, it is crucial to implement various measures, such as using disease-free planting material, practicing crop rotation, soil solarization, and employing resistant banana varieties, as well as soil treatments to reduce the pathogen's survival in the soil. Proper sanitation and good agricultural practices are essential to break the cycle of Tanjore wilt and protect banana crops.

Management

Tanjore wilt, also known as Fusarium wilt or Panama disease, is a devastating fungal disease that affects banana plants, particularly the Cavendish variety. Managing Tanjore wilt requires a combination of preventive and control measures to minimize its impact on banana plantations. Here are some management practices for Tanjore wilt:

Plant Resistant Varieties: Use banana varieties that are resistant to Tanjore wilt. While the Cavendish variety is highly susceptible, there are other resistant cultivars available that can be planted.

Quarantine and Sanitation: Implement strict quarantine measures to prevent the introduction of infected plant material onto your plantation. Ensure that all equipment, tools, and personnel are free from contamination. Remove and destroy infected plants immediately to prevent the disease from spreading.

Soil Management: Rotate crops and avoid planting bananas in the same soil for consecutive seasons. This can help reduce the buildup of the pathogen in the soil.

Fungicide Treatment: In areas with a history of Tanjore wilt, consider applying fungicides that are effective against the Fusarium wilt pathogen. Consult with agricultural experts or local extension offices for guidance on appropriate fungicides and application methods.

Use of Biological Controls: Explore the use of beneficial microorganisms, such as *Trichoderma* spp., which can help suppress the growth of the Fusarium pathogen in the soil.

Proper Irrigation: Manage irrigation carefully to avoid overwatering, which can create favorable conditions for the disease. Ensure good drainage to prevent waterlogging.

Avoid Soil Contamination: Prevent soil contamination by implementing proper sanitation practices, such as cleaning boots and tools when moving between fields.

Resistant Rootstocks: Consider using resistant rootstocks, which can provide some degree of protection to the susceptible scion variety. Grafted plants with resistant rootstocks can be an effective strategy.

Crop Rotation: If possible, rotate banana crops with other non-host crops for several years to break the disease cycle.

Integrated Pest Management (IPM): Implement an IPM approach, which combines various strategies such as biological control, cultural practices, and chemical control in a coordinated manner to manage the disease.

Regular Monitoring: Regularly monitor your banana plants for signs of Tanjore wilt, such as wilting, yellowing of leaves, and brown discoloration of vascular tissue. Early detection can help prevent the disease from spreading.

Training and Education: Ensure that farmworkers are trained to recognize the symptoms of Tanjore wilt and understand the importance of proper sanitation and disease management practices.

Consult Experts: Seek advice and guidance from agricultural experts, local agricultural extension services, or agricultural universities for the latest information and best practices in managing Tanjore wilt in your specific region.

Remember that managing Tanjore wilt requires a multifaceted approach, and it may vary depending on the specific conditions of your plantation. Implementing a combination of these management practices can help reduce the impact of the disease and protect your banana crop.

Conclusion

The management of coconut diseases requires a multi-faceted approach that incorporates prevention, early detection, and a combination of cultural, biological, and chemical control methods. By adopting these strategies and staying informed about the latest research and innovations, coconut growers can protect their crops and ensure the long-term viability of the coconut industry.

References

Alshawwa, I. A., Elsharif, A. A., & Abu-Naser, S. S. (2019). An Expert System for Coconut Diseases Diagnosis.

Bhaskaran, R. (2000). Management of basal stem rot disease of coconut caused by Ganodermalucidum. In Ganoderma diseases of perennial crops (pp. 121-128). Wallingford UK: CABI.

Carvalho, R. R., Souza, P. E., Warwick, D., Pozza, E.A., & Carvalho Filho, J.L.D. (2013). Spatial and temporal analysis of stem bleeding disease in coconut palm in the state of sergipe, Brazil. Anais da Academia Brasileira de Ciências, 85, 1567-1576.

Dollet, M., De Franqueville, H., &Ducamp, M. (2012). Bud rot and other major diseases of coconut, a potential threat to oil palm.

Harrison, N. A., & Jones, P. (2003). Diseases of Coconut. Diseases of tropical fruit crops, 197-225.

Kandan, A., Bhaskaran, R., &Samiyappan, R. (2010). Ganoderma–a basal stem rot disease of coconut palm in south Asia and Asia pacific regions. Archives of Phytopathology and Plant Protection, 43(15), 1445-1449.

Nelson, S. (2005). Stem bleeding of coconut palm.

Sanderson, F. R., Pilotti, C. A., & Bridge, P. D. (2000). Basidiospores: their influence on our thinking regarding a control strategy for basal stem rot of oil palm. In Ganoderma diseases of perennial crops (pp. 113-119). Wallingford UK: CABI.

San Juan Bachiller, N. (2004). Distribution and progression of phytophthora bud rot disease of coconut in selected areas in the Philippines. Diversity and Management of Phytophthora in Southeast Asia, 124-131.

Thevenin, J. M., Kharie, S., Motulo, H. F. J., &Renard, J. L. (1995). Chemical control of coconut bud rot caused by Phytophthora in Indonesia.

19

Diseases of Betelvine and Their Management

Ankit Alok[1], Harshit singh[2], Om Prakash[3] and Shruti Sinha[4]

[1,2]Department of Plant Pathology, School of Crop Protection, College of Post Graduate Studies in Agricultural Sciences, Central Agricultural University (Imphal) Umiam, Meghalaya
[3,4]Department of Plant Pathology, College of Agriculture Iroishemba, Central Agricultural University (Imphal) Manipur

Abstract

Betelvine (Piper betle L.), often referred to as the "natural green gold," holds significant commercial value as an important horticultural crop within the Piperaceae family. This plant holds economic, medicinal, and cultural significance worldwide. However, the successful cultivation of betelvine faces various challenges, both from biotic and abiotic factors, including root and aerial diseases. Throughout its growth, betel vine is highly susceptible to diseases like Leaf spot, Leaf rot, and powdery mildew, which can lead to substantial losses for farmers. These diseases can manifest aggressively and, if left uncontrolled, have the potential to cause extensive damage, including the complete destruction of entire betelvine plantations, often without early warning signs. This chapter aims to investigate and identify various diseases affecting betel vine plants and explore strategies for their effective management.

Keyword: *Betelvine, Diseases, Losses, Management*

Introduction

The betel vine, scientifically identified as *Piper betle* L., belongs to the Piperaceae family, which encompasses plants such as pepper and kava. Paan leaves, widely cultivated in countries like Thailand, Philippines, India, Malaysia, Taiwan, Sri Lanka, and various Southeast Asian nations, are commonly used as a popular mouth freshener after meals. This traditional practice is predominantly observed

DOI: 10.1201/9781042014460-19

in South Asia and among specific Asian diaspora communities worldwide, where betel leaves are commonly chewed as part of a betel quid or paan, often combined with areca nut or tobacco.

In the cultural traditions of Odisha, offering a bundle of betel leaves symbolizes respect and marks the beginning of auspicious events. Taxonomically, betelvine falls under the genus Piper in the Division Magnoliophyta, Class Magnoliopsida, Order Piperales, and Family Piperaceae.

The betel vine is a perennial evergreen climber characterized by its glossy heart-shaped leaves and white catkins that blossom in the spring. It is a unisexual plant and is categorized into two types, non-pungentandodorous, based on the shape, size, and flavor of its leaves. The leaves of the plant are simple with a pointed tip and are typically smooth and glossy. They vary in colour from light green to dark green and have long stalks with 2-3 pairs of secondary veins. The plant's branches usually swell at the nodes and are entirely smooth. Female spikes are cylindrical, while male spikes hang down. Female spikes measuring 2.5-5.0 cm in length.

In the moist areas of East India, female betel plants frequently yield flowers and fruit. Over millennia, nature has offered a diverse array of medicinal compounds, and numerous contemporary medications trace their origins to natural reservoirs. Betel leaves harbor a broad spectrum of bioactive compounds, and they have a history of application in traditional medicine. Abundant in minerals, vitamins, enzymes, proteins, and essential oils, these leaves are a nutritionally valuable resource. They also contain valuable therapeutic components used in the treatment of brain, liver, and cardiac disorders.

Betel vines are perennial, dioecious, and evergreen climbers that flourish in conservatories, locally referred to as Baroj, requiring shaded and humid environments crucial for their optimal growth. Nonetheless, the shaded and humid setting, while conducive to betel vine growth, also creates ideal circumstances for various diseases triggered by bacteria, fungi, viruses, and nematodes. These ailments can significantly hinder plant development and lead to substantial agricultural losses. In India, prevalent diseases affecting betel vines encompass phytophthora leaf and foot rot caused by *P. palmivora* or *P. parasinca*, anthracnose caused by *Colletotrichum capsicum*, bacterial leaf spot caused by *Xanthomonas campesitis* pv. *Bellicola*, basal rot induced by *Sclerotium rolfsii*, and root knot caused by nematodes (*Meloidogyne incognita*). These factors pose significant challenges to betel vine cultivation across the nation.

1. Foot Rot and Leaf Rot

Causal organism: *Phytophthora parasitica var, piperina*

In 1927, Dastur first documented cases of foot rot and leaf rot diseases affecting betelvine (*Piper betle* L.) in Durg, attributing the cause to *P. parasitica* var. *piperina*. This ailment has since been observed in betelvine cultivation across numerous countries, including Indonesia, Myanmar, Sri Lanka, Bangladesh, and others. In 1963, Waterhouse identified the pathogen as *Phytophthora nicottanac* var. *paraonca*. Maiti and Sen, in 1978, later identified it as *P. palmivora*, while Marimuthu, in 1991, referred to the pathogen as *P. palmivora*. In more recent literature, *P. palmivora* has been reclassified as *P. capeier*. (Mehrotra and Aggarwal, 2003).

The Midnapore and Nadia districts of West Bengal, along with several other Indian states, have experienced the most severe instances of foot and leaf rot. The disease incidence in these areas can vary from 30% to 100% for both foot rot and leaf rot, frequently leading to near-total crop failure (Maiti and Sen, 1978).

Symptoms

In 1935, Dastur extensively described the symptoms of foot rot, also known as wet rot disease, which often leads to vine wilting. Infected plants display yellowing leaves and withering shoots, eventually drying to a pale brown hue. The disease initially targets the fine young roots, progressively spreading to older roots and eventually impacting the base or collar region of the plant. In affected plants, the entire underground section undergoes extensive decay, with the pathogen decomposing the soft tissues of older roots and internodal sections of cuttings, leaving only the fibrous part intact.

As for leaf rot, it manifests as circular black or brownish water-soaked spots on the leaves. These spots rapidly enlarge, merging with one another and covering a significant area on the leaf blade, resulting in rotting during continuous wet weather. The central decayed part eventually drops out, forming a hole with regular edges. These symptoms can emerge on any part of the leaves, particularly during summer hailstorms. However, the disease progresses more slowly in dry conditions.

Leaf rot typically occurs during the rainy season, from June to August, and is influenced by external factors. It thrives in conditions of low temperature (15.6°C to 27.4°C), high humidity (91.0% to 99.3%), and diffused light inside the conservatory ("baroj"). These conditions are conducive not only for betel vine growth but also for the pathogen causing the disease. According to Huq (2011), the highest infection rates were observed during the second week of August when the average temperature, humidity, and rainfall were recorded at 20.6°C, 94.6%, and 15.4 mm, respectively.

Management

Tiwari and Mehrotra (1968) found that immersing cuttings in a suspension of *Trichoderma viride* cells was an effective method for reducing the disease. Treatments involving soil drenching and spraying with Bordeaux mixture were also carried out. Significantly, immersing cuttings in a streptomycin solution and applying Bordeaux Mixture (1%) twice a month proved to be a complete preventive measure for foot rot. Furthermore, the application of Fosetyl-Al and

Bordeaux mixture demonstrated efficacy in controlling Phytophthora leaf rot in betelvine.

2. Leaf Spot and Stem Anthracnose

Causal organism: *Collectotrichum piperis, C. capsica*

This disease is of utmost significance for the crop, capable of causing yield losses ranging from 10% to 60%. It manifests in a highly aggressive form, and without proper control measures, it can lead to the complete destruction of entire betel vine plantations while also diminishing the crop's market value (Singh and Joshi, 1971).

Symptoms

Typically, each leaf develops multiple spots of varying shapes and sizes, characterized by a central coloration ranging from light to dark brown or black. These spots are surrounded by a diffuse, yellowish halo. In contrast, stem lesions are narrow, elongated, and dark brown, measuring approximately 7.25 x 1.50-2.30 mm. These lesions tend to merge and eventually encircle the stem. Circular, black lesions may also emerge on the stem, rapidly expanding and encircling it, leading to wilting and drying of the affected parts. It's noteworthy that young leaves and tender shoots are more prone to infection compared to mature ones.

Management

Infected samples should be gathered and then incinerated or completely eliminated right in the field. Cultivating resistant varieties, such as BARI Pan-1 and BARI Pan-2, is another effective strategy. After leaf plucking, it's advisable to apply a spray containing either 0.2% Ziram or 0.5% Bordeaux mixture. Additionally, Tilt 250 EC can be sprayed at a rate of 0.5 ml per liter of water at intervals of 10 days, repeated 2 to 3 times. Another option is a foliar spray of Chlorothalonil at a rate of 2.5 g per liter.

3. Sclerotium Foot Rot and Wilt

Causal organism: - *Sclerotium rolfsii*

The first documented case of Sclerotinia wilt in betel vine was reported by Singh and Chand in 1972, revealing significant losses of 40-62% in affected plants. This disease ranks among the most destructive, significantly diminishing the production of betel leaves. The responsible agent, *Sclerotium rolfsii,* is a pathogenic fungus that resides in the soil and poses a threat to various economically important crops in tropical and subtropical regions globally. It boasts a broad range of host plants and is often referred to as an almost universally pathogenic organism, capable of maintaining its lifecycle even under adverse conditions through the formation of sclerotia.

The fungus itself produces mycelium that appears white to gray and is extensively branched. Sclerotia, which are spherical, smooth, and shiny, can be observed as brown, mustard-like structures on infected stems and in the soil near the betel vine plants.

Symptoms

The disease has the potential to impact betel vine plants at every stage of their growth, rendering them vulnerable from the initial stages onward.. Typically, the infection initiates at the collar region of the plant. Whitish, cottony mycelium becomes visible on both the stem and roots. The affected stem tissue begins to rot at the point of attack, leading to leaf drop and eventual withering, culminating in the complete drying up of the plant.

Initially, the fungal mycelium becomes evident near the base of the vines, in close proximity to the soil line. From this point, the pathogen advances upward, enveloping the stem with a cottony-white mass of mycelia. Following this, water-soaked, gray lesions emerge on the vines, transitioning to a brown hue and resulting in the overall decay of the entire plant system. In the collar zone of the plant, numerous small sclerotia, resembling mustard seeds in color and size, develop. Once the pathogen establishes itself, it undergoes swift growth, generating mycelia and sclerotia. Ultimately, the infected vines, typically aged 3-4 weeks, collapse and succumb to the disease.

Management

An ammonia solution of 0.5% proves to be an effective control method when applied to the soil. Managing wilt or root rot can be achieved through an integrated approach to disease management. This method includes subjecting the soil to solarization in the summer, followed by the use of Trichoderma at a rate of 10 kg/ha (applied both to the soil and as a foliar treatment), along with 200 kg/ha of neem cake and NPK in a ratio of 100:50:50 kg/ha. It's noteworthy that deep irrigation proves to be notably more effective, not only in managing the fungus causing wilt and nematodes but also in positively influencing various growth factors.

In the case of collar rot disease, Mopsin has shown to be highly effective in field conditions, providing 100% disease control when used at a 0.1% concentration. For basal rot disease, Rhizolex 50 WP, applied at 0.4% concentration through soil drenching, has proven to be effective.

Several fungicides, such as Metalaxyl + Mancozeb (0.1%), Difenoconazole (0.05%), Propiconazole (0.05%), Hexaconazole (0.05%), and Metiram (0.1%), have exhibited efficacy in managing *Sclerotium rolfsii*..

In the realm of biological control, *Trichoderma lignorum*, *Gliocladium virem*, and *T. viride* have shown the highest effectiveness against *Sclerotium rolfsii*. Furthermore, in the case of botanical treatments, plant extracts from *Allium sativum*, *Azadirachta indica*, and *Catharanthus roseus* (all at a concentration of 10%) have shown promise in combating *S. rolfsii*.

4. Powdery Mildew

Causal organism: *Oidium piperis*

Powdery mildew has been causing notable damage to betel vines grown in Bassein and Kelva Mahim, near Bombay. Interestingly, this problem hasn't been noted in other parts of the province where betel vine cultivation is prevalent. The onset of this disease aligns with the cold season and tends to subside as hot weather approaches. Mature and well-established plantations are more prone to this ailment compared to recently established gardens. The specific *Oidium* species responsible for the disease was first identified by Stevenson (1926) in Ceylon and later confirmed by Mitra. (1930) in Burma, and Narasimhan (1933) in Mysore.

Symptoms

The disease is characterized by the presence of raised yellow spots on the leaves with an irregular shape. These spots correspond in size and shape to the white powdery patches of mildew found on the underside of the leaves. Occasionally, these patches may also be present on the upper surface of the leaves. Initially small, they gradually enlarge and merge together. Their diameter can vary from a few millimeters to as much as 40 millimeters, and they are coated with a sparse, dusty layer of the fungus. In cases of severe infection, these patches become densely covered, imparting a grayish appearance.

When young leaves are severely affected, they stop growing and may become deformed. The surfaces of these diseased leaves develop cracks, and their edges curl inward. Such leaves appear pale and become brittle, often falling off with the slightest touch. It's important to note that no other part of the betel vine is affected by this particular disease.

Management

To control the disease, a management approach involved applying wettable sulfur as a dusting agent on the leaves and promptly removing infected leaves when they were ready for harvest. Among the betel vine cultivars, male clones such as Kapoori, Tellaku, Vellairettala, Ambadi Badam, and Kulgedu were observed to exhibit resistance to powdery mildew.

5. Bacterial Leaf Spot

Causal organism: *Xanthomonas axonopodis pv, betlicola*

The initial identification of a bacterial leaf spot affecting betelvine was reported by Patel *et al.* (1951) in India, and they designated the pathogen as *Xantomonas betlicola*. However, Vauterin *et al.* (1995) later reassigned the pathogen to *Xanthomonas axonopodis* pv. *betlicola*. Initially perceived as a minor ailment, subsequent reports in 1971 from the Jabalpur region of Madhya Pradesh by Singh and Chand (1971) and in 1978 from Kerala. Nadugala and Amarasinghe (2009) documented an incidence of this disease of up to 60% in major betel cultivating areas, which can escalate to around 75% during the rainy seasons. Moreover, variations in phenotypic traits and virulence were noted among various isolates of *Xanthomonas axonopodis* pv. *betlicola*.

Symptomology

The first sign of this disease, commonly heightened during the monsoon season, is the development of small, water-soaked spots on the undersides of the leaves, situated between the veins. With the advancement of the disease, these spots become apparent on the upper surfaces of the leaves, manifesting as dark, rounded, or angular areas encircled by yellow zones. The centers of these spots initially adopt a mottled brown hue, ultimately transforming into black and undergoing decay.

In the initial phases of infection, individual spots may have a diameter of up to one-fifth of an inch. Over time, these spots merge, forming large, irregular dead areas on the leaves. Occasionally, the central necrotic tissue of older spots falls away. In severely affected vines, diseased leaves turn yellow and drop off. Under damp conditions, a sticky substance oozes from the undersides of the affected leaves, containing a multitude of bacteria. In more severe cases, the stem of the plant may also be targeted, and in extreme instances, the entire vines may succumb to the disease.

Management

Leaf spot disease in betel vine has been linked to a complex involving both fungus and bacterium. In such instances, C. capsici is consistently identified alongside the bacterium X. a. pv. betlicola. Effective control measures for this disease complex include the application of chemicals such as 0.5% Bordeaux mixture or 0.1% copper oxychloride (Yadav *et al*., 1993).

In field conditions, the optimal strategy for managing the leaf spot complex involves a combination of sanitation and the use of B. subtilis, resulting in a 9.7% percent disease index (PDI). This was followed by sanitation alone, which achieved an 11.1% PDI, and sanitation combined with *P. hydropiper*. According to Tripathi in 1984, in field evaluations, streptomycin proved to be the most effective treatment, achieving 80-90% disease control, followed by menadione (78-80%) and resorcinol (68-72%).

The occurrence of bacterial leaf spot decreased significantly with an increase in soil phosphorus levels ranging from 25 to 125 kg/ha, as reported by Wasnikar *et al.* (1993). Balasubrahmanyam *et al.* (1988) indicated that the combination of Copper oxychloride 50 WP at a concentration of 0.25% with 500 ppm streptocycline proved highly efficient in combating bacterial leaf spots. Alternatively, utilizing betel vine cultivars resistant to bacterial leaf spot is another effective approach.

6. Root-Knot

Causal organism: *Meloidogyne incognita*

The primary pest causing substantial damage to betel vine orchards is the root knot nematode, identified as *Meloidogyne incognita*. This nematode infestation is documented to result in yield reductions ranging from 26% to 38%. It is noteworthy that the population of root-knot nematodes peaks between March and June within betel orchards.

Both *Sesbania grandiflora* and *S. sesban* have been identified as reservoirs for these nematodes. Root-knot infestations in betel vine plants can arise from several factors, including the application of untreated green manure, the presence of contaminated tools, and occasionally the cultivation of betel vine varieties that are already infested with *Meloidogyne incognita.*

Symptoms

The primary symptoms of root knot nematode infestation include a reduction in the size of betel leaves, chlorosis (yellowing of the leaves), and the appearance of black spots near the middle of the leaves. In the case of root knot disease, the nematodes induce the formation of tumor-like growths on the tips of the plant roots, known as root galls.

Interestingly, nematodes have been identified as playing a crucial role in the development of wilt and root rot diseases induced by fungi and bacteria. It has been observed that nematodes can make betel vines more susceptible to *Phytophthora palmivora* infections, which enter the plant through wounds caused by *Meloidogyne incognita* (Marimuthu, 1991; Jonathan *et al*., 2000). Sitaramaiah and Devi (1994) observed a positive correlation between the populations of *M. incognita* and the occurrence of wilt disease. Significantly more damage was observed when *Xanthomonas campestris* pv. *betlicola* was combined with *M. incognita* and *Sclerotium rolfsii* (Acharya et al., 1987).

Nematode infestation disrupts the absorption of minerals due to blockages in xylem vessels and damage caused by other style-bearing nematodes, resulting in rootlet holes. This leads to the absorption of soil-borne fungal spores along with minerals due to pressure, ultimately weakening the plants (Maiti and Sen, 1979).

Management of Root-Knot Diseases

The application of 100 gauge black and white polythene mulching for a period of 15 days before planting has been demonstrated to reduce the population of plant-parasitic nematodes in India, according to findings reported by Shivakumar and Marimuthu in 1987. Another effective method involves applying 1 ton/ha of neem oil cake and 2 tons/ha of sawdust, which significantly reduces nematode populations, root galls, and increases the number of harvested leaves, as demonstrated by Jagdale *et al.* in 1985 and Acharya and Padhi in 1988a. Furthermore, amending the planting beds with 2.5 tons of chopped and shade-dried leaves from *Calotropis gigantea* was found to reduce nematode populations by 60%, as shown in Shivakumar and Marimuthu's study in 1986. During experiments with varying levels of Potash, it was noted that 75 kg/ha was the most efficient in decreasing larval populations and the number of root galls. Additionally, a combination of neem cake (0.5 t/ha), NPK fertilizer (150:100:50), and Carbofuran at 0.75 kg a.i./ha also resulted in a decline in nematode populations and the root gall index. The practice of mixed cropping using *Tagetes sp.* Sadaphuli and neem (*Melia azadirachta*) in betelvine gardens has been identified as an effective means of suppressing nematode populations in the soil. Among these approaches, the application of neem cake was found to be the most successful in reducing root galls and increasing the yield of marketable leaves, according to Acharya and Padhi in 1988a.More recently, *Paecilomyces lilacinur* has emerged as an effective biocontrol agent, demonstrating its efficacy both in pots and under field conditions, as documented by Bhatt and Vadhera in 2004.

An alternative method entails the application of *Trichoderma viride*, which is cultured on linseed oil cake, to the soil at 60, 90, and 120-day intervals post-planting. This approach has demonstrated significant efficacy in diminishing root-knot nematode infestations in betelvine, as indicated in the investigation conducted by Bhatt *et al.* in 2002. Regrettably, none of the evaluated cultivars exhibited resistance to root-knot nematodes. Among the cultivars examined, Sanchi and Birkoli exhibited moderate resistance, while the remaining cultivars displayed susceptibility to *M. incognita*, as reported by Acharya and Padhi in 1988b.

In terms of chemical nematicides, Aldicarb is identified as the most effective option for nematode management, as indicated by Bhatt and Vadhera in 2004. Additionally, Carbofuran, applied at a rate of 1.5 kg ai/ha, has proven to be effective in reducing the larval population of root-knot nematodes.

Conclusion

The presence of a shaded and moist environment creates favorable conditions for the development of various diseases caused by fungi, bacteria, viruses, and nematodes, which can significantly hinder plant growth and result in substantial losses for farmers. Major diseases such as leaf and foot rot, anthracnose, bacterial leaf spot, basal rot (collar rot), root-knot disease, and various pests pose significant challenges to betelvine cultivation worldwide, including in India. Betel leaves are commonly used for chewing, often with areca nuts in their raw form.

The excessive or frequent application of chemical pesticides in orchards to manage these diseases can have residual effects. Therefore, to address both environmental and human health concerns, as well as effectively manage betelvine diseases in the future, it will be essential to adopt a balanced approach. This strategy should include carefully combining chemical treatments with environmentally friendly management practices or integrated approaches. Additionally, the use of emerging tools for pathogen identification will be crucial in ensuring proper disease management.

References

Acharya, A., & Padhi, N. N. (1988). Effect of neem oilcake and sawdust against root-knot nematode, Meloidogyne incognita on betelvine (Piper betle L.). Indian Journal of Nematology, 18(1), 105-106.

Acharya, A., Dash, S. C., & Padhi, N. N. (1987). Pathogenic association of Meloidogyne incognita with Sclerotium rolfsii and Xanthomonas betlicola on betelvine. Indian Journal of Nematology, 17(2), 196-198.

Acharya, A., Padhi, N. N., Swain, P. K., & Dash, S. C. (1988). Occurrence of nematodes on betelvine in Orissa. Indian Journal of Nematology, 18(2).

Balasubrahmanyam, V. R., Chaurasia, R. S., Tripathi, R. D., & Johri, J. K. (1988). Evaluation of some fungicides and antibiotics against fungal and bacterial pathogens of betelvine (Piper betle L.). International Journal of Pest Management, 34(3), 315-317.

Bhatt, G. D. (2002). Management strategies for individual knowledge and organizational knowledge. Journal of knowledge management, 6(1), 31-39.

Bhatt, J., & Vadhera, I. (2004). Nematodes of betelvine and their management–A review. Agricultural Reviews, 25(3), 231-234.

Dastur, J. F. (1935, May). Diseases of pan (Piper betle) in the Central Provinces. In Proceedings of the Indian Academy of Sciences-Section B (Vol. 1, pp. 778-815). Springer India.

Huq, M. I. (2011). Studies on the epidemiology of leaf rot and leaf spot diseases of betel vine (Piper betle L.). Bangladesh Journal of Scientific and Industrial Research, 46(4), 519-522.

Jagdale, G. B., Pawar, A. B., & Darekar, K. S. (1985). Pathogenicity of Meloidogyne incognita on betelvine (Piper betle L.). Indian Journal of Nematology, 15(2), 244-244.

Jane, N. S., Deshmukh, A. P., & Joshi, M. S. (2014). Review of study of different diseases on betelvine plant and control measure. Int J Appl Innov Eng Manag, 3(3), 560-563.

Jonathan, E. I., Arulmozhiyan, R., Muthusamy, S., & Manuel, W. W. (2000). Field application of Paecilomyces lilacinus for the control of Meloidogyne incognita on betelvine, Piper betle. Nematologia Mediterranea, 28(2), 131-134.

Maiti, S., & Sen, C. (1978). Leaf rot and foot rot of Piper betle caused by Phytophthora palmivora [flavouring and perfume plants, India]. Note. Indian Phytopathology.

Maiti S, Sen C. Fungal diseases of betelvine, PANS. 1979; 25:150-157.

Marimuthu, T. (1991). Fungal nematode wilt complex in betelvine (Piper betel, L.). Pl. Disease Res, 6, 85-88.

Mehrotra, R. S., & Aggarwal, A. (2003). Rots, Damping offs, Downy mildews and White Rusts. In. Plant Pathology. Tata McGraw-Hill Publishing Company Limited. New Delhi, India.

Mitra, M. (1930). India: new diseases reported during the year. International Bulletin of Plant Protection, 4(7).

Narasimhan, M. J. "Annual Report of the Mycological Section for the year 1931–32,"Rept. Mys. Agr. Dept., for the year ending 30th June 1932, 32–35.

Parvin, R., Aminuzzaman, F. M., Islam, M. S., Hasan, N., Begum, K., & Roni, M. Z. K. (2016). In vitro efficacy of some fungicides, plant extracts and bio-agents against sclerotium rolfsii and control of foot and root rot disease of betel vine. Int. J. Bus. Soc. Sci. Res, 4(3), 211-220.

Ragunathan, C. (1926). Bacterial leaf spot of Betel. Ceylon Dept. of Agric-. Leaflet., (39).

Sitaramaiah, K., & Devi, G. P. (1994). Nematodes, soil microorganisms in relation to Phytophthora wilt disease in betelvine. Indian Journal of Nematology, 24(1), 16-25.

Sivakumar, M., & Marimuthu, T. (1986). Efficacy of different organic amendments against pbytonematodes associated with betelvine (Piper betle. L). Indian Journal of Nematology, 16(2), 278-279.

Sivakumar, M., & Marimuthu, T. (1987). Preliminary studies on the effect of solarisation on phytonematodes of betelvine. Indian Journal of Nematology, 17(1), 58-59.

Stevenson, J. A. (1926). Foreign Plant Diseases. Foreign Plant Diseases

Tiwari, D. P., & Mehrotra, R. S. (1968). Rhizosphere and Rhizoplane studies of Piper betle L. with special reference to biological control of root-rot disease. Bull. Indian Phytopathol. Soc., 4, 79-89.

Tripathi, R. D., Johri, J. K., & Balasubrahmanyam, V. R. (1984). Evaluation of chemicals inhibiting the bacterial leaf spot pathogen of betelvine. International Journal of Pest Management, 30(4), 440-443.

Tutu, A. K. S., Guddun, S. S. K., Kumari, P., & Dey, S. K. D. (2022). An overview of Betel vine (Piper betle L.): Nutritional, pharmacological and economical promising natural reservoir. Advances in Horticultural Science, 36(1), 63-80.

Uppal, B. N., Kamat, M. N., & Patel, M. K. (1946, December). Powdery mildew of betel vines. In Proceedings/Indian Academy of Sciences (Vol. 24, No. 6, pp. 255-259). New Delhi: Springer India.

Wasnikar, A. R., Khatik, S. K., Nayak, M. L., Vishwakarma, S. K., & Punekar, L. K. (1993). Effect of phosphorus on bacterial leaf spot disease incidence, and chemical composition and storage quality of Piper betle leaves. Phytoparasitica, 21, 75-78.

Waterhouse, G. M. (1963). Key to the species of Phytophthora de Bary (No. 92). Commonwealth Mycological Institute

Yadav, B. P., Ojha, K. L., & Prasad, Y. (1993, September). Cultivation of betelvine in Bihar. In National Symposium on Betel Vine Production Technology (pp. 22-23).

20

Diseases of Mulberry and Their Management

[1]*Harshit Singh,* [2]*Snata Kaushik,* [3]*Ankit Alok,* [4]*Chada Anu Reddy*

[1,2,3]*Department of Plant Pathology, School of Crop Protection, College of Post Graduate Studies in Agricultural Sciences, Central Agricultural University, (Imphal), Umiam, Meghalaya*
[4]*Department of Entomology, School of Crop Protection College of Post Graduate Studies in Agricultural Sciences, Central Agricultural University (Imphal), Umiam, Meghalaya*

Abstract

Mulberry plants are widely used for rearing silkworms and producing raw silk. However, the occurrence of diseases in these plants can significantly reduce crop yield and quality, and even lead to plant death. Among the diseases affecting mulberry plants, leaf spot disease caused by the fungus Mycosphaerella moricola and powdery mildew caused by the ascomycete Phyllactinia (syn. P. corylea) are major foliar diseases worldwide. Leaf spot disease alone can cause losses of up to 20% in mulberry production. Fungi are the primary cause of most plant diseases, including powdery mildew, fusarium wilt, root rot, leaf spot, rust, root rots, stem cankers, bacterial blights, and root-knot nematodes. As such, increasing the productivity of mulberry leaves is a major challenge for agronomists, breeders, and plant protection specialists in the field. Effective management of these diseases involves a range of strategies, including the use of fungicides, pruning, and burning diseased plants. By adopting these measures, farmers and researchers can help to ensure the health and productivity of mulberry crops and maintain the global supply of raw silk.

Keywords: *Mulberry, Leaf rust, Fusarium solani, cankers, leaf spot, Silkworm*

DOI: 10.1201/9781042014460-20

Introduction

Mulberry is a very hardy and fast growing perennial plant belonging to the genus *Morus* of the family *Moraceae*. The leaf of mulberry is used for feeding and rearing of the silkworm, for the production of silk yarn. It is estimated that mulberry silk worm (*Bombyx mori*) contributes around 90 % of the total global raw silk production and it is a very attractive economic activities mostly to the rural people. In addition to the utilization of mulberry leaves as silkworm feed, it is being used for many other purposes due to its high nutritive value and delicious taste is getting importance as valuable food stuff as well as environmental importance of mulberry taking into account several aspects of its role as animal feed, foodstuff, phyto-medicine and remediation of polluted soils and atmosphere etc. The mulberry plant boasts impressive versatility, as both the bark and wood serve valuable roles in the creation of paper and sports equipment. Furthermore, recent analyses of the plant have uncovered numerous pharmacological compounds with notable medicinal benefits. It's evident that the mulberry plant holds great potential for further exploration and advancement in the realm of medicine (Ghosh *et al., 2017*). Improving the productivity of mulberry leaf is a significant challenge for agronomists, plant protection specialists, and breeders. Mulberry root rot, caused by the fungus *Rosellinia necatrix* (Hart Berl.) from the Agaricaceae family, is a common disease in several countries, including Italy, Japan, China, India, Eastern Europe, the North Caucasus, Azerbaijan, Georgia, and Tajikistan. (Sayfullaev*, et al.,* 2017).

1. Diseases of Mulberry

Leaf spot of mulberry

Mulberry Leaf Spot caused by *Mycosphaerella moricola*, thrives in wet weather conditions. Although the disease can lead to premature defoliation in severely affected trees, it has little long-term impact on the vigor of mature trees. Symptoms of the disease include small, dark brown and irregular spots on leaves with pale green-yellow halos that appear in the spring. Over time, these spots grow larger and merge, forming a characteristic 'shot hole' with yellow patches around the brownish spot, eventually making the leaf wither. Leaf spot disease can cause an estimated loss of 10% to 20% of leaves and typically affects leaves that are 18 days old. Fungi are responsible for the majority of diseases affecting plants. Some of the most destructive fungal diseases such as Powdery mildew, Fusarium wilt, Root rot, Leaf spot, Rust, stem cankers, and bacterial blights, as well as soil-borne root-knot nematodes. The lesions also produce tiny brown-black spore-producing fruiting bodies called acervuli, and defoliation can occur after severe attacks. Other fungi, including *Cercospora moricola*, *C. missouriensis*, and *Cercosporella* spp., can also cause leaf spots on mulberry leaves during rainy seasons, with *Cercosporella* spp. potentially causing defoliation in older trees. It was found that the pathogen's germ tubes only entered the leaves through stomata and did not form any appressoria. The production of secondary conidia was observed in young leaves.

Managment

If you notice leaves falling prematurely from a tree that is affected by a disease, it is important to collect and dispose of them properly. This can prevent the disease from recurring in the future. The disease thrives in wet weather, so it may be less problematic during dry summers. Healthy and strong trees are less vulnerable to attacks. To keep young trees healthy, they should be fed in spring and watered regularly during dry periods. Certain types of Morus alba are less prone to mulberry leaf spot. To control leaf spots on ornamental plants, the fungicide tebuconazole with trifloxystrobin is recommended, while tebuconazole and triticonazole are suitable for controlling other diseases. For optimal growth in the UK, it needs a warm and sheltered location.

2. Powdery Mildew

Powdery mildew is a common disease caused by the ascomycete *Phyllactinia* spp. (also known as *P. corylea*) that affects mulberry trees worldwide, which are used for silkworm feed. This disease is most severe in the autumn and spring in most areas, and it can extend into the monsoon in hilly regions of India. The disease is characterized by the development of white dust-like colonies on the lower surfaces of leaves (fig. 1) after the germination of wind-borne conidia. Foliage loss due to the disease can reach up to 20% in mulberry trees and can also reduce the quality of foliage, ultimately impacting silkworm cocoon production. (Manimegalai and Chandramohan, 2007). The surface morphology of conidia of *Phyllactinia corylea*, a fungus that causes powdery mildew in mulberry trees, is clavate (fig. 2) and has evenly distributed spine-like protrusions. When inoculated on a leaf, the conidia germinate within 4 hours and produce a single germ tube from a little behind the distal end of the conidium. The hyphae produce special branches, stomatopodia, which enter the leaf through stomata. The disease becomes apparent at the conidial stage, where the hyphae form a mycelial mass within 72 hours after inoculation. The conidiophores appear straight at early stages but become spirally coiled when the conidia mature, which may aid in their detachment. The article concludes that the disease can have significant impacts on mulberry trees and ultimately reduce silkworm cocoon production.

Fig. 1: Symptoms of Powdery mildew on mulberry leaf

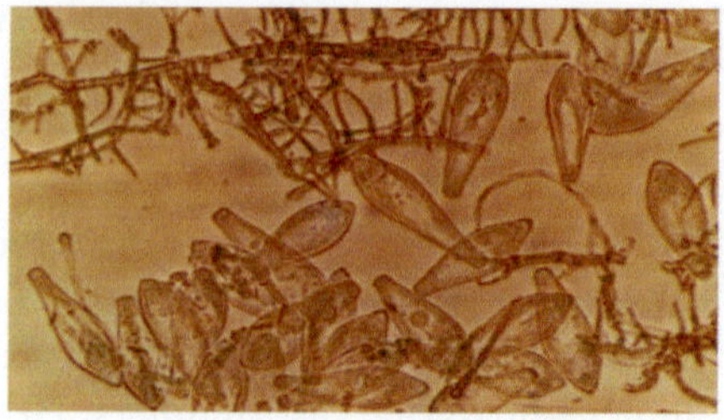

Fig. 2: Spores of *Phyllactinia corylea*

Management

The disease is managed by using fungicides such as wettable sulfur, methyl-benzimidazole-carbamate, or ethylene-bisbenzimidazole carbamate, which are applied on the leaves. However, the mulberry garden is maintained over a long period with frequent pruning of the plant for easy harvesting of leaves for silkworm feed. This makes the use of chemical fungicides not practical from a biosafety standpoint and is also not considered economical for sericulture. One way to minimize the use of fungicides for managing powdery mildew is to focus on developing cultivars that are more resistant to the disease. This approach could help reduce the need for chemical treatments, which can be both costly and potentially harmful to the environment.

3. Fussarium Wilt

Mulberry leaf plays a crucial role in sericulture since the amount and quality of the leaves produced per unit area directly affect cocoon harvest. However, frequent harvesting of leaves can strip the soil of nutrients and leave the plant susceptible to soil-borne diseases such as wilt and root rot. *Fusarium solani* is a common pathogen responsible for wilt in mulberry and is a severe and economically significant disease that results in significant yield losses. The pathogen can thrive in soil and the disease spreads quickly through contaminated soil, irrigation, saplings, and farm tools. Fusarium species can be found in various environments, including soils, living plants, plant debris, and other organic substrates. They can exist as free-living saprophytes or as pathogens and/or endophytes causing various diseases such as wilt, stem rots, root rots, stem canker, and fruit rots in temperate and tropical areas. The symptoms of Fusarium wilt include peeling bark, a distinct odor, and sudden death of plants. The disease spreads quickly through irrigation water, soil, and garden tools, affecting nearby mulberry trees in the same or adjacent rows. (Sanoamuang and Saksirirat, 1984).

Management

The efficacy of three antagonists (*Trichoderma viride*, *Pseudomonas fluorescens*, and *Bacillus subtilis*) and six fungicides (carbendazim, mancozeb, zineb, copper oxychloride, tebuconazole, and carbendazim + mancozeb mixture) was tested against Fusarium wilt pathogen. The results indicated that Trichoderma and bacterial bioagents significantly reduced the growth of the pathogen. The mixture of carbendazim and mancozeb was the most effective fungicide, completely inhibiting the growth of the pathogen at a concentration of 0.1% (Narayanan *et al.*, 2015).

4. Leaf rust of Mulberry

Leaf rust disease is caused by *Cerotelium fici,* belongs to the order Uredinales of class Basidiomycetes. In India, the pathogen produces numerous pin head sized circular to oval brownish to black eruptive lesions on the lower surface of leaves. Genus *Cerotelium* is an obligate microcyclic rust fungus, exists primarily as mycelium, Uredia and Uredospore (fig. 3). Uredospores are oval to round and uninucleate produced singly on uredophores in Uredia. In favourable conditions with high humidity Uredospores germinate through stomata. The hypae grow inter-cellular in the host tissue and sending haustoria into the host cells to draw the nutrients. Uredospores disperse through water droplets and wind currents and spread the disease. The presence of leaf rust, caused by *Cerotelium fici*, is a major issue for mulberry cultivars in India. A study was conducted over three years to determine how various epidemiological factors affect disease development in seven promising mulberry cultivars - Local, K2, S13, S34, S36, S54, and MR2 - during different seasons. Results showed that disease development was greatly influenced by crop age and weather factors such as temperature, relative humidity, rainfall, and rainy days. Disease severity increased significantly with crop age in all cultivars. The relationship between crop age and disease development was best explained by a non-linear quadratic function, with a coefficient determination (R2) of over 90% in both rainy and winter seasons (Srikantaswamy *et al.*, 2006).

Management

Timely utilization of leaves without delaying the leaf harvest especially during winter months and wider spacing in plantation have been found to reduce the disease incidence. In addition, spraying of fungicides like Dinocap and Carbendazim at 0.2 percent concentration helps in controlling the disease with a safe period of 7 days.

5. Sooty Mould Disease

Mulberry cultivars are susceptible to a fungal disease called Sooty mold, which is more common during the winter season from August to December. This disease can cause crop losses of up to 10-15%, and the symptoms include the development of a thick black coating on the upper surface of the leaves. The disease spreads through the presence of white flies in the mulberry field that produces a honey-like substance, which attracts molds. The favorable conditions for Sooty mold development are a temperature range of 20-24°C and high relative humidity above 70%.

Management

To prevent the growth of saprophytic fungi, you can use a 0.2% solution of Indofil-M45 spray. To control infestations of white flies, you can do a foliar spray of 0.02% monocrotophos twice after pruning, on the 15th and 30th day.

6. Root-knot Nematode

Root knot is a disease caused by a nematode called *Meloidogyne incognita*, which is prevalent throughout the year and commonly found in irrigated sandy soils. Disease can cause crop losses of up to 20%, and severely affected mulberry plants show stunted growth with low water moisture in leaves, followed by yellowing of leaf margins. The main indicator of this disease is the formation of knots or galls on roots, which are spherical and vary in size (fig. 4). Young galls are small and yellowish-white, while old galls are big and pale brown. The primary spread of the disease is through contaminated soil, farm implements, and run-off irrigation. Planting infected saplings with other susceptible crops can increase disease intensity, and some susceptible weeds in and around mulberry gardens can act as secondary sources of infection. The root-knot disease thrives in temperatures between 27-30 °C, soil moisture of less than 40%, and pH of 5-7.

Fig. 4: Root knot symptom of mulberry.

Source: W. Saksirirat

Management

To manage root-knot disease caused by *Meloidogyne incognita*, it is recommended to use neem oil cake at a rate of 800 kg/acre/year. This should be applied in four split doses during intercultural operations, after pruning the plant, or after leaf harvest. This can be done by creating trenches of 10-15 cm depth near the root zone of the plant and covering them with soil before irrigating.

7. Complex Root Rot of Mulberry

Mulberry trees are susceptible to leaf and stem blight, wilt, and root rot diseases that can cause the death of the plant. The roots of the affected plant can be easily peeled off, which is an indication of the disease (fig. 5). To control these diseases, eco-friendly measures are recommended. Studies have shown that several *Trichoderma* spp. and bacterial isolates can be effective against these diseases. For instance, *Trichoderma viridae* and *Trichoderma harzianum* are effective against root rot pathogen and root-knot nematode, respectively. Additionally, applying a combination of *T. harzianum*, *T. viride*, farmyard manure, and effective microorganisms in the soil at an interval of 30 days three times can result in a high degree of mulberry survival of up to 72% (Beevi and Qadri, 2010).

Fig. 5: Symptom of mulberry root rot.

Source: W. Saksirira

8. Bacterial Blight of Mulberry

Yellowish-brown bacterial blight caused by *Pseudomonas syringae* pv. *mori* was observed, leading to defoliation and easy breakage of branches. Effective management of this blight involves burning diseased plants and pruning before the rainy season. Treatment with Bordeaux mixture, antibiotics such as hydrochloride and Streptomycin sulfate, has been found to significantly reduce the disease incidence.

Fig. 6: Symptom of bacterial blight of mulberry

Source: Chuprayoon *et al.* (1994)

Conclusions

Mulberry plants are prone to various diseases caused by fungi, which can significantly reduce crop yield and quality. Leaf spot disease and powdery mildew are among the major foliar diseases affecting mulberry plants worldwide. Effective management of these diseases involves several strategies, including the use of fungicides, pruning, and burning diseased plants. To maintain the health and productivity of mulberry crops and ensure the global supply of raw silk, it is essential for farmers and researchers to adopt these measures and continuously work towards developing new and innovative ways to prevent and control these diseases.

References

Ashmita, G., Gangopadhyay, Debnirmalya, Chowdhury and Tanmay. (2017). Economical and

Beevi, N.D. and S.M.H. Qadri. (2010). Biological control of mulberry root rot disease (Fusarium spp.) with antagonistic microorganisms. J. Biopesticides, 3(1): 90-92.

Chattopadhyay, S., Ali, K.A., and Doss, S.G. (2010). Evaluation of mulberry germplasm for resistance to powdery mildew in the field and greenhouse. J. Gen. Plant. Pathol., 76: 87–93.

Chuprayoon, S., Singh, S., and P. Sipayakhun. (1994). Diseases, insect pests of mulberry and their control. Funny Publishing, Partnership Ltd., Bangkok, Thailand.

Dandin, S.B., Jaysawal, J., and Giridhar, K. (2001). Handbook of sericulture technologies. Central Silk Board, Bangalore.

Datta, R.K. and Nanavati, N. (2005). Global silk industry: a complete source book. Universal Publishers, Boca Raton.

Environmental Importance of Mulberry: A Review. 3(10): 067-072.

Govindaiah and Gupta, V.P. (2005). Foliar disease of mulberry and their management. Mulberry crop protection. Central Silk Board, Bangalore, India, 145–177.

Kumar, V., Babu, A.M., Sharma, D.D., and Datta, R.K. (1998). Penetration and infection process of Phyllactinia corylea on mulberry leaf causing powdery mildew. Asexual Stage. Journal of Phytopathology, 146(10): 469-472.

Manimegalai, S. and Chandramohan, N. (2007). Leaf quality of mildew affected leaves and their effect on mortality and economic characters of silkworm, Bombyx mori L. Sericologia, 47: 87–92.

Narayanan, P., Vanitha, S., Rajalakshmi, J., Parthasarathy, S., Arunkumar, K., Nagendran, K., and Karthikeyan, G. (2015). Efficacy of biocontrol agents and fungicides in management of mulberry wilt caused by Fusarium solani. Journal of Biological Control, 29(2): 107-114.

Saksirirat, W. and Sivilai S. (2014). Mulberry Diseases and Their Control. Int. J. Wild Silkmoth and Silk, 18, 17-22.

Sanoamuang, N. and Saksirirat, W. (1984). Complex root rot of mulberry. KKU-ACNARP 1983. Technical Progress Report. Faculty of Agriculture, Khon Kaen University.

Sayfullaev, A.A., Normatov, B.I., Bobomurodov, M.K., and Normatov, R.J. (2021). Types Of Fungal Diseases Of Mulberry. Academicia Globe, 2(6).

Srikantaswamy, K., Gupta, V.P., and Rekha, M. (2006). Influence of some factors on the epidemiology of leaf rust (Cerotelium fici) disease in mulberry (Morus alba L.). Archives of Phytopathology and Plant Protection, 2(39): 129-140.

Summerell, B.A., Salleh, B., and Leslie, J.F. (2003). A utilitarian approach to Fusarium identification. Plant Disease, 87: 117-128.

21

Diseases of Chrysanthemum and Their Management

***Shashank Shekhar*[1*]*, Suraj Kumar Patel*[1]*, Arvind Yadav*[3]*
and Rajneesh Pal*[2]**

[1]Department of Plant Pathology, Sardar Vallabhbhai Patel University of Agriculture and Technology, Meerut, Uttar Pradesh
[2]Department of Entomology, Sardar Vallabhbhai Patel University of Agriculture and Technology, Meerut, Uttar Pradesh
[3]Department of Plant Pathology, Acharya Narendra Dev University of Agriculture & Technology, Ayodhya, Uttar Pradesh

Abstract

In terms of global flower crops, chrysanthemum is ranked second only to rose. It is recognized as the sixth potent flower crop in India that is significant for commerce. It's adaptable; you can use it to make garlands, plant it in pots, and cut blossoms for arrangements. Small-flowered kinds are produced in India for religious offerings, wreaths, garland, and veni, while large-flowered forms are planted for exhibition. In addition to being a beautiful flower, chrysanthemums are rich in vitamins, minerals, and nutrients. The most significant chemical extracts from Chrysanthemum include betaine, choline, flavonoids, and vitamin B1. Elite cultivars carry a significant risk of spreading bacterial, viral, and fungal diseases since they are usually grown vegetatively. Chrysanthemums are attacked by many insect pests and diseases. The major diseases and their symptoms, etiology, spread, and management are discussed.

Keywords: *Chrysanthemum, Disease, Symptoms, Pathogen, Management*

Chrysanthemum

Botanical Name	*Chrysanthemum morifolium*
Family	Asteraceae
Chromosome Number(2n)	18

DOI: 10.1201/9781042014460-21

Introduction

Chrysanthemum, commonly known as "mums" or "chrysanths," is a genus of plants that bloom and belong to the Asteraceae family. These plants are native to Asia and northeastern Europe, with China being the primary centre of diversity for the genus. Chrysanthemums are well-known for their attractive and colorful flowers, and they are popular ornamental plants in gardens, floral arrangements, and as cut flowers.

Chrysanthemum morifolium is a herbaceous perennial flowering crop. It is the second most popular cut flower worldwide, right behind roses, and a significant decorative plant for commerce (Silva *et. al.,* 2013). Certain species' flowers, such as those of Chrysanthemum morifolium and Chrysanthemum indicum, are rich in nutrients and biologically active substances and are used to make medicinal tea and cosmetics, respectively (Lin *et. al.,* 2010).

When it comes to chrysanthemums, flower color is one of the most important characteristics in terms of both biology and economics. It affects the customer's decision to purchase cut flowers, potted flowers, and garden flowers (Posadas *et. al.* 2006). Flower color enhances the plant's visual appeal and is essential for attracting pollinators for reproduction, protecting the plant from photooxidative damage, and supplying resilience to both biotic and abiotic challenges (Samantha *et. al.,* 2011). Intraditional Chinese medicine, some species of chrysanthemum have been used for their potential medicinal properties. For example, Chrysanthemum morifolium is believed to have various health benefits.In some parts of Asia, particularly China and Japan, chrysanthemums are celebrated during special festivals. For example, Japan has a Chrysanthemum Festival known as "Chrysanthemum Day" or "Festival of Happiness."

Diseases of Chrysanthemum

1. Powdery Mildew

Powdery mildew is a common disease caused by a fungus called *Podosphaera* spp. that appears as a white,powdery substance on the leaves, stems, and sometimes flowers of chrysanthemum plants. It can cause leaves to distort and yellow, leading to reduced plant vigor and flower quality.

Symptoms

The development of a white or grayish powdery substance on the surfaces of the leaves, stems, and occasionally flowers is the most characteristic sign of powdery mildew. This powdery substance consists of fungal spores and mycelium. Infected chrysanthemum leaves may show discoloration, starting as yellow or pale green spots or patches. As the disease progresses, these areas may turn brown or necrotic. The affected leaves may exhibit curling, distortion, or cupping. This distortion can result from the fungal growth on the leaf surface. Severe powdery mildew infections can lead to stunted growth and reduced plant vigor. Chrysanthemums may not grow as tall or produce as many flowers as healthy plants. In some cases, heavily infected leaves may drop prematurely, further weakening the plant. If the infection spreads to the flowers, it can cause deformities in the petals and reduce the overall quality of the blooms (Reddy and Reddy, 2016)

Healthy and Infected leaves of chrysanthemum

Pathogen

The fungus *Podosphaera spp.* has a number of species of powdery mildew infections. Many different types of plants, including many agricultural crops and decorative plants, are susceptible to the widespread plant disease known as powdery mildew. The term "powdery mildew" refers to the distinctive white or grayish powdery growth that develops on diseased plant surfaces, such as leaves, stems, and occasionally flowers.

Disease Cycle

The disease cycle of powdery mildew, including the one that affects chrysanthemum plants, is relatively complex and involves several stages. Powdery mildew fungi do not require water for germination or infection, and they can thrive in dry conditions. The life cycle of powdery mildew begins in the fall, when the fungus produces overwintering structures known as cleistothecia. These are small, hard structures that contain fungal spores called ascospores. Cleistothecia can survive in the soil and on plant waste., or on infected plant material. When the environmental conditions become favourable in the spring, the cleistothecia release ascospores into the air. These ascospores are carried by wind currents and can land on chrysanthemum plants. The ascospores germinate on the surface of chrysanthemum leaves, and the resulting fungal structures (haustoria) penetrate the plant's epidermal cells. Once inside the plant, the fungus begins to grow and reproduce.

Secondary Infection (Asexual Reproduction): As the fungus colonises the plant, it produces specialised structures called conidiophores, which bear conidia (asexual spores). These conidia are the powdery substance seen on the plant's surface. Disease spreads easily when conidia are transferred by air currents to other areas of the same plant or to other neighboring plants.Powdery mildew fungus is able to complete multiple generations of asexual reproduction within a single growing season. Each cycle of conidia production, infection, and colonisation can occur relatively quickly, especially in warm and dry conditions. As the growing season draws to a close, the fungus may transition to sexual reproduction. If conditions are suitable, cleistothecia may form on infected plant tissue. Inside the cleistothecia, sexual spores (ascospores) are produced, completing the life cycle.

Etiology

Powdery mildew infections prefer humid conditions between 68 and 81°F. In warm, dry conditions, new spores proliferate and swiftly disperse the disease. In Minnesota, powdery mildew symptoms start to show up in the middle to end of summer. Because they are more vulnerable, older leaves will be the first to get powdery mildew. Spores formed in leaf spots are spread to other leaves by the wind. When conditions are right, powdery mildew can spread quickly and frequently cover all of the leaves. Although cucumber or melon fruit can occasionally become infected, leaves and vines are the primary sites of powdery mildew infections. Squash fruit is not immediately contaminated by powdery mildew. Fewer and smaller fruits are produced by infected plants (Schuh and Grabowski, 2022)

Management

- Plant types that are resistant to powdery mildew, either completely or partially.
- Promote adequate airflow around plants by planting them at the right distance apart, staking them, and controlling weeds.
- Planting with the right spacing is important (Schuh and Grabowski, 2022).
- Potential organic fungicides can be used (Bradshaw, 2015).
- The most effective treatment for powdery mildew was sesame oil. Ammonium oleate also had promising results (Bradshaw, 2017).

2. Chrysanthemum White Rust

Chrysanthemum white rust (*Puccinia horiana*), which originated in South-east Asia and has now spread to various areas in Western Europe, has become a significant issue for farmers of this flower. By spores and contaminated plant matter, this rust is quickly propagated, and chemical management is time-consuming. Some EPPO nations are still able to stop its creation (Water, 1981).

Symptoms

Chrysanthemum white rust is identified by light green to yellow spots that can grow up to 5 mm in diameter on the leaf's upper surface. These areas become brown and necrotic with age. Raised pustules ranging in color from beige to pink arise on the underside of the leaves beneath the spots. Pustules become white with age. Even though they can develop on any green tissue, including the petals, pustules are most commonly observed on immature leaves and flower bracts. Usually, the symptoms appear in chilly, rainy weather.

White rust symptoms on chrysanthemum leaves

Pathogen

P. horiana is an obligatory biotrophic pathogen that mostly affects developing plants' fresh tissue; however it can also infect older leaves (Firman and Martin, 1968). Nowadays, *P. horiana* is almost always present in areas where chrysanthemums are grown or exported, and most nations control it as a quarantine organism (Anonymous, 2004).

Disease Cycle

Chrysanthemum white rust has only two spore stages in its life cycle. The pathogen reproduces only on living chrysanthemum leaf. When teliospores, which develop on the underside of damaged leaves and germinate in the pustules, do so, basidiospores are produced. Because basidiospores are airborne, they can infect new plants when they come into contact with damp chrysanthemum leaves. Both brown rust and white rust of chrysanthemums are caused by the same meteorological factors: cold temperatures, damp weather, or excessive humidity (Agriculture and Natural Resources, University of California).

Etiology

P. horiana is autoecious rust. The unicellular basidiospores created by the in situ germination of the bicellular teliospores are spread by air currents. There are no known further spores. Both teliospores and basidiospores appear to require high humidity and a thin layer of moisture in order to germinate. When basidiospores achieve maturity, they can begin to germinate as soon as they reach 4 to 23 degrees Celsius. At this temperature, discharge starts within three hours. Teliospores can germinate as soon as they reach maturity. Basidiospores may germinate across a wide temperature range, and at 17 to 24 degrees Celsius, either leaf surface may be pierced in 2 hours (Water, 1981).

Management

- Oxycarboxin, triforine, benodanil, triadimefon, diclobutrazol, dibitertanol, and propiconazole are among the active components that have been reported to be beneficial.
- Systemic fungicides such as Azoxystrobin is effective
- Myclobutanil, hexaconazole, and propiconazole all had good eradicative activity (Dickens, 1991).
- *P. Horiana*can be controlled by *Verticillium lecanii,* which is also utilized to biologically control aphids on glasshouse chrysanthemums (Srivastava et. al., 1985).
- Resistant varieties such as Renine, Guilderland, Statesman, Ready and Carillon should be used (Rademaker and de Jong, 1985).

3. Fusarium Wilt

One of the most common and harmful diseases of Chrysanthemum is Fusarium wilt, which is brought on by *Fusarium oxysporium f. sp. Chrysanthemi* and results in infection and loss from the nursery through the flowering stage. According to Locke *et al.,* (1985), the disease is most severe in warm areas. Fusarium wilt of chrysanthemums is challenging to manage because of the pathogen's persistence in the soil and the lack of resistant cultivars (Garibaldi *et. al.*2009).

Symptoms

The *Fusarium oxysporum f. sp. Chrysanthemi* slowed growth, inability to form typical buds and flowers, yellowing, drooping, and loss of turgidity in the leaves are all signs of diseased plants.

Symptoms of fusarium wilt in Chrysanthemum

The invasion and multiplication of the pathogen into vascular tissues often results in disintegration and discoloration in the roots and lower regions of the stem. As infection worsens, the vascular tissues become obstructed and unable to transport needed nutrients and water, which finally causes plant death.When infection progresses, the collar region of plants also shows signs of creamy-white fungal growth (Ghosh and Singh, 1982).

Pathogen

Fusarium oxysporum f. sp. *chrysanthemi* is harmful to chrysanthemum flowers (*Chrysanthemum* spp.). Fusarium oxysporum is a widespread soilborne fungus that can infect a variety of plants, causing a disease known as Fusarium wilt or Fusarium wilt disease.This fungal pathogen typically enters the plant through its roots and then spreads throughout the vascular system, disrupting the flow of water and nutrients. This disruption can lead to wilting, yellowing of leaves, and ultimately, the death of the infected chrysanthemum plant.

Disease Cycle

A soilborne fungus that may live in the soil for up to 10 years without a host is what causes fusarium wilt. Ideal soil conditions include warm temperatures, an acidic pH of 5.0–5.5, and high humidity. The midway to end of the growth season may be when infection symptoms first appear due to the environmental conditions needed for disease development. The fungus enters the plant from the roots and spreads throughout the whole plant by colonizing the vascular tissue. Plant sections above ground undergo yellowing, withering, and eventual death due to pathogen-induced blockages in the vascular tissue. Chlamydospores, which the fungus can generate, can persist for years in plant detritus or soil.

Etiology

Fusarium oxysporum is the fungus that causes fusarium wilt. Fusarium wilt also affects pepper, eggplant, and tomatoes. It can also live on weeds like crabgrass, mallow, and pigweed. The fungus is more severe in acidic soil and grows best in warmer climates (ideal soil temperature is 82°F). Most frequently, the pathogen enters through root wounds made by nematode feeding or culture. Although it is uncommon in commercial seed, fusarium wilt can occur in seeds. The fungus can be propagated on equipment tainted with infested soil or introduced on infected transplants (Schuh *et al.*, 2021).

Management

- **Crop rotations for 3-5 years**: Reducing the inoculum (spores) in the soil can be achieved by rotating away from chrysanthemum crops. While disease eradication is not a given, this will lessen the severity and frequency of infection in the crop that follows. Since weeds like crabgrass, pigweed, and mallow can serve as hosts for the Fusarium wilt infection, regular upkeep of the infected region is required throughout this time.
- **Plant only robust transplants:** To reduce the chance of introducing disease, carefully examine transplants for signs of illness prior to planting.

- Clean tractors and other equipment after each field to prevent the spread of infected soil on tools and equipment. Cleaning should be done often, especially if you're going from an infected area to a clean one.
- **Use sterile potting media:** To reduce the possibility of the disease spreading, it is advised to use sterile soilless potting material when beginning your own seedlings.
- **Control soil pH:** Adding lime to the soil to bring it up to a neutral pH range of 6.5 to 7.0 will lessen the pathogen's persistence.
- **Steer clear of too much nitrogen:** Plants that are more susceptible to Fusarium wilt in the soil have higher nitrogen and lower potassium levels. Use a soil test to ascertain the nutrient levels and refrain from overapplying high nitrogen fertilizers.

4. Leaf Spot

A vast geographic range of Chrysanthemum species, including the western hemisphere and the continents of Asia and Europe, are affected by the Septoria leaf spot disease. Since 1891, it has been documented from the majority of the United States' chrysanthemum-growing regions.The spotting of the leaves tends to decrease the amount and quality of the blossoms and diminishes the cut flowers' quality and marketability.

Symptoms

Septoria causes irregular, dark brown to black leaf patches that are occasionally encircled by a thin yellow halo.The spots typically start on the lower leaves and spread up the plant as the disease advances. The leaf dots, which cover one-third of the leaf surface, range in size from extremely small to enormous. The infected leaves become yellow, die, and remain on the plant as the spots get larger and more numerous (Magie and Overman, 1966).

Symptoms of leaf spot in chrysanthemum

Pathogen

The fungus *Septoria chrysanthemi*Sacc is responsible for this disease (Waddell, 1959). Because it grows best in warm, humid environments, it is more common during times of high humidity and precipitation.

Disease Cycle

The fungus hibernates in the soil or on plant detritus that has become infected. The fungus produces spores in the spring when the weather is favorable and they are dispersed by wind, rain, or splashing water. These spores land on chrysanthemum leaves, where they infect the plant. The disease cycle continues as new spores are produced on infected leaves, leading to further infection.

Etiology

Mild temperatures (68-89 F) and high humidity or moisture are ideal for the growth of the fungus. It is spread through the formation of numerous minute pycnidia, each of which contains countless spores, which are sprayed onto uninfected plants, especially during rainy seasons or when plants are watered in greenhouse environments. The stomata of leaves serve as the fungus' entrance. The fungus can survive for up to four months in leaf detritus and produce spores that can infect healthy plants (Waddell, 1959).

Management

- The most effective method of controlling this leaf spot is to apply weekly treatments of fungicidal sprays such as zineb (1/2 lb 75% WP) + captan (3/4 lb 50% WP) or zineb alone (1 lb 75% WP) in 100 gal water. Every day while it's raining or whenever disease symptoms appear, the spraying should be done.
- Since the lower leaves are the first to become infected, it is essential to thoroughly cover all leaves with the fungicide, whether it is applied as a dust or a spray, to ensure effective protection.
- During and after the agricultural season, cleanliness practices can also assist control this infection by removing unhealthy leaves and plants. Good sanitation by removing and disposing of infected plant material, especially in the fall, to reduce the overwintering source of the fungus (Waddel and Weber, 1963).

5. Bacterial Blight

Bacterial blight of chrysanthemum, also known as *Pseudomonas cichorii* blight, is a destructive bacterial disease that affects chrysanthemum plants. This disease is caused by the bacterium *Pseudomonas cichorii*. It can lead to severe damage to chrysanthemum plants and reduce their overall health and vigor.

Symptoms

Bacterial blight is indicated by dark brown patches and blotches on leaves that extend beyond plant leaves, water-soaked lesions on stems, darkening and dying of buds, and terminal mortality of stems. There was brown to black rot at the root of the afflicted plants. With an increase in spots, the leaves completely dried out. The leaf markings have a diameter of one to two millimeters.

Pathogen

Pseudomonas cichorii is a bacterium that thrives in warm, humid conditions. Chrysanthemum plants may become infected by it when there is heavy humidity, rain, or natural openings.

Disease Cycle

The disease cycle typically begins when *Pseudomonas cichorii* is introduced to the chrysanthemum plant. This can happen through contaminated plant material, tools, or by wind-driven rain. The bacteria may also survive on plant debris or in the soil from previous infections. Once on the plant, the bacteria need to attach themselves to the plant's surface and enter the plant tissues. *Pseudomonas cichorii* often enters the plant through wounds, natural openings (like stomata), or through the roots. Once inside the plant, the bacteria multiply and spread within the plant's vascular system. This leads to the development of infection within the plant tissues. As the bacteria multiply, they produce toxins and enzymes that damage the plant cells, leading to characteristic symptoms of bacterial blight, such as wilting, leaf spots, and stem cankers. As the infection progresses, visible symptoms of bacterial blight become apparent. These symptoms could include wilting, black streaks on stems, water-soaked sores, and yellowing of the leaves. These symptoms can vary depending on the chrysanthemum cultivar and environmental conditions. The bacteria can spread to neighbouring plants through various means. The virus can proliferate more easily when it comes to rainwater, irrigation, and even mechanical contact (such as pruning tools). Aphids are among the insects that can spread the germs from one plant to another.During adverse conditions, such as winter or when host plants are not available, *Pseudomonas cichorii* can survive in plant debris or in the soil. It may also survive on infected seeds or plant material, serving as a source of infection for the next growing season.

Etiology

P. cichorii is not very host specific and has a wide host range. This creates a significant potential for the development of epidemics under favoured environmental conditions. Stomata or wounds are the entry points for infection. When plants are subjected to a lot of free moisture, such when they're misted and *P. cichorii* inoculum is available, wounding is not necessary for infection of the majority of hosts. It is possible for lettuce to contract the pathogen through its epidermal hairs in an environment that is favourable to it (Shirata *et al.*, 1982).

Management

- The most crucial aspect of disease management is sanitation. Plant debris that has become diseased should be removed and destroyed.
- Reduce splashing and lengthen the time that leaves are wet by watering early in the day or using subirrigation.
- Lower relative humidity by heating and ventilating, using fans to improve air circulation, and spacing plants appropriately.
- Avoid handling plants when their foliage is damp. Bactericide products are only partially successful in controlling bacterial infection; copper products are registered for this purpose.

6. Crown Gall

The bacterium *Agrobacterium tumefaciens*, formerly known as *Rhizobium radiobacter*, is the cause of the plant disease crown gall. This bacterium infects a wide range of plants, including chrysanthemums. The development of tumor-like growths or galls on the roots, stems, and even the crown of afflicted plants is the hallmark of crown gall.

Symptoms

In the basal or root crown zone, galls are frequently seen at or just below the soil surface and can range in size from 1.5 to 2 inches. They may be found on plant roots more often than on the stems and leaves of aerial plants. Galls typically have a spherical shape and an uneven, rough surface. On the surface of the plant, they initially appear as tiny protuberances. Galls that are still actively growing are soft and have a light green or almost white colour. Galls mature and becoming black and woody. When a gall is at the base of an unrooted cutting, it might be challenging to distinguish it from callus growth because some galls have a rather smooth surface.

Crown gall symptoms in chrysanthemum

Pathogen

The group of Agrobacterium contains both tumorigenic and nontumorigenic strains. A tumor-inducing Ti plasmid that spreads genes to the host plant and promotes gall formation is only present in tumorigenic strains. The species *Agrobacterium radiobacter* occasionally includes the nontumorigenic strains.

Disease Cycle

The pathogen is thought to live in soil by colonising plant rhizospheres and existing as a saprophyte. Infected plants are a significant source of inoculum as well. Plant waste, soil, water, plant propagation materials, equipment, storage sheds, etc. can all spread the infection. Budding, grafting, and planting procedures are examples of practises that might spread the illness in nurseries and farms.When inoculum comes into contact with wounds—whether they are equipment-induced (like growth cracks) or created by nature (like ploughing or cutting roots during planting and harvesting for the market), infection can result. When moisture is available and the temperature is not too high, bacteria thrive in wounds. For an infection to start, wounds need to be wet for a number of hours (Agrios, 2005).

Etiology

Gram-negative soil bacteria *Agrobacterium tumefaciens* is well-known for its unusual capacity to insert a portion of its DNA, known as the Ti (tumor-inducing) plasmid, into the host plant's genome.This bacterium plays a crucial role in the formation of crown gall tumors in various plant species. *A. tumefaciens* carries a set of virulence (vir) genes on the Ti plasmid. These genes are responsible for several critical functions, including the formation of the T-DNA transfer machinery, the production of signalling molecules (such as acetosyringone), and the regulation of the infection process.

Management

- The most crucial management technique for crown gall is anitation. Work surfaces, containers, and instruments should all be regularly cleaned and disinfected with a commercial cleaner.

- Use only top-notch plants. Avoid damaging plants, especially when they are damp or at the soil's surface. Dig up and remove diseased plants. Before planting, solarizing the soil in hot climates can help lower the number of bacteria that cause crown gall.
- Agrobacterium tumefaciens K-84, a biological control agent that creates an antibiotic that decreases or eliminates infection, can be used to dip cuttings in order to avoid infection (Agriculture and Natural Resources, University of California).

References

Agriculture and Natural Resources, University of California © Friday, 01-Dec-2017 16:13:51 PST Regents of the University of California

Agrios, G. N. 2005. Plant Pathology, 5th ed. Academic Press, San Diego, CA.

Anonymous. 2004. Diagnostic protocols for regulated pests: Puccinia horiana. EPPO Bull. 34:209-211.

Bradshaw, M. 2015. Potential Organic Fungicides for the Control of Powdery Mildew on Chrysanthemum x morifolium (Doctoral dissertation).

Dickens JSW, 1991. Evaluation of some newer fungicides, in comparison with propiconazole, against chrysanthemum white rust (Puccinia horiana). Tests of Agrochemicals and Cultivars 12. Annals of Applied Biology, 118(supplement):32-33.

Firman, I. D., and Martin, P. H. 1968. White rust of chrysanthemums. Ann. Appl. Biol. 62:429-442.

Garibaldi A, Bertetti D, Gullino ML. 2009 – Susceptibility of chrysanthemum and paris daisy varieties to several isolates of Fusarium oxysporumf.sp. chrysanthemi. Commum. Agri. Biol. Scien. 74(3), 651–657.

Ghosh RN, Singh BS. 1982 – A new wilt of Chrysanthemum in India. Indian Phytopathology 35, 338–340.

Hong, Y.; Bai, X.; Sun, W.; Jia, F. The numerical classification of chrysanthemum flower color phenotype. Acta Hortic. Sin. 2012, 39, 1330–1340.

Lin, L.-Z.; Harnly, J.M. Identification of the phenolic components of chrysanthemum flower (Chrysanthemum morifolium Ramat). Food Chem. Food Chem. 2010, 120, 319–326.

Locke JC, Marois JJ, Papavizas GC. 1985 – Biological control of Fusarium wilt of greenhousegrown Chrysanthemums. Plant Dis., 69, 167–169.

Magie, R. 0. and A. J. Overman. 1966. Chrysanthemum diseases in Florida. Univ. Fla. Agr. Exp. Sta. Bull. 637A. 49 p.

Marissa Schuh, horticulture IPM Extension educator, and Michelle Grabowski Reviewed by Natalie Hoidal, Extension educator. 2022, University of Minnesota Extension

Posadas, B.C.; Coker, C.H.; Knight, P.R.; Fain, G. Consumer survey of selected garden chrysanthemum cultivars in Mississippi. HortTechnology 2006, 16, 539–543.

Rademaker W, Jong Jde, 1985. Japanese rust: how susceptible or resistant is the chrysanthemum? Vakbladvoor de Bloemisterij, 40(45):49.

Reddy, P. P., & Reddy, P. P. (2016). Chrysanthemum. Sustainable Crop Protection under Protected Cultivation, 363-374.

Samanta, A.; Das, G.; Das, S. Roles of flavonoids in plants. Carbon 2011, 100, 12–35.

Shirata A, Ohata K, Serizawa S, Tsuchiya Y, 1982. Relationship between the lesion development by Pseudomonas cichorii and growth stage and leaf position of lettuce and its infection mechanism. Bulletin of the National Institute of Agricultural Sciences, C, No. 36:61-73

Srivastava AK, Defago G, Kern H, 1985. Hyperparasitism of Puccinia horiana and other microcyclic rusts. PhytopathologischeZeitschrift, 114(1):73-78
Teixeira da Silva, J.A.; Shinoyama, H.; Aida, R.; Matsushita, Y.; Raj, S.K.; Chen, F. Chrysanthemum biotechnology: Quo vadis? Crit. Rev. Plant Sci. 2013, 32, 21–52.
Waddell, H. T. 1959. Parasitism of Septoria obesa Syd. and J5. chrysanthemellaSacc. on the cultivated Chrysanthemum. Ph.D. Thesis. Univ. Fla., Gainesville. 83 p.
Waddell, H. T. and G. F. Weber. 1963. Physiology and pathology of Septoria species on Chrysanthemum. Mycologia 55(4):442-452.
Water, J. K. 1981. Chrysanthemum White Rust 1. EPPO Bulletin, 11(3), 239-242.

22

Diseases of Rose and Their Management

Snata Kaushik[1], Vaibhav Pratap Singh[2], Harshit Singh[1], and Hans Austin K. H.[1]

[1]Department of Plant Pathology, School of Crop Protection, College of Post Graduate Studies in Agricultural Sciences, (Central Agricultural University Imphal), Umiam, Meghalaya
[2]Department of Plant Protection, Aligarh Muslim University, Aligarh Uttar Pradesh

Abstract

Roses, renowned for their exquisite beauty and captivating fragrance, stand as one of the most cherished flowering plants globally, adorning gardens, residences, and businesses alike. Despite their popularity, roses are not impervious to various diseases, which can compromise their vitality, diminish flowering potential, and mar their aesthetic appeal. Among the prevalent rose diseases are black spot, Botrytis blight, downy mildew, powdery mildew, and rust, stemming from diverse pathogens such as fungi, bacteria, nematodes, and viruses. Black spot, a fungal affliction, manifests as black spots with fringed margins on leaves, while also affecting canes with purple-red blotches that darken over time. This disease intensifies in wet weather conditions. Botrytis blight, another fungal ailment, induces grayish-brown, fuzzy growth on various plant parts, including flowers, buds, canes, and growing tips. It can also infiltrate pruning cuts and wounds, particularly thriving in cool, cloudy, and humid weather. Downy mildew, resembling a fungal-like ailment, triggers purple to red, angular spots or blotches on leaves. Meanwhile, rust, a fungal disease, instigates the formation of orange-red pustules on leaves, stems, and buds, flourishing in warm and humid climates.The impact of these diseases on rose health are substantial, ranging from defoliation and stunted growth to the potential demise of the plant in severe cases. Consequently, it is imperative to implement proactive measures for preventing and effectively managing rose diseases to safeguard the longevity and aesthetic appeal of these beloved flowering plants.

Keywords: *Rose diseases, Botrytis blight, downy mildew, powdery mildew, manage diseases.*

DOI: 10.1201/9781042014460-22

Introduction

Roses, widely cultivated as cut flowers in greenhouses, as entire plants in nurseries, or within home gardens, encounter common issues across these diverse growing environments, including black spot, Botrytis blight, downy mildew, powdery mildew, and rust. Specific challenges like downy mildew and rose rosette gained prominence in the 2010s. The susceptibility of roses to various diseases varies by geographical region; for instance, rose replant disease is a concern in Europe but not in North America. Greenhouses and tropical climates, as seen in Central and South America, face heightened issues with powdery mildew, downy mildew, rust, and Botrytis blight. Disease management proves costly and can constrain rose production, prompting innovations like the adoption of hydroponics in Colombia, leadingto improved pest control, enhanced rose quality, and increased productivity. In Colombian rose cultivation, environmental control in greenhouses is common, employing plastic containers with nutrient film technique (NFT) subirrigation. Utilizing white plastic mulch or raw rice husks aids in moisture retention and weed control. The flowering time for roses varies by cultivar, typically taking 70-80 calendar days from pruning to the first flower crop. Successful disease management plans should align with the overarching goal of rose cultivation, recognizing differing tolerances to diseases among various rose groups. Growers can choose from a range of disease management tactics tailored to their specific circumstances.

1. Armillaria Root Rot

Symptoms

With a global presence, Armillaria Root Rot is an occasional concern influenced by the growing area's history. Above-ground symptoms are broad and lack specificity. Below the bark of infected plants, diagnostic mycelial fans are noticeable at the root crown. To examine, dig down about 15 cm below the soil line and carefully strip thin layers of bark from the root collar. These mycelial fans, thick white layersof the fungus, adhere to the inner root bark and/or the underlying wood. The fungus also generates rhizomorphs, resembling black, shoestring-like structures, occasionally found within the bark and extending into the surrounding soil. While these rhizomorphs may resemble roots externally, they exhibit a distinct structure when cross-sectioned.

Management

Cultural practices promoting vigorous plant growth, facilitated by favorable climates and sound cultural practices, enhance tolerance. Adequate deep watering is crucial, and caution should be exercised against surface watering, especially around the root crown and trunk root area. If utilizing drip irrigation, relocating drip-line emitters away from the root crown and between plants is recommended. Removal ofseverely infected bushes is advised, ensuring the meticulous extraction of small roots from the soil. A successful strategy involves permanently removing soil from around the root crown and the main root area, a method proven effective in managing infected bushes in tree fruits in California and Australia (Schnabel *et al.*, 2012).

2. Black Mold (*Thielaviopsis thielavioides*)

Symptoms and Signs

Characterized by a white, granular fungal growth, Black Mold swiftly extends across cut and bruised areas of stock and scion, transforming into a black appearance. In terms of biology and epidemiology, the fungus thrives on the tips of "Manetti" cuttings (*Rosa chinensis* var. *manetti*) and on wounds resulting from disbudding. This leads to inadequate callus formation and, consequently, impaired root development (Milbrath, 1946). It emerges beneath bud shields on field-budded "Manetti," hindering callus formation and causing the bud to perish. Rose roots may also suffer damage if plants remain in storage for an extended period before planting or if they receive excessive watering post-planting. The primary source of infection is fungal spores present in the dust and debris of storage facilities where plants undergo handling, and cuttings are prepared for planting.

Management

To address this issue, strict sanitation measures are crucial in greenhouses, storage bins, and storage houses. Thoroughly cleaning walls and floors to eliminate all dust and debris precedes the disinfection of storage houses and bins. Employing uncontaminated stock for planting is imperative in effective management.

3. Black Spot (*Diplocarpon rosae*, F.A. Wolf 1912)

Symptoms

Characterized by circular black spots, often with fringed or feathery margins on the leaf, Black Spot results in yellowing and defoliation, particularly in susceptible cultivars. A common symptom includes bare stems with few leaves near the top. In wet conditions, the spots may intensify, merging into large irregular areas covering a significant leaf surface. Severe infections may lead to similar lesions onstems, posing a risk of confusion with other fungal leaf-spotting diseases. The fungus overwinters on living or dead plant tissue from the previous season, primarily affecting newly emerging leaves. Spores, produced on old plant material, are dispersed onto young plant tissue through rain or watering. Once conidia land on a leaf, a minimum of 9 hours of leaf wetness is required for spore infection. Fungal fruiting bodies (acervuli) form in spots within 11–30 days, producing a new set of spores that spread to healthy plant portions. The infection cycle can restart within 10–18 days. Rainy periods between 10 oC and 26 oC encourage black spot development. Leaves infected with black spot release the plant hormone ethylene, causing early defoliation and a bare appearance. Dropped leaves around the plant base serve asa spore reservoir, potentially re-infecting the plant during subsequent rains or irrigation. The fungus's genetic variability means roses resistant in one region may be susceptible in another, and resistance may diminish over time due to changes in the local fungal population.

Management

Efficient management involves selecting cultivars known for resistance in the specific area. For instance, the Carefree Spirit cultivar survived two years in All-American Rose Selections (AARS) tests without fungicide. Practices include avoiding dense plantings and shaded areas, preventing overhead watering that prolongs plant wetness, and burning leaves in a rose planting at the season's end. Pruning infected canes back to two buds, and removing diseased canes before bud break are essential. Fungicide applications should be concentrated at the season's start during bud break, with frequent application in wet weather and occasional treatment in drier periods (1- to 2-week intervals). Managing the disease becomes challenging once initiated. While numerous fungicides are registered for this disease on roses in the USA, only one or two are required at a given time. Alternating or mixing materials with different action modes is recommended to prevent the development of resistant fungal populations.

4. Botrytis Bud and Twig Blight (*Botrytis cinerea*)

Symptoms

Infected buds may exhibit failure to open, drooping, and the development of smooth, slightly sunken, grayish-black lesions. A grayish-brown mycelial growth might emerge on the entire bud, extending into the stem. Petal tips could turn brown, or numerous small flecks, circular brown spots, or blister-like patches maymanifest on petals. Infected stubs or pruning wounds may progress into cane blights or cankers, potentially leading to stem collapse.

Management

Vital is the prompt detection of initial disease signs. Dispose of infected plant debris and tissues from and around flowers, stems, and leaves. Mitigate greenhouse moisture through temperature elevation, venting, and/or enhanced air circulation. Ensure adequate plant spacing for optimal air movement. Prefer drip over overhead irrigation. Utilize calcium nitrate, gypsum, or a comparable calcium source top-dressed or incorporated into the growing media. While fungicides in foliar spray form are an option, emphasis shouldbe placed on cultural controls. To avert the emergence of resistant fungal populations, employ alternate or tank-mix materials with distinct action modes.

5. Cercospora Leaf Spot (*Pseudocercospora puderi*)

Symptoms

Numerous small maroon to purple oval lesions emerge randomly on the leaf surface. These spots initially exhibit a tan to gray center, while their margins maintain a maroon to dark purple hue. Leaves heavily covered in spots turn yellow and prematurely fall off. Typically, the loss of leaves initiates at the cane base, progressing upward through the plant canopy towards the shoot tips. In the southeastern USA, symptoms worsen throughout the summer and early fall, leading to premature leaf shedding. While leaves are most commonly affected, symptoms can also manifest on stems, pedicels, fruits, and bracts.

Management

Prevent dense plantings and shaded areas. Steer clear of overhead watering that prolongs plant moisture. Collect and incinerate all leaves at the season's end. FRAC fungicide groups M5, 3, and 11 have demonstrated effectiveness (Hagan and Akridge, 2005).

6. Downy Mildew (*Peronospora sparsa*)

Symptoms

Irregular spots on leaves ranging from purplish-red to dark brown characterize downy mildew. Leaflets may exhibit yellowing but contain "green island" areas. Fungal growth is often constrained by major leaf veins, resulting in angular lesions as they expand. A grayish spore mass may appear on the leaf underside in persistent humidity, though diagnostic sporulation is not abundant. Severe cases may leadto defoliation, and small spots or long purplish areas on canes can cause twig death. This condition may be mistaken for black spot, rust, or nutrient deficiency in the early stages, and stem lesions may exhibit white sporulation in advanced stages.

Management

To mitigate the disease, reduce greenhouse humidity by increasing temperatures and venting strategically. Clearing leaves and pruning old flowers and stems, followed by burning, burying, or landfill disposal, helps control its spread. Choose cuttings from disease-free plants, scout for symptoms carefully, and remove and destroy any affected tissue. Even if one leaflet shows symptoms, entire leaves should be removed. While hot water soaks have been effective, caution is advised due to potential damage to cuttings. Silicon applications in Colombia have demonstrated efficacy in strengthening tissue and minimizing damage.

7. Powdery Mildew (*Podosphaera pannosa*)

Symptoms

White, powdery fungal growth appears on young leaves, shoots, and buds in cases of powdery mildew. During scouting, it is crucial to inspect the underside of leaves under good lighting conditions, where the infection may initiate. Distorted growth, especially in young tissues, is a common symptom, and severe cases can lead to stunted foliage.

Management

Select rose cultivars known for powdery mildew resistance in the local area. The cultivar Carefree Spirit, surviving AARS tests for two years without fungicide, is noteworthy. Isolate susceptible cultivarsand provide ample spacing for optimal air circulation between plants. Prune canes during dormancy to promote an open habit and improve air circulation. Remove and dispose of diseased canes, and rake and destroy fallen leaves. Employ a high-pressure water hose to thoroughly wet leaf and cane surfaces in theearly afternoon for quick plant drying; this may need to be done two or three times a week. In Western Oregon, this approach reduced powdery mildew. Briefly exposing plants to red light during the dark interval is somewhat effective in greenhouse production against powdery mildew. Fungicides, including a dormant spray of lime sulfur and foliar applications of potassium salts, can be applied. Apply foliar fungicides from early spring when young growth emerges, with frequency depending on the product. Rotate or mix fungicides with different modes of action to prevent resistant fungal populations.

8. Rust (*Phragmidium mucronatum*)

Symptoms/signs

Small orange pustules (aecia) emerge early in spring on both leaf surfaces, enlarging later and becoming more abundant on the lower leaf surfaces. Mottled and chlorotic areas may appear on upper leaf surfaces opposite the spots (uredinia) on the lower surfaces. In late summer and fall, the small pustules turn black (telia), housing the winter spore stages of the rust. Stems may occasionally become infected.

Management

During the dormant season, remove all dead leaves and prune out infected and dead wood. Opt for resistant cultivars and consider early removal of infected leaves, particularly in smaller plantings. Focus applicationson wet weather, and if permitted by the label, wetting agents can enhance the effectiveness of many fungicides. Several fungicides used for black spot control can be beneficial for managing rust.

9. Spot Anthracnose (*Elsinoe rosarum*)

Symptoms

Initially, red spots varying from brown to dark purple appear on the upper leaf surfaces, with spots up to 0.5 cm in diameter scattered, grouped, and occasionally overlapping. Leaves may exhibit chlorosis or yellowing. Subsequently, the spots turn ashen white with a dark red margin, and the affected tissue may detach from the lower leaf surface, leaving a thin papery membrane or falling out entirely. Symptoms extend to stems, hips, and pedicels, posing a potential confusion with black spot or Cercospora leaf spot.

Management

Efforts should focus on removing and eliminating infected plant parts. Adequate airflow should be ensured by spacing or pruning bushes. Although no specific chemical control has been reported, growers use fungicides in the FRAC group 3, commonly employed for black spot, powdery mildew, or rust, to manage this disease.

10. Verticillium Wilt (*Verticillium dahliae*)

Symptoms

Initially, leaves near the growing point of young canes wilt, and lower leaves yellow. In some cases, mature leaves show necrosis between veins while the veins retain their green color. Defoliation progresses from the base to the tip of canes, and in severe cases, permanent wilt, defoliation, and death may occur. If only a few canes are affected, they may either grow normally or experience dieback in the subsequent season. Symptoms are generally more pronounced in the greenhouse than in the field, and the characteristic vascular discoloration seen in other plants is not evident in roses.

Management

Utilize resistant rootstocks and avoid planting in old vegetable fields. In the greenhouse, employ steam- sterilized soil or a sterile soilless potting mix and/ or hydroponics. Remove and discard symptomatic or dead branches, preferably before leaf fall to prevent new inoculum incorporation into the ground. Clean pruning equipment after use and use conifer-based mulch, as conifers resist infection. Exercise caution with general yard waste mulch, as it may contain infected plants and fungus inoculum.

11. Bacterial Cane Blight (*Pseudomonas syringae*)

Symptoms

Symptoms typically initiate at the base of a vegetative bud, leaf scars, or wounds, manifesting as reddish- brown areas on the bark, which later darken to a purple-black, becoming necrotic. This necrosis expands around and along the cane, potentially affecting a significant portion or the entire length of the cane. Vegetative buds on affected parts of the cane dry up and turn brown. The necrotic bark areas exhibit a glossysurface, with the tissue beneath the epidermis appearing brown to dark brown and moist in the early stages. It is often mistaken for winter injury.

Management

In the absence of specific research on controlling this issue, suggested cultural practices include removing and disposing of infected stems and disinfecting pruning shears before cutting additional stems. Copper- based products are recommended before the onset of fall rains and again when half the leaves have fallen.

Bacterial Leaf Spot and Blast (*Pseudomonas syringae* pv. *morsprunorum*) Symptoms

Dark brown, sunken spots emerge on leaves, flower stalks, and calyx parts, leading to the death of flower buds without opening. 1-year-old stems may exhibit black streaks. The disease has been identified only four times in the PNW from 1962 to 1992 and is most prevalent in cool, wet spring weather.

12. Crown Gall (*Agrobacterium radiobacter*)

Symptoms

Galls often appear at or just below the soil surface in the basal or root crown region, occasionally on roots, and less frequently on above-ground plant parts like stems and leaves. These galls typically exhibit a rounded shape with a rough, irregular surface. Initial growth manifests as small protrusions with a light green or nearly white, soft tissue. With age, galls become dark and woody, occasionally shedding outer portions. Distinguishing between gall and callus growth can be challenging, especially if the gall is situated at the plant base or graft/bud union. Infected plants may exhibit stunted growth, reduced vigor, poor foliage, and fewer blossoms. Symptoms might not surface for more than a year if infection occurs at temperatures below 15°C (59°F).

Management

Employ only pathogen-free nursery stock, inspecting new plants and avoiding planting roses with galls. Minimize plant wounds, especially during planting, and use plants with resistant rootstocks. Prune off any galls on above-ground plant parts, frequently disinfecting pruning shears. Remove and destroy severely affected plants. Preplant soil solarization, proven effective for cherry nursery stock in Western Oregon, involves covering rototilled, irrigated ground with clear plastic from mid-July to mid- September. Biological control using *Agrobacterium radiobacter* strain 84, successful in Australia, New Zealand, and Spain, may not be as effective in the USA. Strain K1026, a genetically modified version, can reduce the potential for resistant crown gall bacteria but does not control latent infections or existing galls. Application should occur shortly after wounding for effectiveness.

13. Rose Leaf Curl

Symptoms

Spring leaves are small, leaflets detach easily, leaf epinasty, necrosis of shoot tips, and yellow vein flecking occur. Shoots are pointed with a broad base. Plants may recover in summer but show symptoms again in fall. Biology and epidemiology. Little is known about this disease other than natural spread appears to be slow.

Management

Opt for stock that is clean and/or certified as tested and free of all known viruses during purchase. Remove and dispose of infected plants. Conduct a heat treatment on scion stock plants, exposing them to a 4-week period at 38°C (100°F) before grafting

14. Rose Mosaic [(*Prunus necrotic ringspot virus* (PNRSV)

Symptoms

Symptoms may range widely depending on the time of year, temperature, and type of virus(es) infecting the plant. Characteristic symptoms include chlorotic line patterns (zigzag pattern), ringspots, and mottles in leaves sometime in the growing season. There may also be yellow net and yellow mosaic symptoms. Symptoms often are evident in spring and early summer but may not be on leaves produced in summer. Vein banding may be on leaves in long hot periods. Flower distortion, reduction in flower production, flower size, stem caliper at the graft union, winter survival, and early leaf drop and increase susceptibility to cold injury have all been reported. Some infected cultivars may not show any symptoms at all. Biology and epidemiology. Several viruses are associated with the range of symptoms of rose mosaic, including Prunus necrotic ringspot virus (PNRSV), Apple mosaic virus (ApMV), Arabis mosaic virus (ArMV), and Strawberry latent ringspot virus (SLRSV). The disease does not spread naturally and has no known insect vector, but grafting transfers it to healthy plants. Transmission of ArMV and/or SLRSV could occur by Dagger nematodes, but this has not been extensively studied. Viruses can be in the rootstock or scion or both and may not show symptoms.

Management

Purchase stock that is clean and/or certified as tested and free of all known viruses. Remove and destroy infected plants. However, the disease will not spread unless you propagate from or onto an infected bush.Heat-treat scion stock plants for four weeks at 38°C (100°F) before grafting.

15. Rose Yellow Mosaic (*Rose Yellow Mosaic Virus*, RoYMV)

Symptoms

Symptoms associated with RoYMV infection include yellow mosaic, ring mosaic, premature leafs enescence, and necrotic dark brown rings on canes. Symptoms first appeared in new growth 4–6 weeks post inoculation. Necrotic cane symptoms are observed only in the cultivar Ballerina. Symptoms persist throughout the season (Lockhart et al., 2008). Biology and epidemiology. Potyviruses such as RoYMV are known to be aphid transmitted.

Management

Purchase stock that is clean and/or certified as tested and free of all known viruses. Remove and destroy infected plants.

16. Nematode Diseases (*Meloidogyne hapla*, *Pratylenchus Penetrans*, *P. vulnus*, and *Xiphinema diversicaudatum*)

Nematodes impacting plant growth are categorized based on their feeding behavior. The type of sample required for diagnosis, either roots, soil, or both, depends on the nematode species. Root-knot nematodes (*Meloidogyne* spp.) are sedentary endoparasites that burrow into roots, establishing permanent feeding sites. These feeding sites lead to the development of giant cells, hyperplasia of cortical and vascular parenchyma, and the slowing down of meristematic activity in root tips (Fig. 12). Root-lesion nematodes (*Pratylenchus* spp.) are migratory endoparasites that tunnel within roots, feed internally, and can freely move back into the soil and onto new roots. Dagger nematodes (*Xiphinema* spp.) are migratory ectoparasites that feed externally on roots, moving from cell to cell and piercing them to feed without entering the root tissue. The galls formed at these feeding sites result from the hyperplastic response of cortical cells.

Management

To avoid nematode-related issues in floral production, use soilless media or hydroponics to eliminate soil from the production system. When soil is necessary, steam treatment is crucial for eradicating nematodes. While challenging, steam treatment of soil in planting beds is also feasible. Soil solarization proves useful in reducing nematode populations in the top foot of soil, facilitating the establishment of plants in infested soil. Thermal therapy post-infestation of plants is possible but intricate. Pretreat plants at 38°C for 24 hours, followed by immersion in hot water at 48°C for 35 minutes. Note that this treatment may cause damage to rose roots. The use of resistant rootstocks is complex due to varied reactions to different nematodes in different locations. Preplant fumigation can be effective in the field, but the availability of registered nematicides is diminishing each year.

References

Bagsic, I., Linde, M. and Debener, T. (2015). Genetic diversity and pathogenicity of Sphaceloma rosarum (teleomorph Elsinoë rosarum) causing spot anthracnose on roses. Plant Pathol., 12478.

Bowen, K.L. and Roark, R.S. (2001). Management of black spot of rose with winter fungicide treatment. Plant Dis., 85: 393–398.

Corvallis Rivera, P.A. and Engel, E.A. (2010). Presence of rose spring dwarf-associated virus in Chile: partial genome sequence and detection in roses and their colonizing aphids. Virus Genes, 41(2): 295–297.

Farr, D.F. and Rossman, A.Y. (2015). Fungal databases, systematic mycology and microbiology laboratory, ARS, USDA.

Golino, D.A., Sim, S.T., Cunningham, M. and Rowhani, A. (2011). Evidence of root graft transmission of two rose mosaic viruses, Prunus necrotic ringspot virus and Apple mosaic virus in rose roots tocks. Phytopathology, 101: S62.

H.R. (2010). Continuous lighting reduces conidial production and germinability in the rose powdery mildew pathosystem. Plant Dis., 94: 339–34.

Hagan, A.K. and Akridge, J.R. (2005). Chemical control of cercospora leaf spot on Fuchsia Meidiland shrub rose. Alabama Coop erative Extension, 587.

Hammett, K.R.W. (1971). Symptom differences between rose wilt virus and Vertidllium wilt of roses. Plant Dis., 55: 916–920.

Hausbeck, M.K. and Moorman, G.W. (1996). Managing Botrytis in greenhouse-grown flower crops.Plant Dis., 80: 1212–1219.

Horst, R.K. and Cloyd, R.A. (2007). Compendium of rose diseases and pests, 2nd edn. APS Press.

Lockhart, B., Zlesak, D. and Fetzer, J. (2008). Identification and partial characterization of six new viruses of cultivated roses in the USA.

Milbrath, J.A. (1946). Control of black mold fungus Chalaropsis thielavioides Peyr. on Manetti rose, vol. 8, Oregon State University technical bulletin.

Mohan, S.K and Bijman, V.P. (2010). Bacterial cane blight of rose caused by Pseudomonas syringae. Acta Hortic., 870: 109–113.

Philley, G., Hagen, A.K. and Chase, A.R. (2001). Rose diseases. In: Jones RK, Benson DM. Diseases of woody ornamentals and trees in nurseries. American Phytopathological Society, 76.

Pscheidt, J.W. and Ocamb, C.M. (2016). Pacific Northwest plant disease management handboo k. Or egon State Unive rsity,

S alem, N. , Golino, D.A., Falk, B. W. and Rowhani, A. (2008). Identification and partial characterization of a new Luteovirus associated with rose spring dwarf disease. Plant Dis., 92:508–512.

Schnabel, G., Agudelo, P., Henderson, G.W. and Rollins, P.A. (2012). Aboveground root collar excavation of peach trees for Armillaria root rot management. Plant Dis., 96:681–686.

St. Paul Intrama, S. (1968). Coniothyrium rose canker in Oregon. PhD thesis, Oregon State University.p.95.

St. Paul Pie, K. and De Leeuw, G.T.N. (1991). Histopathology of the initial stages of the interaction between rose flowers and Botrytis cinerea. Eur. J. Plant Pathol., 97: 335–344.

Suthaparan, A., Stensvand, A., Torre, S., Herrero, M.L., Pettersen, R.I., Gadoury, D.M. and Gislerod,

Index